Architectures of Hurry—Mobilities, Cities and Modernity

'Hurry' is an intrinsic component of modernity. It exists not only in tandem with modern constructions of mobility, speed, rhythm, and time–space compression, but also with infrastructures, technologies, practices, and emotions associated with the experience of the 'mobilizing modern'. 'Hurry' is not simply speed. It may result in congestion, slowing-down, or inaction in the face of over-stimulus. Speeding-up is often competitive: faster traffic on better roads made it harder for pedestrians to cross, or for horse-drawn vehicles and cyclists to share the carriageway with motorized vehicles. Focusing on the cultural and material manifestations of 'hurry', the book's contributors analyse the complexities, tensions, and contradictions inherent in the impulse to higher rates of circulation in modernizing cities.

The collection includes, but also goes beyond, accounts of new forms of mobility (bicycles, buses, underground trains) and infrastructure (street layouts and surfaces, business exchanges, and hotels) to show how modernity's 'architectures of hurry' have been experienced, represented, and practised since the mid nineteenth century. Ten case studies explore different expressions of 'hurry' across cities and urban regions in Asia, Europe, and North and South America, and substantial introductory and concluding chapters situate 'hurry' in the wider context of modernity and mobility studies and reflect on the future of 'hurry' in an ever-accelerating world. This diverse collection will be relevant to researchers, scholars, and practitioners in the fields of planning, cultural and historical geography, urban history, and urban sociology.

Phillip Gordon Mackintosh is Associate Professor of Geography, Brock University. He is the author of *Newspaper City: The Liberal Press and Toronto's Street Surfaces, 1860–1935* (2017), numerous publications on turn-of-the-twentieth-century cultures of urban reform and city planning, historical cycling, and urban historical geographies of class, gender, and race, and is co-editor of *The World of Niagara Wine* (2013).

Richard Dennis is Emeritus Professor of Geography, University College London (UCL). His books include *Cities in Modernity* (2008) and *English Industrial Cities of the Nineteenth Century* (1984). He has written for numerous publications on housing, public transport, and imaginative literature in nineteenth- and twentieth-century London and Toronto. He is an editorial committee member of *The London Journal*, and previously was associate editor of *Journal of Urban History* and series editor of *Cambridge Studies in Historical Geography.*

Deryck W. Holdsworth is Emeritus Professor of Geography, Pennsylvania State University. He is the co-author of *Homeplace: The Making of the Canadian Dwelling over Three Centuries* (1998) and was co-editor of the *Historical Atlas of Canada, Volume III, Addressing the Twentieth Century* (1990). He has authored numerous journal articles on office buildings, folk and industrial housing, and insights from historical hotel guest registers.

Routledge Research in Historical Geography

Series Edited by Simon Naylor (School of Geographical and Earth Sciences, University of Glasgow, UK) and Laura Cameron (Department of Geography, Queen's University, Canada)

This series offers a forum for original and innovative research, exploring a wide range of topics encompassed by the sub-discipline of historical geography and cognate fields in the humanities and social sciences. Titles within the series adopt a global geographical scope and historical studies of geographical issues that are grounded in detailed inquiries of primary source materials. The series also supports historiographical and theoretical overviews, and edited collections of essays on historical-geographical themes. This series is aimed at upper-level undergraduates, research students, and academics.

For a full list of titles in this series, please visit www.routledge.com/Routledge-Research-in-Historical-Geography/book-series/RRHGS

Historical Geographies of Prisons: Unlocking the Usable Carceral Past
Edited by Karen Morin and Dominique Moran

Historical Geographies of Anarchism: Early Critical Geographers and Present-Day Scientific Challenges
Edited by Federico Ferretti, Gerónimo Barrera de la Torre, Anthony Ince and Francisco Toro

Cultural Histories, Memories and Extreme Weather: A Historical Geography Perspective
Edited by Georgina H. Endfield and Lucy Veale

Commemorative Spaces of the First World War: Historical Geographies at the Centenary
Edited by James Wallis and David C. Harvey

Architectures of Hurry—Mobilities, Cities and Modernity
Edited by Phillip Gordon Mackintosh, Richard Dennis and Deryck W. Holdsworth

Architectures of Hurry—Mobilities, Cities and Modernity

Edited by Phillip Gordon Mackintosh,
Richard Dennis and
Deryck W. Holdsworth

Routledge
Taylor & Francis Group
LONDON AND NEW YORK

First published 2018
by Routledge
4 Park Square, Milton Park, Abingdon, Oxon OX14 4RN

and by Routledge
605 Third Avenue, New York, NY 10017

Routledge is an imprint of the Taylor & Francis Group, an informa business

British Library Cataloguing-in-Publication Data
A catalogue record for this book is available from the British Library

Library of Congress Cataloging-in-Publication Data
A catalog record for this book has been requested

ISBN: 978-1-138-72984-1 (hbk)
ISBN: 978-1-315-18960-4 (ebk)

Typeset in Times New Roman
by Swales & Willis Ltd, Exeter, Devon, UK

Contents

Illustrations

Figures

Tables

Contributors

Christina E. Dando is Professor of Geography, University of Nebraska Omaha. Her research interests include the impacts of media and technology on human perception and interaction with the environment, and exploring the intersections of landscape, media, and gender. She is the author of *Women and Cartography in the Progressive Era* (Routledge, 2018).

Jason Finch researches and teaches modern urban literatures at Åbo Akademi University in Finland. He is the author of, most recently, *Deep Locational Criticism* (John Benjamins, 2016) and is a co-founder of the Association for Literary Urban Studies. Current research focuses on the literary history of the London slum, and on presentations of the urbanity of provincial English cities (especially Birmingham, Liverpool and Bradford).

Boyang Gao was an undergraduate at Peking University and completed her doctoral studies at Beijing's Chinese Academy of Sciences. She is currently Associate Professor, Department of Urban and Real Estate Management, Central University of Finance and Economics, Beijing. Her research focuses on the economic transformation of China's megacities, including Beijing, Chongqing, and Tianjin.

Glen Norcliffe is Professor Emeritus and Senior Scholar, Geography, York University, Toronto. His research explores the cultural political economy of industry in contemporary and historical settings, including issues of globalization, impacts of neoliberalism, the production of sport, and development in peripheral regions. His most recent book is *Critical Geographies of Cycling: History, Political Economy and Culture* (Routledge, 2015).

Sherry Olson is Professor of Geography, McGill University, and a member of the Centre interuniversitaire d'études québécoises (CIÉQ Laval-UQTR); co-author with Patricia Thornton of *Peopling the North American City, Montreal 1840–1900* (McGill-Queen's, 2011), author of an environmental history of Baltimore, and of numerous papers on the two cities – their horses, diseases, fires, riots, and other vexations.

Colin G. Pooley is Emeritus Professor of Social and Historical Geography, Environment Centre, Lancaster University, UK. His research focuses on the social geography of Britain and continental Europe since circa 1800, with recent projects focused on residential migration, travel to work, and other aspects of everyday mobility. His most recent book is *Mobility, Migration and Transport: Historical Perspectives* (Palgrave, 2017).

Marilyn E. Pooley is a historical geographer. She was formerly a Teaching Associate in the Environment Centre at Lancaster University, UK, and in retirement is researching (with Colin Pooley) everyday mobility in nineteenth- and twentieth-century Britain using life writing.

Mary Anne Poutanen is a historian and member of the Montreal History Group and the Centre de recherche interdisciplinaire en études montrèalaises (McGill University). Author of *Beyond Brutal Passions: Prostitution in Early Nineteenth-Century Montreal* (McGill-Queen's, 2015) and co-author of *A Meeting of the People: School Boards and Protestant Communities in Quebec, 1801–1998* (McGill-Queen's, 2004), she teaches interdisciplinary studies at McGill and history at Concordia University.

David Rooney is Keeper of Technologies and Engineering at the Science Museum, London, and an Honorary Research Associate in the Department of Geography at Royal Holloway, University of London. He has wide research interests in histories and geographies of urban infrastructure, from time distribution networks to traffic systems, and is currently writing on the politics of congestion in twentieth-century London.

Dhan Zunino Singh, PhD, History (University of London), works on the cultural history of urban mobilities. He is a research assistant at the National Scientific and Technical Research Council (CONICET), National University of Quilmes, Argentina, and an associate editor of the *Journal of Transport History*. He lectures postgraduate seminars on mobilities at Buenos Aires University.

Acknowledgements

In this era of wavering historical geography education, the editors happily acknowledge the influence of the students of our historical geography classes who, over the years, have helped us sort out many of the organizing ideas of this book: Deryck's GEOG 427 Urban Historical Geography and Phil's GEOG 3P41 Modern City in Historical Perspective. Richard appreciates the enticements of former doctoral student, Sam Merrill (now at Umeå University), and colleague Carlos Lopez-Galviz (Lancaster University), who lured him into transport and mobility studies just when he was thinking of slowing down.

Richard also thanks the London Transport Museum, and especially its director, Sam Mullins, for successive invitations to talk about the social and cultural history of London's trains and buses. To this end, he was grateful for opportunities to talk about architecture(s) of hurry and 'more haste, less speed' in conferences in Munich, Odense and KCL, and in lectures and seminars at Leicester, Brock, Queen's (Kingston), and McGill.

Deryck would like to thank the late Gunter Gad, University of Toronto (his long-time collaborator on the emergent office district) for discussions on the Bourse in its many forms; Tony King, who has always encouraged him to think globally about the social production of the built environment; and Gretta Tritch Roman, for many conversations about New York and Chicago commodity exchanges.

Phil cherishes the opportunity to work closely with the many talented and insightful contributors to *Architectures of Hurry* – and especially with Richard and Deryck. He also thanks his colleague and 'sounding board', Michael Driedger.

We all are grateful for the enthusiastic reception from the audience attending the special sessions on 'Architectures of Hurry: Mobilities and Modernity in Urban Environments' that we offered at the Sixteenth International Conference on Historical Geography, at the Royal Geographical Society, London, 2015. Listening closely all day was Faye Leerink, an acquisitions editor for Routledge, who championed the book from the beginning; thanks also to Ruth Anderson, the editorial assistant who shepherded it through the publication process. We have a special fondness for the eager support of Simon Naylor and Laura Cameron, co-editors of Routledge's Research in Historical Geography Series: their 'delight' in the prospect of the book inspired us. The volume would not exist without the remarkable thinking, writing, and especially the deadline-meeting cooperation of our contributors: Boyang, Christina, Colin, David, Dhan, Glen, Jason, Marilyn, Mary Anne, and Sherry. They made the process a pleasure. Thanks also to Harry Mackintosh for his assistance with images. Lastly, Deryck and Phil thank Richard for his close reading of E.M. Forster – and suggesting the ICHG sessions in the first place!

1 Architectures of hurry

An introductory essay

Richard Dennis, Phillip Gordon Mackintosh and Deryck W. Holdsworth

Nature does not hurry, yet everything is accomplished.
(Lao Tzu, Chinese philosopher, sixth century BC)

Though I am always in haste, I am never in a hurry.
(John Wesley, English cleric, eighteenth century)

No man [sic] who is in a hurry is quite civilized.
(Will Durant, American philosopher and historian, early twentieth century)

Don't rush. Be quick, but don't hurry.
(Earl Monroe, American basketball player, born 1944)

In his survey of *Keywords*, Raymond Williams famously observed of 'Community' that 'it seems never to be used unfavourably'.[1] Had he been writing about 'Hurry' as a keyword, he would probably have reached the opposite conclusion. Writers may be ambivalent about 'Speed', they may use 'Haste' both positively and negatively – Wesley's godly haste (above) is matched by the proverb warning that those who 'marry in haste, repent at leisure' – but few of them have a good word for 'Hurry'. This collection of essays takes its title (albeit pluralized to imply a diversity of meanings) from E.M. Forster's disdainful reference to 'the architecture of hurry' in his novel, *Howards End* (1910). But Forster is by no means the only novelist to denigrate 'hurry'.

Writing in journalistic mode for an American popular magazine, the prolific English novelist Arnold Bennett commented that, 'The first-rate organizer is never in a hurry'. He went on, in the gender-biased spirit of his times (1929), to claim that:

> Women on the contrary are too often in a hurry. In addition to expecting too much from human nature, they expect too much from clocks [and] they have a superstitious notion that, when it is convenient to them, the hands of clocks will cease to move.[2]

In so doing, he was recognizing the subjective, emotional and less-than-rational character of 'hurry', in contrast to the physical laws that govern and measure speed, but he was also questioning the place of 'hurry' in modernity. Bennett could hardly be described as 'modernist' in literary style, but he was quintessentially 'modern' in his productivity and efficiency as an author of at least 40 novels and short-story collections, 21 books of non-fiction, 11 plays and countless magazine articles.[3]

Yet 'Hurry' as much as speed is intrinsic to modernity. We cannot conceive of modern society and modern capitalism without invoking, not just speed, but the desire for speeding-up, the fears and anxieties associated with acceleration, and the physical, psychological, and emotional consequences of hurry. We are accustomed to exploring processes of rationalization, liberalization, and regulation of the movement of populations and things as central features of the modern urban-industrial world. The subsequent requirement *cum* desire for people, goods, money, information and even weather to 'be somewhere', at some 'when', birthed extraordinary cultures of interaction, expedited and warranted by transportation and communications innovations. Bicycles, buses, trains, streetcars, automobiles and ferries whisked everyday people from homes perhaps to go to school, open a shop, clock in at a factory, rise in an elevator to a skyscraper office, or attend an evening or weekend social – all mediated by modernity's preoccupation with time and space compression. Yet ever-changing transportation and communications infrastructures simply stimulate more connectedness, more mobility, more distant and near interactions – and more hurry – to perpetuate still more ironies of modernity, the clash of its conveniences and securities with its ambiguities and absurdities. Long ago, Marshall Berman demonstrated these conflicting tendencies in his brilliant dissections of Baudelaire's Paris, Dostoevsky's St Petersburg and Robert Moses' New York, just as David Harvey trenchantly deconstructed the modern contradictions of late capitalism at every scale, from the global and imperial to the streets of Baltimore and Paris.[4] Hurry epitomizes these ironies and contradictions. In seeking to speed up, we often end up doing things more slowly (as another proverb, 'more haste, less speed', reminds us); if we do succeed in physically speeding up, we often fail to anticipate the negative psychological and emotional side-effects: order and discipline provoke resistance, de Certeau's 'tactics' and 'transgressions', and a resort to 'short cuts'.[5]

Hurry and mobility

Recent scholarship in the social sciences and humanities has produced substantial literatures on modernity and speed, and on modernity and mobility.[6] Studies of mobility first of all focused on flows of people and commodities; more recently on information, ideas, policies and practices. There were explorations of the friction of distance, of differentiation by class and gender, of time–space compression and distanciation and their political and economic correlates.[7] Meanwhile, new strands of social and cultural history and human geography have focused on the senses and the emotions. How were speed and mobility experienced?[8] Variously, this research has been related to physics, physiology, neurology, psychology and aesthetics. When we travel, do we experience our own moving through the landscape, or does the landscape move past us? Of course, it is both, and this is the two-pronged point about mobilities and modernity: their capacity to marry technologies and emotions to fashion exhilaration. So, moderns hurry corporeally through technologically mediated landscapes, even as landscapes move through them, modern human bodies and minds subject to the science and sentiment of motion. Thus, there have been discourses on the thrill of acceleration, ever-increasing speed, and on the anxieties associated with excessive speed, mobility and displacement. Mimi Sheller and John Urry have taken these notions to their logical conclusion, averring that speedy mobility is the enemy of public civility; mobility in modernized landscapes daily produces tragedy, as speed, mass and inertia converge in distinctly modern and antisocial ways.[9] This collection, however, focuses not on mobility *per se*, nor on speed as an objective attribute of modernization, but on 'hurry', a necessarily relative phenomenon. You can travel at high speed without hurrying, and you can travel quite slowly but still be 'in a hurry'.

Tim Cresswell differentiates between 'movement', which he aligns with space and location, and 'mobility', socially produced movement laden with meaning, which he associates with place.[10] In the same way, *we might suggest that we speed through space, but we hurry from place to place.*

In his agenda-setting *Mobilities* (2007), Urry identified five interdependent 'mobilities' that contribute to the production of social life: First, and the most obvious form of physical mobility, the corporeal movement of people, extending from daily routines of movement to or from work through to once-in-a-lifetime migrations chosen by intercontinental settlers or forced upon exiles or refugees; second, the physical movement of objects, such as manufactured commodities, moving through the production cycle from raw materials to finished product, and then traded, deployed, re-traded, recycled or abandoned; third, imaginative travel through diverse print and visual media that allow us to experience places otherwise inaccessible; fourth, communicative travel through the transmission of messages, letters, telegrams, phone calls and, now, various forms of electronic communication; and fifth (and perhaps less relevant to the historical focus of *this* collection), virtual travel. Urry argued that there are increasingly complex *systems* that make mobility both possible and – through their guarantee of repetition – predictable.[11]

In the nineteenth century, which Urry described as 'the century of "public mobilization" through new times, spaces and sociabilities of public movement', mobility systems included national postal services, commercial electric telegraph networks, railways and steamship companies offering scheduled services advertised in timetables, and other sources of trustworthy information such as guidebooks and package tours, organized by expert agencies such as Thomas Cook.[12] These networks were sustained by a fixed infrastructure of rails, roads, pipes, cables and their terminal or access points: railway and bus stations, telephone exchanges and kiosks, sewage works, hotels, post offices and postboxes, warehouses and department stores, a veritable *architecture* to facilitate the mobility of people, things and ideas.

George Revill also discusses the significance of infrastructure, but from a methodological perspective, as a way in which historians and historical geographers can engage with the 'mobilities turn'.[13] He notes that, in mobility studies of the past, often our only forms of evidence are apparently mute buildings, structures and machines of transport archaeology and second-hand accounts in print and visual media. This may be unduly pessimistic, as effectively first-hand accounts in diaries, novels and witness statements can replicate the ethnographic and participatory sources employed by researchers of contemporary mobilities (as several chapters in this collection demonstrate). But Revill's point is that, even in restricting our attention to assemblages and the agency of infrastructure, we can move beyond traditional 'transport history' into 'mobility history'.

Mobility is about more than connectedness; it is also about the speed and frequency of intermittent connections. Today, when we rely so heavily on continually available and updatable electronic communications, we may think of newspapers or letters as impossibly slow ways of spreading information. But the Victorian press and post office were not the sad remnants that we experience today. Nowadays, our daily newspapers often take several days to publish reviews of live events such as plays and concerts, and, if we are unlucky and our newsagent receives copies printed the previous evening, we miss out on the latest sports results or, indeed, any news that breaks after about 6 p.m.[14] The assumption seems to be that the printed paper doesn't really matter. For the latest news, we go online. But, through the nineteenth and even into the late twentieth century, major newspapers published several editions daily, frequently changing their headlines and layout to accommodate 'breaking news', and an early morning edition would publish full reports and reviews of concerts, speeches (including sermons) and sporting events that had happened only a few hours previously. There was also space for 'stop press', breaking news that could be incorporated into the printing of the current edition while a new edition was being prepared. 'Stop press' gave readers the latest cricket or football scores in games that were still in progress when the paper was being printed and would still be in progress when the paper was being purchased on the streets. Contrasting with today's at most once-a-day deliveries, multiple postal collections and deliveries allowed for an exchange of correspondence over a period of a few hours, even between correspondents who lived in different towns. So although, in general, 'progress' has involved speeding-up, some forms of mobility have slowed down over time, irrespective of the slowing-down attributable to congestion or safety concerns.

Nevertheless, increasing speed – faster forms of transportation, and faster transfers of information and capital – is associated with increasing levels of mobility. The more quickly we can go, the more often we will go. The faster we transmit information, the more information is transmitted. Journeys once not worth making (because they took so long that you would need to head home shortly after arriving) are now taken for granted. So, the increasing separation of home and work, the increasing spatial range of shopping and leisure activities, the ability of businesses to market their products over larger territories, all depend on the time it takes to commute, to travel to friends or leisure spaces, to deliver goods to customers. We are accustomed to thinking about 'the annihilation of space by time', and to the clichéd repetition that 'time is money', both phrases that originated in the nineteenth century, as much in popular advertising as in theoretical works of political economy.[15]

Speed in business makes more likely transactions that were rare in earlier periods: the ability to borrow or invest, irrespective of the remoteness of the lender or the object of investment; the speeding-up in the circulation of capital, reducing the 'turnover time' for realizing profit.[16] There were successful exceptions, especially in eighteenth-century transatlantic trade that could entail credit periods of nine months or more, but these were transactions between merchants buying and selling commodities on credit rather than speculative investments.[17] Credit, derived from the Latin *credo*, implies a degree of knowledge that warrants belief and trust, whereas investments in distant, relatively unknown enterprises known only at second or third hand involve much higher levels of risk – the uncertainty of speculation, but also the potential for fraud and duplicity among actors at a distance who do not know one another personally. Hence the need for discipline and regulation: speed and mobility as central elements in *modern* society. The discipline of the timetable is matched by the discipline of the individual user, consumer, employee or manager, especially the need for time discipline and punctuality. The impact of clocks, watches, and more directly disciplinary chronometers such as time recorders, factory sirens, or whistles to signal train departures was classically epitomized by the White Rabbit in *Alice's Adventures in Wonderland* (1865), muttering 'Oh dear! Oh dear I shall be late!', taking a watch out of his waistcoat pocket, looking at it and then hurrying on. More formally, the American neurologist, George Beard, discussed how clocks and watches obliged us to keep time with a new precision, making us nervous about being even a few moments late. Ever more accurate, and ever more synchronized, timekeeping was reflected in systems of pneumatic or electrically transmitted pulses to ensure that all clocks across a business, city or state showed exactly the same time, and by national and international agreements to standardize time and establish time zones, all habituating moderns to a hurried life of 'clock time'.[18]

However, as John Tomlinson contends, speed could also escape discipline: the thrill of speed – the unruly speed of Mr Toad at the wheel of his motor car, the irresistibility of 'scorching' (the desire for 'swift motion', as Frances Willard [president of the Women's Christian Temperance Union] wrote of her bicycling experience), the provocation of Marinetti's Futurist Manifesto, speed as a weapon or, at the very least, to quote Tomlinson, speed as '[s]ubversive and impetuous, conjoining hedonism with a peculiar sort of existential heroism'.[19] While in some ways speed was being democratized by modernity – after all, passengers in third-class carriages travelled at the same speed as those in first-class carriages in the same train – it was also being privatized and eroticized, not least in the anticipation of the crash. Jeffrey Schnapp explores the history of writing and thinking about vehicle crashes, drawing particular attention to the different perspectives of drivers, passengers and observers, especially voyeurs.[20]

Speed could be a mixed blessing, even for those who avoided addiction to its experiential aspects. As Urry commented, '[t]he mechanization of movement through the railway initiates the valuation of speed'.[21] Technologies that improve speed, mobile or communicative, are valued regardless of their cost, their environmental impact or their tendency to increase social inequality; high speed equals high status. This kind of thinking assumes that journey time is wasted time, that nothing we might do en route – think, read, write, talk, eat, sleep, observe – has any value compared with the saving of time, and that there is no value in waiting, anticipating the moment of arrival, the delivery of a message or the receipt of goods. Speed means *now*, instant gratification, the suborning of the quiet rewards of unhurried patience. Speed is not just a component of modernity, but a determinant of monetization. No wonder a key text on speed and, more generally, the 'pace of life', is Simmel's *The Philosophy of Money* (1900).[22]

For our interests in this book, Simmel's essay on 'The Metropolis and Mental Life' (1903) is an even more fundamental text, where he discusses 'the intensification of nervous stimulation', 'the unexpectedness of onrushing impressions', 'the tempo and multiplicity of economic, occupational, and social life' and the consequent development of attitudes of reserve, apparent insensitivity, the blasé attitude, as coping mechanisms among the citizens of big cities.[23] Observers of life in New York and Berlin suggested that the intensified tempo and competition inherent in the modern metropolis resulted in, at the least, irritability, nervousness, irrational responses, fear and, more clinically, neurasthenia, neuralgia, nervous dyspepsia, tooth decay, premature baldness, heart disease, cancer, kidney disease, criminal behaviour, insanity and suicide. The novelist William Dean Howells, who, when he first arrived in New York in the late 1880s, celebrated the steam-hauled elevated trains as 'the most ideal way of getting about in the world' and 'intoxicating', by 1907 was lamenting the 'wear and tear of the nerves' from trains 'roaring by the open window', an experience reproduced in King Vidor's silent film, *The Crowd* (1928), in which the archetypal young couple, John and Mary Sims, find marital bliss challenged by the elevated trains that thunder past their tenement window.[24] Incidentally, King Vidor shows us that silence can be deafening, and that noise is an intrinsic component of hurry: often unsettling, distracting, confusing or discordant.

Hurry and the rhythm of modern life

Such attitudes and experiences, as much as the material infrastructure of roads, railways, pipes, wires and buildings, constitute a kind of *architecture* – what the English novelist E.M. Forster described in *Howards End* (1910) as an 'architecture of hurry'.[25] Forster's novel reflects his anxious engagement with modernity, but especially with the modern metropolis. It is usually explored by historical sociologists and literary geographers in terms of its epigrammatic summons, 'Only connect'. This may be interpreted in various ways, as an exhortation to 'connect' old and new, male and female, business and intellect, mind and body, town and country. But Forster's text invites many other social scientific engagements, not least his references to the 'civilization of luggage' – his dissatisfaction with the rootlessness of modern society – and the 'architecture of hurry'. Forster begins Chapter XIII of his novel with a series of observations about geographical and social change over the previous two years. London was in a state of 'continual flux':

> [H]er shallows washed more widely against the hills of Surrey and over the fields of Hertfordshire. This famous building had arisen, that was doomed. Today Whitehall has been transformed; it would be the turn of Regent Street tomorrow. And month by month the roads smelt more strongly of petrol, and were more difficult to cross, and human beings heard each other speak with greater difficulty, breathed less of the air, and saw less of the sky.[26]

Forster moves from the established history of the changing built environment – the spread of suburbia way beyond the limits of the county of London into rural Surrey to the south and Hertfordshire (where the former farmhouse 'Howards End' is located) to the north; the rebuilding of Whitehall (especially a new War Office, completed in 1906, and government offices at the south end of the street partly opened in 1908); and the rebuilding of Regent Street (begun in the 1890s, but not completed until the 1920s) – to the experience by its inhabitants of a busier, more crowded, more polluted city. But then he considers the implications for Margaret Schlegel, the older of the two sisters at the heart of the novel:

> In the streets of the city she noted for the first time the architecture of hurry, and heard the language of hurry on the mouths of its inhabitants – clipped words, formless sentences, potted expressions of approval or disgust. Month by month things were stepping livelier, but to what goal?[27]

Forster directs our attention to a physical architecture of hurry as buildings are demolished and replaced by bigger and, presumably, more profitable or more efficient structures, to a changing infrastructure of streets in which pedestrians are subordinated to motor traffic, but also to a less personal experience of social interaction. Today, we could imagine Forster railing against people who traverse the streets attached to their smartphones: earbuds umbilically occluding the ambient sounds of urban life – including its mobilized dangers – oblivious of other pedestrians and obstacles; texting truncated communications with abbreviations and emoticons; consulting social media rather than engaging in face-to-face conversations, technologically mediating their very presence on the street. But, whatever the medium of hurry, Forster is ultimately concerned about its social value, its consequences for urban society.

Forster's Austrian contemporary Robert Musil was equally intrigued by the erosive and impersonal rhythms of modern urban life in his gargantuan and (in its production) very unhurried novel *The Man Without Qualities*, begun in 1921, still unfinished when Musil died in 1942, and set in Vienna in 1913. Consider the novel's opening, in which 'Automobiles *shot out* of deep narrow streets . . . Where more powerful lines of speed *cut across* their [pedestrians'] casual haste, they *clotted up*, then *trickled on faster* and, after a few oscillations, *resumed* their steady rhythm' (our emphasis).[28] Generalizing, the narrator tells us that this city, 'Like all big cities . . . was made up of irregularity, change, forward spurts, failures to keep step, collisions of objects and interests, punctuated by unfathomable silences'. Immediately, we focus in on a specific incident: 'a heavy truck . . . which had braked so sharply that it was now stranded with one wheel on the curb' and a pedestrian who 'lay there as if dead', while passers-by busied themselves 'vaguely wanting to help' but 'really only marking time while waiting for the ambulance to bring someone who would know what to do and have the right to do it'.[29]

The scene promptly switches farther down the same street, to the house where the eponymous 'man without qualities' is 'ticking off on his stopwatch the passing cars, trucks, trolleys, and pedestrians . . . He was gauging their speeds, their angles, all the living forces of mass hurtling past'.[30]

Musil draws our attention to the erratic rhythms of modernity, a theme developed by cultural geographer Tim Edensor, in the wake of Henri Lefebvre's excursions into 'rhythmanalysis', differentiating between eurhythmy (orderly, regular rhythm) and arrhythmia (lack of rhythm). Modern mobility is a mélange of 'speeding up, stopping, slowing down, hurrying and waiting', sometimes by design but more often by the clash of urban rhythms.[31]

The rhythms and routines of movement are matched by the rhythms and routines of speech, although we should not entirely dismiss the value of 'clipped words' and 'potted expressions'. Yi-Fu Tuan differentiates between 'talk' and 'conversation'. Although the latter implies serious exchange and interaction, the former still has its place in maintaining community. So, Tomlinson suggests that mobile-phone chat can play a role equivalent to communal singing in church or pub or on the terraces at a sporting event. The words matter less than the act. 'Clipped words' at least acknowledge the existence of those we hurry past and imply, at the most basic level, that we belong together in this place.[32]

The vocabulary of hurry

Not only the language *of* hurry but the language *about* hurry merits our attention. The idea of hurry was present in human antiquity, as the translation from Lao Tzu at the head of this Introduction indicates. In positing hurry as incompatible with the natural world, Lao Tzu's aphorism exposes the modernity-before-modernity that characterized life in so-called 'pre-modern' cities. But the English word 'hurry' is of much the same vintage as the word 'modernity'. The *Oxford English Dictionary* (*OED*) traces 'hurry' back to the end of the sixteenth century, noting its equivalence to 'hurly' and hence, 'hurly-burly'. Hurry in its earliest and 'now obsolescent' meanings was about 'commotion' or 'mental agitation or disturbance; excitement; perturbation', 'excited, hasty, or impetuous motion; rush', but, by the eighteenth century, it had acquired its current meaning: 'action accelerated by some pressure of circumstances, excitement, or agitation; undue or immoderate haste; the condition of being obliged to act quickly through having little time; eagerness to get something done quickly'.[33]

As the *OED* notes, 'hurry' can be associated with 'haste' and 'rush', reminding us of the motto 'More haste, less speed', which, as Tomlinson asserts, 'warns against a certain type of ill-considered and counterproductive disposal of energy and effort rather than attacking the goal of speed itself'.[34] Being hasty, rushing into things without thinking, being told to 'hurry up', all usually lead to less speed, taking longer to do things properly.[35] Hurry becomes one human reaction to the processes of modernization – trying to adjust, to keep up with the speed of economic, political and technological change. It is rarely 'tactical' or 'transgressive', but merely a spatial practice moulded to the form of modern capitalist society.

'Rush hour' originated in the United States in the 1880s and 1890s, initially concerned with the crowds using public transport during what, less emotively, transport managers prefer to term 'peak hours'. The term migrated to Britain in the early twentieth century, although, as Richard Dennis notes in his chapter in this volume, the Royal Commission on London Traffic (1905) differentiated between 'slack hours' and the much more honest 'crush hours', the crush denoting both overcrowding of passengers on buses and congestion of buses on streets. But 'rush hour' was also used of situations where mobility was confined to the interior of buildings: rush hours in restaurants and hospital kitchens, for example.

Another word in the lexicon of hurry is 'lift'. Here we are thinking less of 'the lift' as the British word for 'elevator', although this is an important aid to speed, facilitating the development of ever taller buildings from the 1870s, and more of getting or giving a 'lift'. The *OED* ascribes first usage to Jonathan Swift (1712), who 'got a lift in a coach to town'. 'Lift' in Swift's sense meant literally to take one's feet off the ground in order to move. More curiously, it is an acknowledgement of the efficacy of mediated mobility. To give a lift is to construct an architecture of hurry, to get a lift, to acquiesce to that construction. A lift in Swift's time signified moving at the speed of a horse-drawn cart, dray or carriage, the greater the speed, the more inconvenient the mobility on unpaved roadways. By the early twentieth century, giving or getting lifts involved speeds regularly exceeding 40 mph on roadways designed to accommodate the pneumatic tyres of the automobilized vehicle doing the lifting. By mid century, air travel in the 'jet age' involved 'lifting off' to achieve airspeeds between 500 and 600 mph, the speed of the first Boeing 707 in the 1950s. Airlines announce the end of high-speed flights as 'wheels down', a certain indication of the end of one hurry and the start of another – to deplane and to wait, often impatiently, for luggage. Architectures of hurry must necessarily involve, not only defeating gravity and friction as well as time and space with 'lift', but also believing that to do so is bound up in the purpose of modernism, and essential to its explanation (as circular as such thinking seems).

Three words less obviously associated with 'hurry' are 'terminal', 'exchange' and 'nomad'. 'Terminal' is of interest here because of the way it has changed its meaning in recent years. In the nineteenth and most of the twentieth centuries, terminal signified the beginning- or end-point of a journey, a place of arrival or departure, a place where individual users were usually in a hurry, and where re-ordering – organizing queues for tickets or, more recently, security checks; differentiating between types of traveller and types of journey; providing distractions (shops, cafes) while travellers waited; separating the business of waiting from that of boarding or alighting – contributed to a reduction in anxiety, if not to any measurable speeding-up. 'Terminal' also became the point in an electrical device where a connection could be made, such as a contact with a battery or a wire forming part of a circuit, but it was still a critical point in a flow, now of electrical energy rather than people. Latterly, the terminal became the screen and keyboard that gave access to information and services through a computer network. As Tomlinson notes, terminals, which were once fixed locations, often with heavily symbolic architecture, are now portable and personal – mobile devices that allow us access as we hurry, but allow others to fix our constantly changing location.[36]

'Exchange' is a much older word, but, as a place of exchange (see Deryck Holdsworth's chapter in this volume), it is contemporaneous with hurry and modernity, entering the language in the late sixteenth century. In the nineteenth century, however, its locational usage broadened from being a place for the transaction of business; it also became a place for changing trains, and a place for connecting telephone subscribers. The former is especially intriguing. 'Liverpool Exchange' opened as a new terminus of the Lancashire & Yorkshire Railway in May 1850. Previously, the terminus had been nearly a mile farther out from Liverpool's city centre, and the name was chosen to emphasize the new station's proximity to the Liverpool Royal Exchange.[37] In the same week in 1850, 'Bradford Exchange' was opened, again by the Lancashire & Yorkshire Railway, in this case in collaboration with the Great Northern Railway. The station could hardly claim to be any more convenient for Bradford's wool exchange than its rival (Bradford Market Street, now Forster Square). Nor were local residents persuaded that it helped in the process of changing trains: there were regular demands in the local press for a 'Central Railway Station', urging haste while affordable vacant sites still existed near the centre of the fast-growing town. Nothing materialized, but the idea of 'exchange' as a place to change trains, or to change between rail and road as effortlessly as possible, was maintained when Bradford Exchange was rebuilt and, in 1983, renamed Bradford Interchange.[38] Other cities acquired 'exchange' stations through the late nineteenth and early twentieth centuries. Apart from Manchester (1884), where the new station could just about claim to be the nearest station to the Manchester Royal Exchange, they were in places that lacked prominent commercial exchange buildings. As a station name, 'Exchange' now signified purely the ability to transfer between trains expeditiously.

By the late nineteenth century, nomadism was associated, not only with transiency, which in turn equated to restlessness, but with a consequent absence of civilization. The English novelist George Gissing invoked nomadism as a modern disease:

> There can be no home without the sense of permanence, and without home there is no civilization – as England will discover when the greater part of her population have become flat-inhabiting nomads.[39]

E.M. Forster, only a few pages on from his allusion to the architecture of hurry, declared that:

> The feudal ownership of land did bring dignity, whereas the modern ownership of movables is reducing us again to a nomadic horde. We are reverting to the civilization of luggage, and historians of the future will note how the middle classes accreted possessions without taking root in the earth, and may find in this the secret of their imaginative poverty.[40]

T.S. Eliot alluded to 'barbarian nomads' encamped in 'mechanised caravans' threatening chaos on a culture that depended on stability and continuity. More recently, and less emotively, Raymond Williams identified 'nomad capitalism' as exploitative and destructive of community. All these views were expressions of what Cresswell refers to as 'sedentarist metaphysics' and all perceived mobility, and especially self-interested, individualistic mobility as antithetical to established community.[41]

There seemed to be two conceptions of nomadism: one as invasive, the other as circular, whether the circular mobility of the rich (Gissing's 'flat-inhabiting nomads' and Forster's 'nomadic horde', restlessly moving between one fashionable resort and another, hurrying to keep up with changing fashions), or that of the poor (constantly moving between one rented dwelling and another but, because of their dependence on support from extended family and friends, never moving far, a process encapsulated in fiction in Gissing's 'slum novel', *The Nether World*, and in social science by Charles Booth's observations of patterns of residential mobility in London's East End).[42] Both were intrinsic to modern urbanism. And both were deplored by contemporary commentators.

The connection of hurry to modernity was noted, more than half a century before Forster, by Matthew Arnold in his poem, 'The Scholar Gipsy' (1853).[43] Arnold treated 'hurry' as a 'disease' of modern life, harking back to a golden age:

> Before this strange disease of modern life,
> With its sick hurry, its divided aims . . .

Yet the subject of the poem was a poor Oxford student who becomes a 'gipsy'; in effect, he espouses nomadism and, of course, is envied by the poem's narrator for doing so. In this case, therefore, hurry was *not* about mobility and transiency per se – nobody could be more transient than the 'gipsy-poet' who is glimpsed here, there and everywhere: a shadowy, ephemeral presence. Rather, it was about a particular kind of modern hurry:

> For what wears out the life of mortal men?
> 'Tis that from change to change their being rolls;
> 'Tis that repeated shocks, again, again,
> Exhaust the energy of strongest souls . . .

Sick hurry was infectious. The 'scholar gipsy' was urged to:

> . . . fly our paths, our feverish contact fly!
> For strong the infection of our mental strife . . .

This is a foreshadowing of today's 'hurry sickness', defined by the popular magazine *Psychology Today* as:

> a behavior pattern characterized by continual rushing and anxiousness; an overwhelming and continual sense of urgency . . . a malaise in which a person feels chronically short of time, and so tends to perform every task faster and to get flustered when encountering any kind of delay.

The term can be attributed to cardiologists Meyer Friedman and R.H. Rosenman (1974).[44] These usages situate hurry in the day-to-day business of having too much to do, a wider view of hurry than the mobility-centred hurry on which we focus in this book. Nevertheless, we can apply Arnold's allusion to the infectiousness of 'feverish contact' to the mobility of traffic and crowds. How do you avoid accelerating on an arterial road when you have a hurrying truck pressing on your tail, urging you onwards? How do you avoid hurrying in a crowd where everybody else is committed to hurry – all 'dodg[ing] in and out like runners on a football field' as if 'trying to reach the bank to have a check cashed before three o'clock', as Richard Harding Davis wrote of late-Victorian New York's Broadway?[45] This is hurrying as a stampede, but hurrying *against* the crowd, or hurrying in a crowd of leisurely dawdlers – trying to walk purposefully along Oxford Street when everybody else is window-shopping – can be equally destructive of personal health. Richard Hornsey notes the attempts of the London 'Safety First' Council to 'retrain walkers to "Keep to the Left"' on sidewalks so that 'those nearest the kerb would always be facing the oncoming traffic and thus less likely to step into its path'.[46] This was part of a campaign to segregate pedestrians and motor vehicles (discussed further in David Rooney's chapter in this volume), but there have also been occasional proposals that fast- and slow-moving pedestrians should be confined to separate lanes. Most recently, a variety of world cities, including Washington, DC, Chongqing and Antwerp, have experimented with separate lanes for texting or mobile-phone use more generally, all born out of the frustration of more hurried walkers.[47]

So there are occasions when we condone hurry and prefer the language of hurry to that of speed. We rarely think of ourselves as 'speeding' up and down escalators, when we choose to walk, keeping to the left (in London) to overtake those who prefer to stand and keep to the right. Indeed, 'speeding' implies that we are going *too* fast, exceeding the speed limit. When, recently, London Underground managers attempted to stop commuters from walking up and down escalators in 'peak hours', claiming that more people could be accommodated if everybody stood rather than walked, they met with resistance from the hurriers who could not accept that their hurry was counterproductive or unnecessary.[48] There were other, more primitive, instincts at work: anxieties about safety and self-preservation (whether from fire, for those who still remember the 1987 King's Cross escalator fire; from the jostling that allows pickpocketing; or from terror, and therefore the desire to pass through underground public space as quickly as possible). Hurry can be a matter of us and them. When we hurry, we do so rationally and purposefully; when others hurry, they are taking risks and putting us at risk.

Speeding-up and slowing-down

Countering hurry and hurry-sickness have been recent attempts to slow down (such as the 'slow professor' movement, an early twenty-first-century emanation of the anti-modern in neo-liberalized universities across the world).[49] But few other than the well-off can afford slowness: most consumers of 'slow cooking' have either the time or the money to decline 'fast food', and organized advocacy of 'slow food' and 'slow cities' tends to be as much about celebrating the local and regional over the global as it is about challenging speed. The slow movement claims not to be anti-speed, but to favour doing everything at the 'right' speed.[50] A popular focus in historical literary and cultural studies of modern urban life has been the *flâneur* and, more recently, the possibility of the *flâneuse*. The essence of *flânerie* is a self-conscious denial of hurry. The dilemma of the flâneur/flâneuse, faced with the rush and bustle of the city street, was typified by Baudelaire's poet, 'crossing the boulevard in a great hurry [*sic*], splashing through the mud, in the midst of a moving chaos, with death galloping at me from every side'.[51] Modern traffic was incompatible with the studied mobility of the leisurely flâneur, whom it condemned to death. And yet, even the flâneur/flâneuse might hurry – an implicit lesson in Mackintosh and Norcliffe's historical geographical study of women, bicycling and 'flâneurie' (also Christina Dando in this volume).[52]

The flâneur/flâneuse occupied an ambiguous place in the social hierarchy, a city person with elite cultural taste but, especially in the figure of the bohemian artist, no money – whom George Gissing regarded as 'unclassed' or even 'born in exile'. For Gissing, urban mobility involved a great deal of walking (see Jason Finch in this volume). He never earned more than a modest clerk's income from his writing and, for the first decade of his career, made ends meet by coaching private pupils, generally in their own homes, usually a long way from his own lodgings. This might confirm the stereotype that the poor live in 'slow' cities, condemned by poverty to walking – or 'carrying the banner', the term Jack London used to describe London's East End homeless walkers-through-the night in 1903 – while the wealthy and educated occupy the 'fast lane', even playing 'fast and loose'.[53] But, as Gissing's Henry Ryecroft observes, there is also a sense in which, if 'time is money', it is also the case that 'money is time': 'What are we doing all our lives but purchasing, or trying to purchase, time?' (an idea elaborated with dystopian urgency in the 2011 Andrew Niccol film, *In Time*, where humans have been genetically engineered to die at the age 25, but time extensions can be purchased, meaning wealth equals time).[54] Money buys you the ability to be fast, but it also buys you the time to think, reflect, relax. In recent years, there have been counter-literatures on the realities of congestion, the absence of much objectively measurable acceleration in the speed of city traffic, but also, as noted above, on the desirability of slowness.

Our focus in this Introduction has been on the association of hurry with modern urbanism, but this is not to deny that hurry as a practice – if not as a word – was present in pre-modern societies, and that rural dwellers can sometimes feel obliged to hurry just as much as urban populations. Returning to the theme of nomadism, we may conceive it as intrinsic to modernization, and hurry as an ancient social and cultural necessity born of nomadism and following herds, hunting and gathering, locating shelter, protecting offspring and avoiding hazards. Early humans would have desired – an emotion probably connected to hurry – convenience, the easier acquisition of nutrition, shelter, tools, and the general methods of subsistence and survival. And if agriculture, then also urbanization, where ancient peoples hurried to construct settlements, create surpluses and their protection and the trade and militaries arising concomitantly. Of the Romans' consequential urban imperialism, Richard Sennett writes, 'Power needed stone' – including stone infrastructure.[55] The Romans appear to have understood hurry as part and parcel of their infrastructure impulse. Infrastructure substantiated Roman civic republicanism, which seems much like hurrying to produce built civility, and imperialism, to which architectures of hurry in support of militarism, war, conquest and wealth are essential. Thus, Roman civility hinged on mobility and its facilitation.[56] Either this implies a disconnection between mobility and modernity or it indicates that ancient Rome was, in many respects, the first modern city.[57]

Scope and structure of the argument

Architectures of Hurry divides into three themes: 'Modes of hurry', 'Local and global infrastructures' and 'Practices of mobility' – and an epilogue: 'Mobilizing hurrysome historical geographies'. 'Modes of hurry' introduces the reader to discussions of familiar urban mobilities: bicycles, buses, subways and cars in different parts of the world and in different eras. 'Local and global infrastructures' points its curiosity at urban geographies of hurry, from street surfaces and sidewalk technologies to office buildings and hotels. 'Practices of mobility' contains two enquiries into nineteenth- and early twentieth-century uses of mobility – as depicted in life-writing about various forms of urban (and rural) mobility. The volume concludes with the editors' reassessment of historic 'hurry' from the vantage of the twenty-first century.

Part I begins with Christina Dando, in Chapter 2, finding architectures of hurry in women's discovery of the bicycle's potential for speed – and liberation – in turn-of-the-twentieth-century America. Dando shows how a culture of 'scorching', or speeding on the bicycle, encouraged women to seek out and even create 'new cycling spaces'. In this process of locating and making cycling geographies, in the brief times of the day women had to escape workday modernity in search of leisure, women cyclists began 'speeding up to slow down', employing the swiftness of their bicycles to hurry respite.

Using the historical London bus as an exemplar, Richard Dennis, in Chapter 3, reveals how forerunners of the iconic red omnibus illustrate many of the components of an architecture of hurry. From their engineering, design, use of onboard communication, and advertising (of routes, times and stops), to passenger education in the use of public transit and public education in the use of streets now inhabited and governed by surface public transit, London buses represented in miniature the broader Western impulse to reorganize urban environments to accommodate the modernist preoccupation with hurry. This produced what Dennis calls a 'language and body-language of hurry on streets and in buses', intertwining gender and class, while imposing a new urban deontology of rights, risks and especially responsibilities on the users of streets in cities determined to suborn pedestrianism to automobility – all producing an imaginary of hurry.

Dhan Zunino Singh's Chapter 4 contemplates architectures of hurry in relationship to the infrastructural generation of urban 'velocity' in Buenos Aires, Argentina. The Argentine capital offers stories of two competing methods of accommodating public transportation at the turn of the twentieth century: the construction of surface and underground transport infrastructure. Recognising that 'Modernisation . . . implied a radical change of urban fabric', authorities in Buenos Aires undertook a programme of street and transportation improvements to connect the central city with the burgeoning suburbs. In the process, they produced an irony: 'improved architectures of hurry contributed to what we now call "traffic generation"'.

Glen Norcliffe and Boyang Gao examine, in Chapter 5, the hurried rise of transportation and its infrastructure in China, from the Qing dynasty to Mao, and then its twenty-first-century repercussions, including the threat to the bicycle, a metaphor of the hurry-slow impulse in China. China has become the world leader of public transit in its proliferating megalopolises – 16 metropolitan areas with populations exceeding 10 million and 22 cities over 5 million, according to the 2010 census.[58] But all this intensive city-building has occurred only in the last half-century, in the wake of Mao's preoccupation with infrastructure and Deng Xiaoping's 'Open Doors agenda' in 1978 'to catch [China] up to other industrialized countries'. Consequently, Norcliffe and Gao give us a China impatient to construct architectures of hurry, but the unintended consequence is the opposite of urban hurry: 'circulatory embolisms'. This, as China's millennials, 'the *bā líng hòu* (born in the 1980s) and the *jiǔ líng hòu* (born in the 1990s)', have begun to resist hurry through their embrace of the bicycle.

Part II, 'Local and global infrastructures', begins with Phillip Gordon Mackintosh's Chapter 6 on surface infrastructure in late-Victorian and Edwardian Toronto, Canada. Contrasting 'architectures of sluggishness' with incipient architectures of hurry at the turn of the twentieth century, Mackintosh depicts modern Toronto as frustrated by an infrastructure paradox: a city council, a city engineering department and Toronto's urban reformers in general who were determined to modernize the city's deficient infrastructure, but nevertheless deferred to the whims and financial control of a property-owning electorate that paid for – and thus influenced – the quality of surface infrastructure in the city. This they did through Toronto's local improvements by-law. Feeling 'besieged by a relentless and expensive modernism', Torontonians used the local improvements petition to fight what they believed were unnecessary road and sidewalk recommendations by city engineers eager to mobilize Toronto. The consequence was mile upon mile of inferior 'organic infrastructure', composed of wood, stone and gravel. The predilections of, and the remarkable tolerance for inconvenience by, property owners literally mired in mud and politics the city's modernist aspirations for an urban environment of hurry.

Broaching the development of architectures of hurry as they concerned 'promiscuous pedestrian crossing' on modern streets, David Rooney's Chapter 7 excavates the conception and construction of pedestrian guard rails in interwar London. In this, the London Metropolitan Police assistant traffic commissioner, Alker Tripp, figures centrally, particularly through his attempt to impose safety regulations on London's 'laissez-faire road landscape'. This revised geography of the street necessitated, in part, the segregation of pedestrian and vehicle traffic on the East India Dock Road in the populous, bustling and, crucially, 'cosmopolitan' East End neighbourhood of Poplar. The consequent suborning of individual liberty to a singular and automobilized understanding of the public good in the streets of Poplar was the consequence of a broader contemporaneous apprehension of the city, its people, and their behaviour, class and skin colour. Not surprisingly, this included the employment of racialization and segregation tropes that shaped, literally, the road infrastructure and its social controls.

Deryck Holdsworth's Chapter 8 explains the transformation of nineteenth-century mercantile exchanges and offices – what he calls 'new architectures of transactional space' – from stationary wood, brick and mortar buildings to architectures of hurry. The myriad trade activities within their walls, by vast numbers of employees busily generating trade and commodity and supply chains around the globe, the networking of indefatigable traders in nearby coffee-houses, the dedicated shipping technologies loading and transporting countless tonnes of goods in connecting ports, all create the sense of the exchange building as a mobile thing, its 'large colony of offices' undulating with 'hundreds of worker bees or ants'. Indeed, Holdsworth gives us a modern urban irony: the stationary exchange building, but as capitalist homunculus, armies of workers and technologies animating – hurrying – its apparently static visage. And then, of course, there is the mutability of these buildings themselves, as businesses expanded, moved from, and/or replaced, their existing premises in the steadfast pursuit of investment and capital. Thus, the 'idea of geographical mobility of business', Holdsworth writes, 'needs to be expanded to include these stationary buildings'.

In Chapter 9, Sherry Olson and Mary Anne Poutanen find architectures of hurry in an unlikely source: Victorian and Edwardian hospitality. Here, they illustrate Montreal's early hoteliers' and guests' imagining of hospitality in a burgeoning context of hurry. Contending the hospitality industry 'drew upon the entire city', Olson and Poutanen link Montreal's flows of people, horses, goods, credit, knowhow and news with an emergent and reciprocal interest in supplying hospitality services – inns, hotels and boarding houses – to an increasing world of hurriers dependent upon strangers to take them in. It was also an environment fraught with 'delays, uncertainties, weather, predation and violence, inefficiencies in the connections between modes of transport (baggage lost or stolen), and failures of communication', all demanding architectures of hurry to right. Hospitality was structurally a world of masculine organization and gender division that included the 'front stage' but especially the 'back stage roles of women' and a persistent emphasis on male-dominated understandings of family in the accommodation of guests in the construction of home away from home.

Jason Finch's Chapter 10 on pedestrianism in George Gissing's fictional memoir, *The Private Papers of Henry Ryecroft* (1903) leads off Part III, 'Practices of mobility'. Finch pulls us into Gissing's ambulatory London, diverted by what he calls the '"frenetic" mobilities of the Victorian city', in which walking was the primary mode of urban mobility. Contrasting the successfully retired Henry Ryecroft with his younger, struggling writer-self, Finch locates in Gissing's life-writing a London posing as architecture of hurry – only a generation earlier, as Gissing remembers his young adult life (as tutor and writer) as he circulated 'in a great hurry' through the famous city's streets. It is these urbanized and capitalized 'architectures of London' that provide Gissing, and us, a malaise-making insight about hurry, mobility and modernity: alas, 'it is just as noble to charge around a dirty and rapacious city grabbing money, as it is to contemplate a beautiful fire in rural seclusion'.

Colin Pooley and Marilyn Pooley associate hurry with the life-writings of British transport users after 1840, exploring sped-up modernity at the level of individual transport users, the focus of Chapter 11. Pooley and Pooley's use of diaries helps them analyse the perceptions of Britons who experienced dramatic changes in, not only mass transportation, but also the public engagement with urban and rural space – mediated by Victorian horse-powered, steam and electric public transport, human-powered individual transport, and twentieth-century petroleum-powered automobility and air travel. They discover that hurry, although important, was not a central preoccupation of convenience-conscious travellers: '[f]or a wide range of reasons . . . older and slower forms of transport flourished alongside new technologies'. Travel, which had become routine for the life-writers studied, defined everyday life. Yet this is probably why hurry mattered, especially kinaesthetically, because it could make ordinary travel fun – or convenient, when quotidian matters interfered with routines and schedules.

Finally, there are many 'architectures of hurry' that we do not discuss in this volume. The only literal architecture on which we focus is that of subway stations, hotels and commercial exchanges. We extend 'architecture' to encompass the infrastructure of rails, streets and street furniture, and the vehicles that use them. We are also interested in time as an architecture, framing personal mobility on roads and railways and work practices in the arenas of commerce and hospitality. But we have not explored 'architectures of hurry' in the domestic sphere, either in the building and layout of homes or in the hurried routines of domestic life.[59] There has been copious recent research (too copious to enumerate here) on spaces designed either to speed up or slow down acts of consumption: bars without seats to encourage rapid 'perpendicular drinking' compared with lounge bars and restaurants; music halls with 'twice nightly' performances, artistes instructed to keep to schedule and script, and tables and comfortable chairs removed from the auditorium; and also on the changing 'architecture' of artistic production: one-volume paperback novels to read on the train, compared with mid Victorian 'three-deckers' borrowed from a circulating library; cinema 'shorts'; musical items brief enough to fit on to phonograph cylinders or early gramophone records; but also public parks with meandering walks suited to ambling, strolling or sauntering. But these sites, too, might benefit from an explicit focus on hurry in their design, construction, operation and patronage.

For the most fundamental 'architecture' of hurry, however, we return to Forster's account of a city in constant flux and the increasingly hurried processes of urban redevelopment as blocks of flats replaced houses, and as those flats, in turn, after a few years 'might be pulled down, and new buildings, of a vastness at present unimaginable, might arise where they had fallen'. Forster exaggerated, of course. But in today's Chinese cities, it is estimated that the lifespan of new buildings is only 25–30 years, and some buildings are demolished when they are only 10 years old.[60] This is 'creative destruction' in extreme proportions, prompted in part by poor-quality, sometimes illegal, construction, but mostly by the logic of advanced capitalism, as time–space compression and the continuing hurry of migrants to big cities create new configurations of spatial relations that, in turn, prompt the revaluation of what land is worth and what should occupy it. Historical geographers – of London, Toronto, Montreal, New York and Buenos Aires, as much as of Shanghai or Beijing – could profitably (*sic*) pay as much attention in the future to the economics underlying hurry as to its cultural dimensions. The whole built environment of modern cities was designed not only for hurry but by hurry.

Notes

1 Raymond Williams, *Keywords: A Vocabulary of Culture and Society* (New York: Oxford University Press, 1983), 76.

2 Arnold Bennett, 'This Modern Living: A Study of the Difficulties Which a Complex Civilization Presents for the Women of Today', *Vanity Fair* 32: 4 (September 1929); for a wider discussion of Bennett's writing on women and domestic efficiency, see John Nash, 'Arnold Bennett and Home Management', *English Literature in Transition, 1880–1920* 59 (2016): 210–33.

3 The tally could be extended if Bennett's, or his publishers', penchant for packaging groups of earlier works under new titles, is acknowledged: *Clayhanger* (1910), *Hilda Lessways* (1911) and *These Twain* (1916) repackaged as *The Clayhanger Family* (1925); *How to Live on 24 Hours a Day* (1910), *Mental Efficiency* (1911) and *Self and Self-Management* (1918) repackaged as *How to Live* (1925).

4 Marshall Berman, *All That Is Solid Melts into Air: The Experience of Modernity* (New York: Simon & Schuster, 1982); David Harvey, *The Urbanization of Capital* (Oxford: Blackwell, 1985); David Harvey, *Paris, Capital of Modernity* (New York and London: Routledge, 2003); David Harvey, *The Enigma of Capital: And the Crises of Capitalism* (London: Profile, 2010).

5 Michel de Certeau, *The Practice of Everyday Life* (Berkeley: University of California Press, 1984); Tim Edensor, 'Rhythm and Arrhythmia', in *The Routledge Handbook of Mobilities*, eds Peter Adey, David Bissell, Kevin Hannam, Peter Merriman and Mimi Sheller (New York: Routledge, 2014), 167.

6 Select literature includes: Jeffrey Schnapp, 'Crash (Speed as Engine of Individuation)', *Modernism/Modernity* 6 (1999): 1–49; Mimi Sheller and John Urry, 'The City and the Car', *International Journal of Urban and Regional Research* 24 (2000): 737–57; Tim Cresswell, *On the Move: Mobility in the Modern Western World* (New York: Routledge, 2006); John Tomlinson, *The Culture of Speed: The Coming of Immediacy* (London: Sage, 2007); John Urry, *Mobilities* (Cambridge: Polity, 2007); Enda Duffy, *The Speed Handbook: Velocity, Pleasure, Modernism* (Durham, NC: Duke University Press, 2009); Hartmut Rosa and William E. Scheuerman, eds, *High-Speed Society: Social Acceleration, Power and Modernity* (University Park, PA: Pennsylvania State University Press, 2009); Peter Merriman, *Mobility, Space and Culture* (London: Routledge, 2012); James Faulconbridge and Alison Hui, 'Traces of a Mobile Field: Ten Years of Mobilities Research', *Mobilities* 11 (2016): 1–14; Adey et al., *Routledge Handbook of Mobilities*; Colin Divall, ed., *Cultural Histories of Sociabilities, Spaces and Mobilities* (Abingdon: Routledge, 2016).

7 For example: Derek Gregory, 'The Friction of Distance? Information Circulation and the Mails in Early Nineteenth-Century England', *Journal of Historical Geography* 13 (1987): 130–54; Tanu Prya Uteng and Tim Cresswell, eds, *Gendered Mobilities* (Abingdon: Routledge, 2008); David Harvey, 'Time–Space Compression and the Postmodern Condition', in *Modernity: Critical Concepts—Volume IV: After Modernity*, ed. Malcolm Waters (New York and London: Routledge, 1999), 98–118.

8 Stephen Kern, *The Culture of Time and Space, 1880–1918* (Cambridge, MA: Harvard University Press, 1983); Schnapp, 'Crash'; Tomlinson, *The Culture of Speed*; Duffy, *The Speed Handbook*.

9 Sheller and Urry, 'The City and the Car'.

10 Cresswell, *On the Move*, 2–4.

11 Urry, *Mobilities*, 47, 13.

12 Ibid., 91, 14.

13 George Revill, 'Histories' in Adey et al., *Routledge Handbook of Mobilities*, 506–16.

14 Guy Bergstrom, 'Understanding the News Cycle at a Newspaper', *the balance*, 8 August 2016 (accessed 7 June 2017), www.thebalance.com/understanding-the-news-cycle-2295933.

15 See especially David Harvey, 'Between Space and Time: Reflections on the Geographical Imagination', *Annals of the Association of American Geographers* 80 (1990): 418–34. The annihilation of time and space has been traced by Michael Freeman, *Railways and the Victorian Imagination* (New Haven, CT: Yale University Press, 1999), 78, to the *Liverpool Railway Companion* (1833); see also Wolfgang Schivelbusch, *The Railway Journey* (Leamington Spa: Berg, 1986), 33–44. In London, successive covers of the *District Railway Map of London* through the 1880s and 1890s featured a train emerging from a tunnel mouth emblazoned with the slogan, 'Time is Money'.

16 David Harvey, *Enigma of Capital*, 42. On the exploitation of milliseconds in twenty-first-century stock trading over fibre-optic cable, see Michael Lewis, *Flash Boys: A Wall Street Revolt* (New York: Norton, 2014).

17 See, for example, *Joshua Johnson's Letterbook, 1771–1774: Letters from a Merchant in London to his Partners in Maryland*, ed. Jacob M. Price (London: London Record Society, 1979), *British History Online* (accessed 12 September 2017) www.british-history.ac.uk/london-record-soc/vol15.

18 George Beard, *American Nervousness* (New York: Putnam, 1881), cited in Cresswell, *On the Move*, 17–18; Mustafa Dikeç and Carlos Lopez Galviz, '"The Modern Atlas": Compressed Air and Cities c.1850–1930', *Journal of Historical Geography* 53 (2016): 11–27; Alexis McCrossen, *Marking Modern Times: A History of Clocks, Watches, and*

Other Timekeepers in American Life (Chicago: University of Chicago Press, 2013), 4; also Michael O'Malley, *Keeping Watch: A History of American Time* (New York: Viking, 1990).

19 Tomlinson, *Culture of Speed*, 9; Frances Willard, *Wheel within a Wheel: How I Learned to Ride the Bicycle* (London: Hutchison, 1895), 50.

20 Jeffrey Schnapp, 'Crash'.

21 Urry, *Mobilities*, 99.

22 Georg Simmel, *The Philosophy of Money*, trans. David Frisby (London: Routledge, 2011[1900]), esp. Chapter 6, 'The Style of Life'.

23 Georg Simmel, 'The Metropolis and Mental Life', in *Classic Essays on the Culture of Cities*, ed. Richard Sennett (Englewood Cliffs, NJ: Prentice Hall, 1969), 47–60.

24 William Dean Howells, *A Hazard of New Fortunes* (Oxford: Oxford University Press, 1990[1890]), 63, 50; Howells, *Through the Eye of the Needle* (New York: Harper, 1907), 10–11, cited in Kern, *Culture of Time and Space*, 126; Richard Dennis, *Cities in Modernity: Representations and Productions of Metropolitan Space 1840–1930* (New York: Cambridge University Press, 2008), 340–3.

25 E.M. Forster, *Howards End* (London: Penguin, 2000[1910]), 93.

26 Ibid., 92.

27 Ibid., 93.

28 Robert Musil, *The Man Without Qualities* (London: Picador, 2017[1930]), 3.

29 Ibid., 4–5.

30 Ibid., 6.

31 Edensor, 'Rhythm and Arrhythmia'; J. Germann Molz, 'Representing Pace in Tourism Mobilities: Staycations, Slow Travel and The Amazing Race', *Journal of Tourism and Cultural Change* 7 (2009), 272.

32 Yi-Fu Tuan, *Cosmos and Hearth: A Cosmopolite's Viewpoint* (Minneapolis: University of Minnesota Press, 1996); Tomlinson, *Culture of Speed*, 119–20.

33 This and subsequent references to the *Oxford English Dictionary* are from OED Online, June 2017, Oxford University Press (accessed 13 September 2017), www.oed.com.libproxy.ucl.ac.uk/.

34 Tomlinson, *Culture of Speed*, 4.

35 But note again the ambiguity of 'haste', which in certain circumstances (see the quotation by Wesley at the beginning of this chapter) can denote purposeful speed, the opposite of idleness.

36 Tomlinson, *Culture of Speed*, 102–5.

37 'Exchange New Station, Liverpool', *Manchester Times*, 15 May 1850.

38 David Joy, *A Regional History of the Railways of Great Britain: Volume VIII South and West Yorkshire* (Newton Abbot: David & Charles, 1975), 75–83; 'Central Railway Station', *Bradford Observer*, 30 October 1856: 7; 'Central Railway Station for Bradford', *Bradford Observer*, 1 March 1860: 7.

39 George Gissing, *The Private Papers of Henry Ryecroft* (London: Constable, 1921[1903]), 'Winter: XV', 237.

40 Forster, *Howards End*, 128.

41 T.S. Eliot, *Notes towards the Definition of Culture* (London: Faber & Faber, 1948), 108; Raymond Williams, 'Mining the Meaning: Keywords in the Miners Strike', in *Resources of Hope*, ed. Raymond Williams (London: Verso, 1989), 124, both cited in Cresswell, *On the Move*, 32–6.

42 George Gissing, *The Nether World* (London: Dent, 1973[1889]): there are references to residential mobility throughout, but see esp. Chapter XV, 130; Charles Booth, *Life and Labour of the People, Volume I* (London: Williams & Norgate, 1889), 27.

43 Matthew Arnold, 'The Scholar Gipsy' (accessed 13 September 2017), www.poetryfoundation.org/poems/43606/the-scholar-gipsy. Thanks to Jeremy Tambling for alerting us to Arnold's use of 'hurry'. We recognize 'gypsy' is now a term of marginalization, but keep Arnold's usage for consistency.

44 Rosemary Sword and Philip Zimbardo, 'Hurry Sickness' (accessed 13 September 2017), www.psychologytoday.com/blog/the-time-cure/201302/hurry-sickness; Meyer Friedman and Ray Rosenman, *Type A Behavior and Your Heart* (New York: Knopf, 1974).

45 Richard Harding Davis, 'Broadway', *Scribner's Magazine* 9 (5 May 1891): 585–605.
46 Richard Hornsey, '"He who Thinks, in Modern Traffic, is Lost": Automation and the Pedestrian Rhythms of Interwar London', in *Geographies of Rhythm: Nature, Place, Mobilities and Bodies*, ed. Tim Edensor (Aldershot: Ashgate, 2010), 99–112.
47 'A fast lane for pedestrians?' *Daily Mail*, 17 July 2014 (accessed 2 October 2017), www.dailymail.co.uk/news/article-2696568/TV-puts-fast-slow-lanes-DC-sidewalk.html; 'Chinese city opens "phone lane" for texting pedestrians', theguardian.com, 15 September, 2014 (accessed 2 October 2017) www.theguardian.com/world/shortcuts/2014/sep/15/china-mobile-phone-lane-distracted-walking-pedestrians; 'Antwerp now has "text lanes" for pedestrians who are glued to their mobile phones', *Independent*, 13 June 2015 (accessed 2 October 2017) www.independent.co.uk/news/world/europe/antwerp-now-has-text-lanes-for-pedestrians-who-are-glued-to-their-mobile-phones-10317952.html.
48 'Why Londoners won't put up with Holborn station's new standing only escalator policy', www.standard.co.uk/lifestyle/london-life/why-londoners-wont-put-up-with-holborn-stations-new-standing-only-escalator-policy-a3123651.html (accessed 13 September 2017).
49 Alan Latham and Derek McCormack, 'Speed and Slowness', in *The Sage Companion to the City*, eds Tim Hall, Phil Hubbard and John Rennie Short (London: Sage, 2008), 301–17; Maggie Berg and Barbara Seeber, *The Slow Professor: Challenging the Culture of Speed in the Academy* (Toronto: University of Toronto Press, 2016).
50 Tomlinson, *Culture of Speed*, 148.
51 Charles Baudelaire, 'Loss of Halo', in *Paris Spleen*, trans. Louise Varése (New York: New Directions, 1970[1869]), 94.
52 Mackintosh and Norcliffe coined the term *flâneurie* to represent the presence of both male and female *flâneurs* at the turn of the twentieth century: Phillip Gordon Mackintosh and Glen Norcliffe, 'Flâneurie on Bicycles: Acquiescence to Women in Public in the 1890s', *Canadian Geographer* 50 (2006), 17–37.
53 Jack London, *The People of the Abyss* (London: Macmillan, 1903), 113.
54 Gissing, *Ryecroft*, 'Winter: XXIV', 262.
55 Richard Sennett, *Flesh and Stone: The Body and the City in Western Civilization* (New York: Norton, 1994), 89.
56 Giuseppe De Luca, 'Infrastructure Financing in Medieval Europe: On and Beyond "Roman Ways"', in *Infrastructure Finance in Europe: Insights Into the History of Water, Transport, and Telecommunications*, eds Youssef Cassis, Giuseppe De Luca and Massimo Florio (Oxford: Oxford University Press, 2016), 42.
57 Mary Beard, *SPQR: A History of Ancient Rome* (London: Profile Books, 2015).
58 'List of Cities in China by Population and Built-up Area' (accessed 7 June 2017), https://en.wikipedia.org/wiki/List_of_cities_in_China_by_population_and_built-up_area; see also, 'More than 100 Chinese Cities now above 1 Million People' (accessed 5 September 2017), www.theguardian.com/cities/2017/mar/20/china-100-cities-populations-bigger-liverpool.
59 But see Richard Dennis, 'The Architecture of Hurry', in *Cityscapes in History: Creating the Urban Experience*, eds Katrina Gulliver and Heléna Tóth (Farnham: Ashgate, 2014), esp. 117–24, where Dennis devotes space to the apartment/flat as an 'architecture of hurry' in late nineteenth- and early twentieth-century London and Toronto.
60 Forster, *Howards End*, 40; Qian Yanfeng, '"Most homes" to be demolished in 20 years', *China Daily* (accessed 2 October 2017), www.chinadaily.com.cn/china/2010-08/07/content_11113982.htm; Wade Shepard, '"Half the houses will be demolished within 20 years": On the disposable cities of China', *City Metric*, 21 October 2015 (accessed 2 October 2017), www.citymetric.com/skylines/half-houses-will-be-demolished-within-20-years-disposable-cities-china-1470.

Part I

Modes of hurry

2 'She scorches now and then'

American women and the construction of 1890s cycling

Christina E. Dando

She has a fair and lovely face,
  A face that wins men;
She rides a bicycle with grace
  And scorches now and then.

She scorches now and then, but in
  No crowded thoroughfare;
In country ways she takes her spin
  Where travelers are rare.

And thus to woman, man or child
  No danger can come nigh
From her, for she's of temper mild
  And wouldn't hurt a fly.[1]

In a few lines, the writer of 'Nellie On Her Bike' conveys that there are spaces where scorching is appropriate and others where it is not. 'Scorching' was to cycle fast and aggressively, a practice largely associated with young men. As women ventured out and began to pick up speed, American society had specific ideas about appropriate spaces for women and their behavior in these spaces. Cycling introduced many American women to the pleasure of flying along on a beautiful day as society warned women about the dangers of uncontrolled swiftness.

This tension between fast riding and controlled, stately, appropriate riding, and between the fast pace of modern life and the new concept of having slower leisure time, is central to the bicycle's architecture of hurry. Besides the social construction of cycling and speed, this architecture also includes the bicycle's material goods, as well as the physical infrastructure needed for cycling. I am interested in the practices of cycling, the ways such practices impacted the landscape and, at times, resulted in the creation of new cycling spaces, in what Peter Soppelsa describes as the co-construction or coproduction of technologies, practices, and geographies.[2] I will begin with a brief history of cycling in America before considering how cycling was socially constructed for women. I will then consider the ways in which the American landscape was transformed in order to facilitate women's hurrying. Gender, cycling, and landscape were bound together in new ways in the 1890s, resulting in transformations, not only in American women's lives, but also in the landscape itself.

'More bicycle mad than ever': the rapid diffusion of bicycling

> This year the United States is more bicycle mad than ever . . . The bicycle has been discerned to be the most marketable commodity of the hour, and every manufacturer whose plant was adaptable enough to include bicycle construction seems to have adjusted himself and his factory to the work of reaping the most inviting harvest that offers. The original bicycle makers make many more bicycles this year than ever; sewing-machine companies make them; arms manufacturers make them; so do machinists who are out of a job; men clever with tools who want to start in business, and everyone else who can. People who cannot make bicycles and are not otherwise occupied, busy themselves in the sale of them.[3]

For most Americans towards the end of the nineteenth century, personal mobility was largely limited to walking, horseback riding, horse-drawn buggies and wagons, and occasionally a train ride. Leisure time was largely limited and expected to be filled with meaningful activities. The introduction of the ordinary bicycle in 1870 offered for men and boys a new leisure time activity that promoted physical outdoor activity. The ordinary featured a large front wheel and small back wheel, with the cyclist precariously perched atop the larger wheel. Given women's dress of the age, with tight corsets and long, voluminous skirts and petticoats, few women attempted the ordinary bicycle. But many men and boys adopted the ordinary and took to the road.

As American men took to the bicycle, it quickly edged into American popular culture. The latest technological advancements were reported in *Scientific American* and in cycling magazines. With the development of the safety bicycle in the late 1880s, with its two equal-sized wheels, more men and women took up cycling. To meet the demand for cycles, manufacturers developed new mass-production techniques, eventually producing 1.2 million bicycles annually.[4] More and more producers entered the market, aggressively marketing their products: the latest in frames and saddles or seats (gender specific), tires, skirt guards, fenders, cyclometers, cycling clothes and shoes, and book and guides.[5] Magazines and newspapers featured articles and short stories on cycling and advertisements for a wide variety of cycling products. A cycling article from 1896 observed: 'The growth of wheeling became rapid, dramatic, almost sensational . . . . The manufacture of the slender steel machine has become a colossal industry'.[6] While the bicycle facilitated the speed of modern life, modern life hurried to profit from the bicycle fad.

As its popularity grew, prices dropped, and used bicycles were increasingly available, making it accessible to a broader segment of the population. By 1896, the majority of Americans could afford to own or rent a bicycle, travel on cycle paths and roads equally, and enjoy the pleasures of cycling on a lovely day.[7]

As we consider the greater mobility made possible by the bicycle, we must recognize that mobility is impacted, not only by social class, but also by gender and race.[8] As Tim Cresswell and Tanu Priya Uteng write: 'How people move (where, how fast, how often etc.) is demonstrably gendered and continues to reproduce gendered power hierarchies'.[9] Bicycling from the start was a gendered practice. Even when technological advances made it possible, women's cycling was socially constructed as different from men's. Bicycle illustrations and advertisements promoted, not only their product, but also a lifestyle, particularly for women (see Figures 2.1, 2.2 and 2.3).

'Her sweetest refuge': women and cycling

> Leaving her desk, her sewing-table, her counter, or school-room, this worker knows, with unerring instinct, where she will soonest find the quickest reaction from her mental and bodily fatigue. Her bicycle is her first thought, her sweetest refuge. Once in her saddle, the world of petty cares runs behind her like the road she travels. It is no use on your wheel to ponder over vexing questions and the irritating worries of the day. While you roll rapidly along, a brand new set of nerves and interests is quickened . . . . One's voluntary inclination is for speed, and speed means accelerated circulation and a sense of mental buoyancy that purifies the brain of its ill humors.[10]

American women took slowly to bicycling, limited initially by the available technology and its costs. Only a handful of women took on the ordinary bicycle. American magazines, newspapers, and books valorized men on the ordinary, associating the high-wheel bicycle with athletic men in good physical condition.[11]

The tricycle offered some women their first taste of mobility. Tricycles were first used in the 1870s, with the tricycling fad peaking in the 1880s.[12] With a tricycle, women could appropriately engage in bicycling, in a modest seated position and in proper dress for the age. By 1888, tricycling by women was widely practiced: 'there is now hardly a town of any importance, where good roads exist, that does not boast one or more lady rider; and in the cities there are scores'.[13] The League of American Wheelmen (LAW) magazine, *The Wheelman*, included articles and stories on tricycling, such as the romance 'Love on Wheels', featuring a man on an ordinary and a woman on a tricycle (1882).[14] But the tricycle's size, weight, and price limited its audience, restricting its adoption. In 1888, Julius Wilcox wryly observed the gendered unfairness of cycling with women limited to 'a clumsy wheelbarrow of a tricycle' – 'offering a woman a stone to eat while men have a soft biscuit'.[15] In particular, the weight of the tricycle limited its speed. Writing in 1894, Elizabeth Robins Pennell advocated for the safety, commenting that '[t]he bicyclist can scorch in triumph along the tiniest footpath, while the tricyclist trudges on foot, pushing her three wheels through mud or sand'.[16]

A change was already afoot. Ladies' safety bicycles were produced by a number of companies, beginning in the late 1880s, featuring two equal-sized wheels, new pneumatic tires for a smoother ride, and a lowered top tube to allow cycling in skirts, resulting in a safer and easier ride.[17] These technological advances led to wider adoption by women: 'By 1896, every third bicycle that was ordered was an open-frame women's model'.[18]

The new safety presented societal challenges. 'Appropriate' women's dress of the late nineteenth century restricted mobility. Cycling companies such as Columbia promoted women's dress designed for cycling, involving a less restrictive corset ('bicycling corset') and either slightly shorter skirts or a divided skirt.[19] Although early feminists had been calling for dress reform for decades, the bicycle offered a sensible reason for reform: 'skirts, while they have not hindered women from climbing to the topmost branches of the higher education, may prove fatal in down-hill coasting'.[20] Despite the practical rationale, the debates over appropriate cycling costumes for women were extensive. What to wear cycling was a major consideration for women. Books and articles on women cycling spent substantial space discussing possible outfits that met the needs of cycling while maintaining decorum.[21] A short story published in *The Woman's Journal* presents a different argument about dress, with the story's heroine arguing that, 'just because of that "dainty waist", I have to stop every few minutes and rest, I can't make any speed, and must get off and lead my wheel up every slight incline'.[22] Her beau admires her corseted figure and does not want her to wear a cycling outfit. After he nearly loses her to another man, he recognizes that she looks dashing in her cycling outfit and agrees to trust her judgment. Although it is not stated, this also suggests she can now cycle more efficiently – and faster.

Books, articles, and advertisements framed bicycling as appropriate for women: they could ride bicycles 'with grace and even modesty', compromising neither femininity nor family life.[23] Maria Ward's *Bicycling for Ladies* (1896) is an example of a women's bicycle manual, covering everything from its basic mechanism, how to mount and ride, what to wear, and how to repair, with illustrations.[24] An article on 'Hints for Cycling Women' in *The Woman's Journal* dictates how to cycle:

> In cycling the elegant woman does not coast, neither does she race. Rapidity of movement she considers neither conducive to grace nor as evincing good style. On the contrary, she sits erect, with elbows well in, gliding along slowly, and with so little body motion that loss of dignity is not thought of in her connection. She does not wear her skirt so short as to attract attention when she dismounts. In fact, in everything connected with the wheel her movements are so quiet and unobtrusive as to excite the admiration of the onlooker instead of the derision so frequently accorded. 'Repose is always elegance', and rapidity on the wheel is quite the reverse.[25]

Elegance was the image a cycling woman was to cultivate, which involved not only dress but also speed and behavior. Illustrations in texts and advertisements visually demonstrated how a lady cycled, with an appropriate costume and behavior (Figure 2.1). Women were a specific audience for advertisers' efforts: 'In particular, women were targeted, not as homemakers, but as active, independent

Figure 2.1 Advertisement featuring a woman cycling through the American landscape. Columbia bicycles were produced by the Pope Manufacturing Company, owned by Albert Pope. Pope was a passionate advocate for cycling and one of the founders of the League of American Wheelmen.

Source: *Scientific American* 78 (April 16, 1898): 256.

people who enjoyed recreational pursuits on an equal footing with men'.[26] With the right cycle and outfit, a woman could embody the advertisements: 'a holistic image of a beautiful rider *and* beautiful bicycle'.[27] Through various mediums, women were being encouraged to cycle while simultaneously being directed how to cycle and where to cycle. While bicycling gave women greater freedom, at the same time, they were to cycle in a lady-like manner and in appropriate settings.

And why were women to cycle? It was offered as a solution/cure to the problems of modern life:

> The great middle class, which makes the core of a nation, lived more and more in towns and cities, and chained itself more closely indoors. Each decade saw the growth of new morbid tendencies in its womankind, which went far toward proving the black prophecies of the croakers. It was the spinning silver wheels which at last whirled women into the open air, giving them strength, confidence, and a realization that to feel the pulse bounding with enjoyment is in itself a worthy end.[28]

Cycling offered an opportunity to recuperate from the 'anxieties of modern life', always equated with cities, through getting away to nature in the rural countryside, engaging in exercise, and, if nothing else, having a change of landscape.[29] Those 'spinning silver wheels . . . whirled women' from cities to the countryside, speeding them from their 'chains' of obligations to the 'open air', free as a bird (Figures 2.1 and 2.3). As Bisland writes in 1896: 'One's voluntary inclination is for speed, and speed means accelerated circulation and a sense of mental buoyancy that purifies the brain of its ill humors'.[30] If women were to be refreshed, there seemed to be a certain amount of speed involved in their leisure time, a dichotomy between the concept of leisure (slowing down) and speeding up, as cyclists were inevitably to do.

Although it was initially framed as a leisure activity, Americans soon learnt the practical use of bicycling. During a street car strike in Philadelphia, women used bicycles as an alternative form of transportation:

> Clerks in stores, typewriters, and the whole great army of employed women rode their wheels to business; women who came to buy left bicycles in the check rooms of the great shops. Never before had the Quaker City realized what a useful servant the wheel can be.[31]

Their use in times of necessity led to their usage on a more regular basis: 'The armies of women clerks in Chicago and Washington who go by wheel to business, show that exercise within bounds need not impair the spick-and-spandy neatness that marks the bread-winning American girl'.[32] And, in New Orleans, '[t]eachers rode bicycles to school to save carfare'.[33] Given that most working women were expected to begin work at a specified time, we have to imagine that there were moments when the bicycle enabled them to hurry to work on time.

As more and more individuals took up cycling, in cities already crowded with traffic, there came the need for laws to regulate their use and their behavior.[34] Early laws prohibited cycling on sidewalks. Soon came other qualifications: 'New Jersey was the first state to require bicycles to be lighted at night and also be equipped with a bell to give a warning signal'.[35] Massachusetts, Michigan, and New Hampshire required cyclists to keep on the right side of the road except when passing, leading to the custom of passing on the left. Cycling guides and handbooks, as well as LAW road books, included sections on the 'Rights of Cyclists' as well as 'Rules of the Road'.[36] Speed limits were introduced and 'set at ten miles per hour on the highways and eight miles in the public parks, but these laws were apparently frequently violated and the phrase "fined for scorching" became a familiar one'.[37]

There were, however, times when it was viewed as acceptable to achieve scorching speeds, such as to keep up with men who cycled faster. A 'Miss Bacon' explains that she 'first began to ride with my father and brothers, but it was not until the introduction of the pneumatic "safety" that I found myself equally mounted with them and able to go at their pace'.[38] Note that it is 'at their pace': with proper equipment Miss Bacon could keep the pace they set. Mary Sargent Hopkins, in an essay on bicycling for a women's magazine, explains that there are two kinds of woman rider: 'those who ride for speed only, and those who ride for health and pleasure'.[39] She goes on to explain that some of the speed riders:

> ride fast from necessity, not choice, with husbands and brothers, and they are obliged to hustle 'willy-nilly' to keep up. They have a pride in this, and their husbands and brothers have a pride in them. Still I do not think this speed is entirely voluntary, for I never heard a woman hurry her husband or complain of his lagging.

Hopkins appears to tolerate women riding fast to keep up with their male companions, seeing this more as 'involuntary' speeding, a matter of 'necessity', than one that women pursued.

Besides keeping up with companions, another acceptable instance for women to speed or scorch was to ride for assistance. In 1886, *The Cycle* magazine published a 'true' account of an emergency that had a woman riding her tricycle late one night to bring a doctor to help an elderly woman.[40] In another case, a Miss Alice York followed a thief:

> who had stolen a watch and pocketbook from her house. She chased him for nearly a mile, and compelled him, at the point of a revolver, to give up the booty. The watch was valued at $150 and the pocketbook contained $50.[41]

More common were fictional stories published in women's magazines where women leap to the aid of injured men and ride for help, such as 'The Old Man and the New Woman', 'Farmer Green's Conversion', 'Rosalind A-wheel'.[42] In such stories, the need for speed is a matter of life and death (and often ends in romance). But note that these stories are from women's magazines, written for a feminine audience and encouraging women to take up cycling, demonstrating that not only could cycling be a pleasant form of exercise, it had practical usages. Although a supposed 'proper' woman would never scorch, in case of emergencies she could achieve great bursts of speed, seemingly effortlessly and without practice, to save the day.

'A passion for scorching': the pleasures of inappropriate cycling

> The feminine scorcher is not an altogether lovely object . . . A woman with her back doubled into a bow-knot, her hat awry, her hair disheveled, and her face scarlet with exertion, is neither fascinating nor attractive; she takes on an anxious, worried look in her eyes, has her muscles developed at the expense of her feminine grace, and her complexion coarsened by the rude contact of wind and weather . . . The woman who has a passion for scorching loses half the pleasure of riding; she rushes along without taking time to contemplate the beauties of nature; the melting hues of summer sunsets, the charm of the smiling landscape, are all lost upon the inveterate scorcher.[43]

All three illustrations in this chapter depict women traveling quickly and seemingly enjoying it (especially Figures 2.1 and 2.3), judging from the smiles on their lovely faces. But one of these illustrations captures behavior that was frowned upon for both men and women, but especially for women (Figure 2.2).

Books and magazine articles warned against scorching and the danger it posed to bystanders: 'As to speed, it must be a matter of common sense, of which the cyclist often shows a sad lacking, but the scorcher is a man, never a woman, mind you; we are cowardly to risk it'![44] Scorching was frowned upon because of the risk of harm it posed to innocent bystanders if the rider lost control. It was also viewed as displaying poor cycling form, resulting in cyclists hunching over their handle-bars as they concentrate on pedaling as fast as they can. Ward, in her guide to women's cycling, makes it clear that 'speed work' and 'scorching' are not to be done on the street or road, but rather suggests they are 'a sport' that 'should be done only on a track', and 'reckless scorching is to be condemned at all times'.[45] In a collection of bicycle poems entitled *Lyra Cyclus*, there are a number of poems about scorching.[46] The majority of the male scorching poems are humorous, addressing the 'scorcher's spine' (curved over) or being caught by police speeding (note the police officer behind the 'Pretty Little Scorcher' in Figure 2.2). Mary Hopkins, in 'How to Ride a Bicycle', describes female scorchers as 'sexless . . . in knickerbockers, sweaters, and peaked caps, with shoulders humped over their handle-bars and head thrust out turtle-wise'.[47] The quote above adds that scorching can lead to 'an anxious, worried look in her eyes', that, in the effort to hurry and cycle, our scorcher ends up anxious, defeating the whole intent of cycling. An editor of *Outing* likened the practice of scorching to intemperance:

> The habitual scorcher is not a rational being; an occasional sprint along a smooth road is enjoyed by the slowest of coaches, your correspondent included. Scorching, or riding "all out", is intemperance just as much as a man who uses alcoholic beverages to excess, and bad effects are apt to result in both cases.[48]

Both scorching and intemperance were viewed as signs of a lack of self-control, of not being able to rein in impulses.

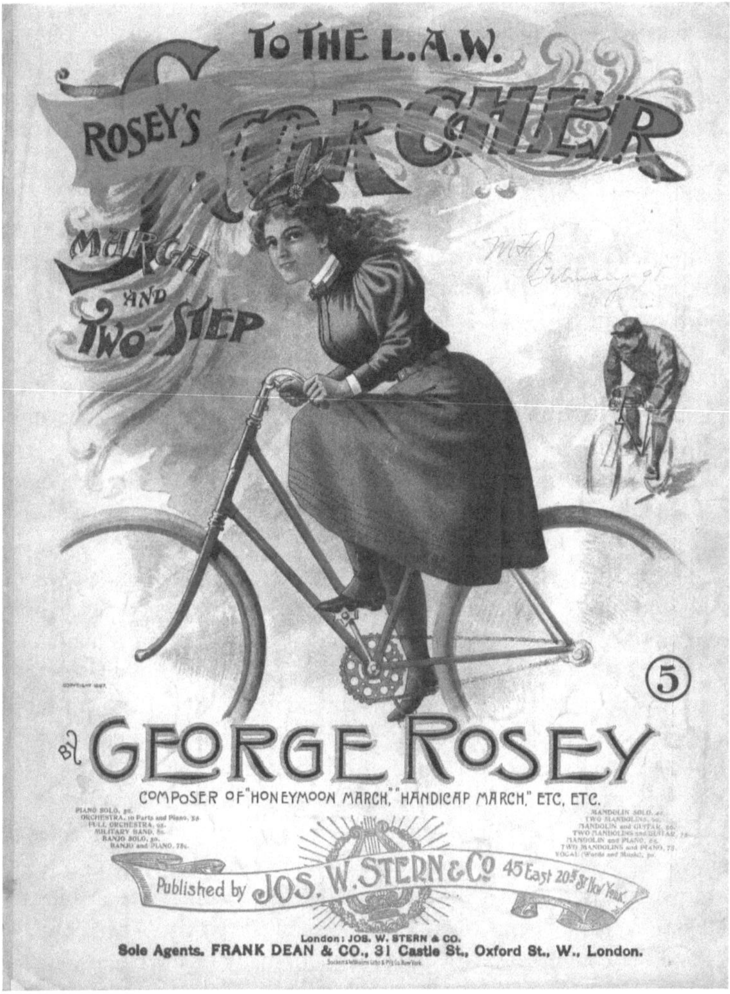

Figure 2.2 Sheet music for 'The Scorcher March and Two-Step' by George Rosey (1897). Rosey also composed music for the song 'The Pretty Little Scorcher' (lyrics by Dave Reed Jr.), whose cover is very similar to this. In the 1890s, singing, like bicycling, was a popular pastime. Composers wrote and published songs reflecting American society at the time, with numerous songs written about bicycling, such as the classic 'Daisy Bell (Bicycle Built for Two)'. An audiofile of Rosey's 'The Scorcher's March' performed by Sousa's band can be found on the Library of Congress website: www.loc.gov/item/00582452/.

Source: Courtesy of The Lilly Library, Indiana University, Bloomington, Indiana. Used by permission of the Edward B. Marks Music Company.

Despite societal warnings about scorching, nearly every bicycle publication, whether for men or women, captures the sheer joy of cycling – of the sensation of flying as you pedal yourself along. Although Maria Ward's guide to cycling is clear in condemning scorching, Figure 2.3 is the cover of her book, capturing a woman obviously enjoying a bit of speed, her feet up as she coasts along, at a quick enough pace that her hat has flown off, a smile on her lips. In written accounts, the imagery of birds and flying is often used to capture the sensation of gliding along on a bicycle, the air whistling past, propelled by your own accord. For example:

> I had the offer of the use of a Triumph Safety from a friend, and determined to try a long journey alone. Of course I had fears of Mrs. Grundy, but so great was the longing to fly through the air on wheels that I determined to defy that tiresome personage.[49]

> When I first saw my Columbia wheel just released from the crate, I was sure I could mount it and ride away instantly across the continent, to the moon, or anywhere. It seems such an easy thing to do, just as it seems that it would be easy to fly when you are watching a bird on the wing. Cycling is the perfection of motion, more like what flying must be than anything that has ever been tried.[50]

> You imagine you are a bird, sailing over flower-covered prairies; you fancy yourself a greyhound bounding after a breathless and frightened jack-rabbit; you even compare yourself with a flash of lightning or a whizzing cannon-ball.[51]

Although none of these examples is clearly scorching, we are left to wonder at what point does quick, pleasurable riding become scorching? When the back humps up and the face is distorted in the effort to achieve the greatest speed? Perhaps the greatest problem society had with scorching was its 'unattractiveness'. As women cycled, they were turned into an image to be consumed, subjected to 'scopic regimes of power', with the gaze turned on them as they cycled.[52]

Yet, there was a fascination with female scorchers, manifesting in illustrations as well as songs and poems, such as the sheet music art for 'The Scorcher' (Figure 2.2) and the poem that begins this essay. In many of the poems, songs, and stories, women scorchers are framed as attractive:

> She rode along the road
> In a costume a la mode,
> And threw a gleam of sunshine on the pike,
> As she gripped the handle bar,
> And beat the trolley car
> And her golden hair was hanging down her bike.[53]

These riders were not just a vision of loveliness: their scorching was the sign of a passionate cyclist. It was viewed as especially alluring when a young male

Figure 2.3 Cover of Maria Ward's *Cycling for Ladies* (1896). *Cycling for Ladies* is an example of the numerous books and articles published for women advocating women's cycling and providing information on everything from purchasing a cycle and selecting a costume to maintaining and repairing a cycle. Ward did not endorse scorching as a practice, yet her cover captures a woman in mid-coast, feet up, her hair and hat suggesting considerable pace.

Source: Image courtesy of the Avery Architectural & Fine Arts Library, Columbia University, New York.

bicyclist finds a female cyclist who likes to scorch as he does: 'Two hearts that scorch as one'.[54] In the short story 'A Dangerous Sidepath', a young man finds his true love in a young woman who likes to scorch as he does, 'down hill'.[55]

'All these charms . . . within easy reach': hurrying to relax

> Who can resist the temptation to fly from the heat and ride away into the depths of the woods, where the boughs of the trees caress each other above the roadway, the buttercups and daisies nod familiarly from the grass, the bees hum drowsily in the patches of sunlight, and, in the shadows, the birds call lazily from nest to nest? When all these charms are within easy reach, how one blesses the light steel roadster which, uncomplainingly, bears the burden of the dust and the heat and carries us on, on, from the hot, paved streets into the smiling home of Mother Nature! Whether it be only for a ride through the Park, or for a long, lovely spin over country roads, the bicycle is always ready, never weary, and brings its rider health and happiness.[56]

An important element to keeping and maintaining a decent speed, as well as having a pleasant cycling experience, was having a suitable road surface to cycle upon and a beautiful landscape to gaze upon. This was a matter of road quality as well as accessibility (on the poor quality of road and street surfaces, see Mackintosh in this volume). Bicycles might be seen as 'geographically liberating', but *where* to cycle was a significant issue for all.[57] The American road network was rudimentary, largely unimproved, and so locating acceptable cycling routes was important. New landscapes might be theoretically available to explore, but the available infrastructure was not necessarily up to the task. Cyclists, especially women, were advised: 'in order to cover distance without dismounting and within a time limit, where the speed attained is an element to be considered, good roads should be chosen'.[58] But what constituted 'good roads'? And where to find them?

Cycling clubs became an important source of information for all cyclists about locations for cycling and road conditions, disseminating the information through their newsletters, road books, and magazines. Many of these cycling clubs were affiliated with the LAW. As cycling clubs and magazines dedicated to cycling came into existence, discussions of good roads/routes for cycling and the need for better roads in general were published, providing directions to cyclists on which roads were better and which were worse, in both urban and rural areas.[59] LAW maps used a code system to indicate the road conditions and the topography – a cartography of hurry – and cyclists could then choose their routes based on the information provided. Guides to cycling directed cyclists to join cycling clubs, if for no other reason than for access to current road conditions and maps.

Early cycling enthusiasts were predominantly in major cities. In the 1870s, major cities had roads and boulevards viewed as 'well suited' for cycling – Boston, New York, and Philadelphia, for example.[60] Grace Denison commented on the roads in major American cities: 'the bosky elegance of Central Park, the new and happy-go-lucky circuit of forty-five miles of the parks and boulevards of Chicago'.[61] A cycling column touts the fine asphalt streets of Washington with 'some splendid hills for coasting, and all around the city are well-kept roads running through beautiful country'.[62] Most streets and boulevards had been constructed for leisurely carriage rides. As cycling expanded, cyclists took to these specific venues, looking for an even road surface, pleasant surroundings, and an opportunity to ride without significant other traffic. In New York City:

> In the early morning and afternoon many society women go out singly or in groups for a quiet spin in the Park . . . A favorite ride is up the Boulevard, along Riverside Avenue to High Bridge, where an impromptu picnic is most rarely enjoyed by the keen appetite which the wheel engenders.[63]

As more Americans took up cycling, these spaces became congested, and scorchers presented a public safety risk, with police levying fines to deter the practice and taking to bicycles to pursue the offenders (Figure 2.2).

These urban rides, of a half-hour or hour away from worries and cares, were seen as the tonic to solve the malaise or 'growth of new morbid tendencies' affecting modern women: 'This tonic of steel is death to mental mildew'.[64] The speed of modernity and its mental and physical demands could be resolved by time spent cycling:

> The woman of affairs has learned that an hour, or even half an hour, may be stolen from the working day, with profit to both woman and affairs . . . An hour of the wheel means sixty minutes of fresh air and wholesome exercise, and at least eight miles of change of scene; it may well be put down to the credit side of the day's reckoning with flesh and spirit.[65]

Within cities, major parks and boulevards provided spaces where women could slip away for a ride, but, over time, the novelty wore off. If you only have 30 minutes, you have limited choices of where to cycle, and, after repeatedly cycling the same spaces, cyclists would be looking for new landscapes with decent roads to explore on their rides.

The demand for cycling spaces led to calls to create a 'dedicated cycling infrastructure', with dry, hard surfaces.[66] The first bicycle path constructed in the United States was in New York City: the Ocean Parkway Bicycle Path, running from Prospect Park to Coney Island, a distance of 5½ miles, opened in June 1894.[67] Simple, inexpensive cycle paths were constructed in upstate New York, Denver, Minneapolis, Portland, and California, often funded by the cyclists themselves from user fees or taxes.[68] In New York State alone, it is estimated that 2,000 miles of bicycle paths were constructed. In the Minneapolis area, paths linked the centers of Minneapolis and St. Paul, 'creating the first intercity bicycle path'.[69]

As cyclists mastered their wheels, they looked for new cycling locations, offering novel challenges and fresh landscapes, outside urban areas. Rural roads in some parts of the country were in decent enough shape to provide pleasant riding conditions, and cycling magazines and LAW road books touted New England roads, described as a 'cycler's paradise'.[70] Cycling guides and magazines directed cyclists to explore the landscape beyond the city: 'there are highways radiating from the heart of the city to the borderland of the country, where one may breathe new inspiration for the world – the world that we persist in having too much with us in the getting and spending efforts that lay waste the powers'.[71] *Outing* magazine had a regular 'Cycling' column in the 1890s where specific routes for weekend rides were laid out with maps. For example, in June 1896, the focus was on the ample cycling opportunities in the Northern Atlantic states, laying out a variety of rides – from Philadelphia to the Jersey Shore, from Providence to Boston, from New York to Albany – with specific directions 'to the lady tourist'.[72] The quote at the start of this section suggests that nature is 'within easy reach', thanks to the bicycle, facilitating urban dwellers' opportunities to recuperate from the pace of modern life. Just as life was 'sped up' through modernity, especially in urban areas, bicycles allowed urbanites to 'hurry up' and get to their leisure – in essence, speeding up to slow down.

But, despite nature being in 'easy reach', outside cities the infrastructure was less developed: 'In 1904, of a total of 2,151,379 miles of rural roads, only 144 miles' offered a suitable surface.[73] Most late-nineteenth-century roads, with the exception of those in major cities, were poorly constructed and poorly maintained. Rural roads might be composed of compacted earth that, when wet, especially in the spring, turned entirely to mud.[74] Jean Rudd, writing about learning to ride her safety cycle comments that, in addition to learning to ride, she had '[a]nother difficulty to encounter. The roads all about my home are heavy country roads – always heavy – either with dust or mud or sun dried sand. And not only heavy, but furrowed with parallel ruts like a corduroy road built lengthwise'.[75] At this point in American history, roads, especially rural roads, were outside the purview of state and federal governments, with their maintenance left to counties and towns, often cared for by farmers suspicious of city cyclists.[76] As a result, travel on rural roads by wagon was relatively slow, with rural life literally being slower than city life (note the farmer plowing behind the cyclist in Figure 2.1).

If the goal of the cyclist was a quick easy ride in the country, then she would need to have knowledge of the roads that would facilitate such a ride, whether through her own experiences or relying on the collective experience/memory of her local cycling club. And women were quick to acquire such knowledge. A Mrs J. Redding was described in 1889 as 'an expert bicycle-rider . . . said to know all the roads around New York for twenty-five miles'.[77] Despite the uneven road quality, rural roads offered a different benefit to women cyclists. In rural areas, there was less traffic than in congested urban areas, allowing women cyclists greater liberties to dress as they wished, cycle as they wished, go where they wished, beyond the judgmental gaze of police, fellow cyclists, and non-cyclists.

Industrialization and urbanization distanced Americans from landscape, changing the landscapes Americans experienced and their interactions with landscape. For urban Americans, landscape became something that they would travel to, in order to experience it, associating landscape with rustic rural scenes or with wilderness.[78] An 1886 guide to Boston schools commented on cycling and emphasized the idyllic virtues of landscape appreciation: 'The benefits derived are the greater owing to the pleasure of viewing the country, thus diverting the mind from any labor or exercise, and resting it by change'.[79] Whereas landscape was presented as another antidote to the rush of modernity, the incessant hurrying, scorching was seen as interfering with the pleasures of riding: 'she rushes along without taking time to contemplate the beauties of nature; the melting hues of summer sunsets, the charm of the smiling landscape, are all lost upon the inveterate scorcher'.[80] Rather, it is by slower, stately riding that women can have the time to take in and appreciate landscape:

> Women are too fond of the beauties which they see all about them on country roads to be turned into mere propellers of a machine over a certain number of miles for the purpose of making a record on a cyclometer.[81]

Only by taking things at their own pace, fast or slow, could cyclists possibly find the rejuvenation they were seeking.

'The immortally blissful scorcher': considering bicycling and the speed of modern life

> Dante, the Infernal geographer, mapped out Hades in seven circles, or wheels. The modern layman is inclined to build Heaven on two . . . The peak of Heaven will no longer be deemed a towering and jagged mountain. It will be represented as an ideally level straightaway track, with no trucks or impediments of any kind, and never an ordinance or a policeman to check the immortally blissful scorcher.[82]

The idyllic image of a woman bicyclist as captured in Figure 2.1 is of a free individual, moving through a beautiful landscape at her own pace, free to choose her own path and her own speed. But was our cyclist truly cycling down a path of her own free will?

Women were essentially directed in all aspects of cycling – especially by other women. Directions tied to mobility, how to cycle, where to cycle, attempted to keep women in their place. Susan Hanson, writing about women and mobility in general, states that, 'women are quite literally kept in their place by being denied access to certain locations at certain times'.[83] Public locations where more people might watch cycling, such as in city parks or on boulevards, were locations where women's practices of cycling might be witnessed and judged.[84] Critics might call out comments about the cyclist in the moment or write a letter to the local newspaper or cycling magazine critiquing the rider. Away from cities, on country roads, were locations where women could cycle away from the public gaze, allowing them greater freedom and opportunities for individual expression. Bicycling opened rural geographies to women where they could cycle and where they could cycle as they chose to, within the physical limitations of the roads themselves.

Cyclists were initially optimistic that there might be a national network of cycling paths, allowing them the entire American landscape at their disposal: 'Everlasting agitation on behalf of cycle paths will in time, and at no distant date, see the country netted with them'.[85] Bicyclists' needs for better cycling roads buttressed the Good Roads movement and the general improvement of American infrastructure. But, by 1900, the cycling fad was fading, and a new obsession was taking hold: automobiling. Ultimately, it was the automobile and the creation of a national highway network that enabled Americans to truly get away in a hurry.

New technologies, new mobilities create new geographies, but these geographies are subject to similar societal restriction, perpetuation, and perhaps at times the creation of new geographies of exclusion, disconnection, inequality, and immobility.[86] The bicycle and the new mobilities it brought American women led to greater freedoms and greater access to American geographies, but not to total freedom. Scopic regimes of power watched, critiqued, and sanctioned women as they moved – and this continues today. In the end, there were great benefits to American women, transforming how they moved, both in their dress and in their spatial mobility. As Susan Hanson writes, 'Mobility is empowering, and because it is empowering, more mobility, especially for women, is a good thing'.[87]

Societal condemnation of scorching came from both male and female cycling advocates, while extolling the virtues of cycling, the pleasures of flying along, and the sense of accomplishment from mastering the bicycle. Women cycling advocates participated in the scopic regimes of power, telling women how to dress and how to ride, how to be 'seen'. Yet they often simultaneously offered ways of resisting, suggesting locations where women could cycle at their own speed and beyond the gaze. Maria Ward states that 'reckless scorching' is to be condemned, but goes on to suggest that non-reckless scorching on a track was acceptable. Nellie found ways to cycle on her own terms, out on a country road, as did the lovers in 'A Dangerous Sidepath' and the cyclist on the cover of Ward's book. But speed is not everything. 'Daisie' encourages her female readers to explore new roads:

> I think we all keep to the good roads too much. We don't strike out into new territory, explore new fields of observation, try unfamiliar paths. I seldom go upon the road that I do not seek to find some new way to reach this or that point, and I have to thank this disposition for much that is rich in my experience.[88]

Jean Rudd, who complained about the 'heavy country roads', seems to echo 'Daisie's' sentiment. When Jean encounters an improved macadamized road, 'it nonplused me. I had become so accustomed to overcoming obstacles that a smooth road was only a perplexity'.[89] Though society may have told women cyclists to seek out good roads and to cycle appropriately, women found ways to challenge themselves and grow, to find their 'bliss', even scorching now and then.

Notes

1 *Boston Courier*, 'Nellie on Her Bike', in *Lyra Cyclus or the Bards and the Bicycle*, ed. Edmond Redmond (Rochester, NY: n/a, 1897), 18.
2 Peter Soppelsa, 'Intersections: Technology, Mobility, and Geography', *Technology and Culture* 52 (2011): 677.
3 Anonymous, 'The Rule of the Bicycle', *Scribner's Magazine* 19 (1896): 783.
4 Ross Petty, 'Peddling the Bicycle in the 1890s: Mass Marketing Shifts Into High Gear', *Journal of Macromarketing* 15 (1995): 32.
5 See Petty, 'Peddling the Bicycle'. With regard to seat design, there was considerable concern over women's seat design and the possibility of women being 'over stimulated' while bicycling. See also Margaret Guroff, 'How We Roll', *Raritan* 35 (2016): 111; Ellen Garvey, 'Reframing the Bicycle: Advertising-Supported Magazines and Scorching Women', *American Quarterly* 47 (1995): 68–9.
6 Anonymous, 'The World Awheel', *Munsey's Magazine* 15, 2 (1896): 131.
7 Anonymous, 'Ravages of the Bicycle Craze', *Scientific American* 74, 23 (20 June 1896): 391.
8 It should be noted that bicycling was overwhelmingly associated with white Americans. Black Americans were (and it could be argued still are) not associated with bicycling. Organizations such as the League of American Wheelmen and the Century Road Club only admitted white cyclists (Century Road Club of America, *Century Road Club Manual* [Terre Haute, IN: Moore & Langen (1898), 111]; Lorenz Finison, *Boston's Cycling Craze, 1880–1900: A Story of Race, Sport, and Society* [Amherst & Boston, MA: University of Massachusetts Press, 2014], 18). However, black Americans found opportunities. *Scribner's Magazine* reported in 1895 that, 'A colored "Professor," so called, who had acquired some reputation as a teacher in New York during the spring months, opened a school in Newport and achieved fame and temporary fortune' (James Townsend, 'The Social Side of Bicycling', *Scribner's Magazine* 17 (1895): 704–8). There were 'colored riders' clubs in New York, Boston, and Philadelphia (Finison, *Boston's Cycling Craze*, 21). And one of the most famous racing cyclists in the country was Major Taylor (Finison, *Boston's Cycling Craze*, 207–8). Black Americans did bicycle, but, as we consider bicycling and consumption of space, and particularly the popular image of cycling, we need to acknowledge that this is very much privileging a specific segment of the American population.
9 Tim Cresswell and Tanu Priya Uteng, 'Gendered Mobilities: Towards a Holistic Understanding', in *Gendered Mobilities*, eds T. Uteng and T. Cresswell (Farnham, UK: Ashgate, 2008), 2.
10 Mary Bisland, 'Woman's Cycle', *Godey's Magazine* (April 1896): 386–7.
11 Garvey, 'Reframing the Bicycle', 67–8.
12 Ross Petty, 'Women and the wheel: the bicycle's impact on women', in *Cycle History: Proceedings of the 7th International Cycle History Conference*, ed. R. van der Plas (San Francisco: Rob van der Plas, 1997), 117. One of the earliest tricycles dates to the second half of the 1600s. The tricycle did not become popular until 1877, and James Starley developed the Coventry Rotary, creating a tricycle craze (Carlton Reid, *Roads Were Not Built for Cars: How Cyclists Were the First to Push for Good Roads & Became the Pioneers of Motoring* [Washington: Island Press, 2015, 10]).
13 Charles Dodge, 'Out-Door Athletics for American Women', *Godey's Magazine* 117 (1888): 209.
14 Anonymous, 'Love on Wheels', *The Wheelman* 1 (1882): 208–11.
15 Julius Wilcox, 'Ladies' Bicycles', *The Bicycling World and L.A.W. Bulletin* 16, 22 (1888): 368.
16 Elizabeth Robins Pennell, 'Cycling', in *Ladies in the Field*, ed. V. Greville (New York: D. Appleton, 1894), 254.

17 Julie Wosk, *Women and the Machine: Representations from the Spinning Wheel to the Electronic Age* (Baltimore & London: Johns Hopkins University Press, 2001), 97; Petty, 'Women and the wheel', 118.
18 Petty, 'Women and the wheel', 42.
19 Sue Macy, *Wheels of Change: How Women Rode the Bicycle to Freedom (With a Few Flat Tires Along the Way)* (Washington, DC: National Geographic Society, 2011), 4–5; Ralph Bogardus, 'The Reorientation of Paradise: Modern Mass Media and Narratives of Desire in the Making of American Consumer Culture', *American Literary History* 10 (1998): 508–23; Petty, 'Women and the wheel', 128.
20 Marguerite Merington, 'Woman and the Bicycle', *Scribner's Magazine* 17 (1895): 703.
21 See, for example: Margaret Le Long, 'Alone and Awheel, From Chicago to San Francisco', *Outing* 31 (1898): 592–6; Maria Ward, *Bicycling for Ladies: The Common Sense of Bicycling* (New York: Brentano's, 1896); Fanny Erskine, *Lady Cycling: What to Wear & How to Ride* (London: British Library, 1897).
22 Ida Harper, 'A Modern Christmas Story', *The Woman's Journal* 26, 52 (December 28, 1895): 414. *The Woman's Journal* was a magazine dedicated to women's rights, especially suffrage. Ida Harper became very active in suffrage and eventually wrote a multi-volume history of the movement.
23 Garvey, 'Reframing the Bicycle', 70.
24 See also Erskine, *Lady Cycling*, and Lillias Davidson, *The Handbook for Lady Cyclists* (London: H. Nisbet, 1896).
25 Anonymous, 'Hints for Cycling Women', *The Woman's Journal* 26, 35 (August 31, 1895): 275.
26 Petty, 'Women and the wheel', 43.
27 Phillip Gordon Mackintosh, 'A Bourgeois Geography of Domestic Cycling: Using Public Space Responsibly in Toronto and Niagara-on-the-Lake, 1890–1900', *Journal of Historical Sociology* 20, 1/2 (2007): 144.
28 'The World Awheel', 158.
29 Catherine Gudis, 'Driving Consumption', *History & Technology* 26 (2010): 372.
30 Bisland, 'Woman's Cycle', 386–7.
31 'The World Awheel', 159.
32 Merington, 'Woman and the Bicycle', 703.
33 Dale Somers, 'A City on Wheels: The Bicycle Era in New Orleans', *Louisiana History: The Journal of the Louisiana Historical Association* 8 (1967): 236.
34 A bicycling column in *The Designer and the Woman's Magazine* describes the 'motley throng' of cyclists who come out on summer evenings: 'one after another of the great army of city toilers speeds up the avenue on a noiseless steed, making swiftly for the leady glades of the Park or the half-deserted roads above the Harlem'. See 'Bicycling: A Spectator's Point of View', *The Designer and the Woman's Magazine* 8 (July 1898): 70.
35 Sidney Aronson, 'The Sociology of the Bicycle', *Social Forces* 30 (1952): 311.
36 Luther Porter, *Cycling for Health and Pleasure: An Indispensable Guide to the Successful Use of the Wheel* (New York: Dodd, Mead, 1895); Anonymous, 'Rule of the Bicycle'; Ward, *Bicycling for Ladies*.
37 Aronson, 'The Sociology of the Bicycle', 311.
38 Sarah Tooley, 'Through the Air on Wheels: Interview with Miss N.G. Bacon', *Woman's Signal* 37 (1894): 168.
39 Mary Hopkins, 'How to Ride the Bicycle, and What to Wear', *The New England Kitchen Magazine* 2 (October 1894): 12.
40 Daisie, 'From a Feminine Point of View', *The Cycle* 1 (September 17, 1886): 415.
41 'The Woman Awheel', *The Woman's Journal* 27, 37 (September 12, 1896): 294.
42 Flora Comstock, 'Rosalind A-wheel', *Godey's Magazine* (April 1896): 389–93; Hattie Dutro, 'The Old Man and the New Woman', *The Woman's Journal* 30 (June 17, 1899): 190; Ruth Raymond, 'Farmer Green's Conversion', *The Woman's Journal* 30 (November 4, 1899): 350.

43 Anonymous, 'The Feminine Scorcher', *Godey's Magazine* 132 (1896): 446.
44 Grace Denison, 'The Evolution of the Lady Cyclist', *Massey's Magazine* 3 (1897): 284.
45 Ward, *Bicycling for Ladies*, 78.
46 Edmond Redmond, *Lyra Cyclus or The Bards and the Bicycle* (Rochester, NY: n/a, 1897).
47 Hopkins, 'How to Ride the Bicycle', 12.
48 The Prowler, 'Cycling Fixtures', *Outing* 22 (June 1893): 60.
49 Tooley, 'Through the Air on Wheels', 168.
50 Jean Rudd, 'My Wheel and I', *Outing* 27 (1895): 124.
51 Helen Follett, 'A Honeymoon on Wheels', *Outing* 29 (October 1896): 3–7.
52 Gudis, 'Driving Consumption', 372–3.
53 *Boston Courier*, 'The Beautiful Scorcher', in Redmond, *Lyra Cyclus*, 113.
54 Earl Eaton, 'Two on a Tandem', in Redmond, *Lyra Cyclus*, 32.
55 John Wood, 'A Dangerous Sidepath', *Outing* 22 (June 1893): 212.
56 Anonymous, 'Thro' Highway and Byway', *The Designer and the Woman's Magazine* 2 (1895): 56.
57 Glen Norcliffe, *The Ride to Modernity: The Bicycle in Canada, 1869–1900* (Toronto: University of Toronto Press, 2001), 23.
58 Ward, *Bicycling for Ladies*, 100.
59 See Christina Dando, 'Riding the Wheel: Selling American Women Mobility and Geographic Knowledge', *Acme: An International E-Journal for Critical Geographies* 6 (2007): 174–210.
60 Charles Pratt, *The American Bicycler: A Manual for the Observer, the Learner, and the Expert* (Boston: Houghton, Osgood, 1879), 131–2.
61 Denison, 'The Evolution of the Lady Cyclist', 282.
62 'On the Wheel', *The Designer and the Woman's Magazine* 2, 1(May 1895): 56.
63 Kathleen Mathew, 'The Bicycle in Society', *The Peterson Magazine* 106 (1895): 602–3.
64 'The World Awheel', 158; Mary Hopkins, 'Outdoor Papers: The Bicycle', *The New England Kitchen Magazine* 1 (September 1894): 310.
65 Merington, 'Woman and the Bicycle', 703.
66 Reid, *Roads Were Not Built for Cars*, 160. Cycling paths, once created, proved a tempting alternative to the poor American roads to other travelers. The editors lament in *The Referee*, a cycling magazine, that the Coney Island Cycle Path was currently in horrible condition: when the adjacent roads became too muddy, wagons and hook-and-ladder trucks used the cycling path instead, leaving it deeply rutted (Anonymous, 'The Coney Island Path Ruined', *The Referee* 14 [1895]: 31). Seattle's 'fabulous cinder-paved bike trail system – which cyclists built and paid for with their own money and sweat equity' was abused by horseback riders, delivery men with their wagons, and farmers who let their cattle trample the path (Knute Berger, 'The War on Bikes Should Be History', *Crosscut.com* [accessed October 11, 2016], http://crosscut.com/2016/seattle-war-on-bikes-should-be-history/; Frank Cameron, *Bicycling in Seattle 1879–1904* [publisher not identified, 1982], 16). America's poor roads were a problem clearly faced by all.
67 Reid, *Roads Were Not Built for Cars*, 163.
68 Ibid., 166; James Longhurst, 'The Sidepath Not Taken: Bicycles, Taxes, and the Rhetoric of the Public Good in the 1890s', *The Journal of Policy History* 25 (2013): 558. Longhurst examines the policy aspects of the sidepath movement, an effort to create physically separate cycling paths.
69 Ross Petty, 'Bicycling in Minneapolis in the Early 20th Century', *Minnesota History* 62 (2010): 90. The American Geographical Society Library has several maps of cycling paths in its collections, such as paths in Albany County and Long Island, New York.
70 The Prowler, 'Cycling', *Outing* 29 (1896): 287.
71 Merington, 'Woman and the Bicycle', 704.
72 The Prowler, 'Cycling Notes', *Outing* 28 (1896): 48.

73 Gary Tobin, 'The Bicycle Boom of the 1890s: The Development of Private Transportation and the Birth of the Modern Tourist', *Journal of Popular Culture* 7, 4 (Spring 1974): 840.
74 Reid, *Roads Were Not Built for Cars*, 147.
75 Rudd, *My Wheel and I*, 127.
76 Reid, *Roads Were Not Built For Cars*, 150.
77 Anonymous, 'Concerning Women', *The Woman's Journal* 20 (August 24, 1889): 265.
78 Aronson, 'The Sociology of the Bicycle', 311.
79 Dexter Smith, *Cyclopedia of Boston and Vicinity* (Boston, MA: Cashin & Smith, 1886), 83.
80 'The Feminine Scorcher', 446.
81 'The World Awheel 1896', 159.
82 Marmaduke Humphrey, 'A Cycle Show in Little', *Godey's Magazine* 132 (April 1896): 367.
83 Susan Hanson, 'Gender and Mobility: New Approaches for Informing Sustainability', *Gender, Place & Culture* 17 (2010): 10.
84 On a more disturbing note, a 'spectator' describes 'The bicycle "masher" who "ogles every passably pretty woman he meets with the libidinous leer of a satyr, and, if the chase promise fair measure of amusement, pursuing his prey until darkness enshrouds her"'. See 'Bicycling: A Spectator's Point of View', 71.
85 Anonymous, 'Bicycling Notes', *The Designer and the Woman's Magazine* 4 (June 1896): 87. We are probably closer today to a national network of cycling paths, many created from former railroad corridors.
86 Peter Merriman, 'Mobility', in *International Encyclopedia of Human Geography: Volume 7*, eds R. Kitchin and N. Thrift (London: Elsevier, 2009), 135.
87 Hanson, 'Gender and Mobility', 9.
88 Daisie, 'From a Feminine Point of View', 415.
89 Rudd, 'My Wheel and I', 127.

3 The London bus

An unlikely architecture of hurry

Richard Dennis

The London bus might seem an unlikely 'architecture of hurry'. Its best-known depiction in twentieth-century popular culture, Flanders and Swann's song, 'A Transport of Delight', characterized the 'big six-wheeler, scarlet-painted, London Transport, diesel-engined, ninety-seven horse-power omnibus' as the antithesis of hurry and a source of congestion in its own right, sticking to a maximum of 20 mph – 'in the middle of the road' and 'in convoys' – thereby frustrating the progress of other road traffic, especially taxicabs.[1] Flanders and Swann's lyrics reflect a rivalry between cab and bus drivers that dates back to the origins of the horse-drawn omnibus in the early nineteenth century, when hansom cabbies and short-stage coach drivers objected vociferously to the new buses, which challenged their monopoly on 'public' (for hire) transportation in the city and, by carrying twenty or more fare-paying passengers in a single vehicle, could undercut cab and stage-coach fares. Yet, far from depicting slowness and congestion, the earliest images and news stories about horse-buses emphasized their speed and the exhilaration, even recklessness, of bus travel.

This chapter therefore examines discourses surrounding, not only the speed of London buses during the first hundred years of their operation, spanning the transition from horse to motor power, but also the excitement, anxieties and frustrations that variously constituted the experience of 'hurry' when travelling by bus. It also traces technological, design and organizational changes that were intended to accelerate travel, some of which, such as the introduction of fixed bus stops, constituted a literal, material architecture of hurry. It offers historical evidence to illustrate Cresswell's thesis that mobility cannot be understood without examining both representation and 'material corporeality', and it also shows how, in the nineteenth century, public transport could be as much an 'enemy of civility' as, for Sheller and Urry, automobility became by the late twentieth century.[2]

Speed

Horse-drawn omnibuses were introduced in London in 1829 by George Shillibeer, who had previously been a coachbuilder in Paris, where omnibuses had already been operating for several years. His first route, from near Paddington to the

City, followed the line of the New Road (now Marylebone Road, Euston Road, Pentonville Road, City Road), which had been laid out in 1756 to allow heavy traffic to bypass the mostly middle-class residential districts north of Oxford Street and Holborn. But, as these districts expanded, so demand was fuelled for improved public transport to convey commuters to and from City workplaces. Shillibeer soon had numerous imitators, prompting a period of chaotic rivalry between operators preceding the introduction of regulation in the mid 1830s. Buses would race one another to be first at the next likely pick-up point (usually an inn) and zigzag across the road to prevent rivals from overtaking.[3]

Contemporary newspapers regularly reported on court cases and accidents attributable to racing and 'furious driving'. In April 1830, for example, *The Times* noted incidents of racing in Broad Street and Coleman Street in the City, as well as on the New Road, and, in January 1831, a 77-year-old man was run over by an omnibus 'passing at a very quick rate' on City Road. He died of his injuries a few days later. An inquest jury 'strongly reprobated the conduct of the omnibus coachmen in general', but returned a verdict of accidental death.[4] The worst cases involved trials for manslaughter, though the lack of authoritative evidence and frequently conflicting accounts by witnesses meant that there were few guilty verdicts. At the Old Bailey, fifteen charges of manslaughter against the drivers of horse-buses between 1833 and 1898 yielded only three convictions. In 1908 and 1909, there were also three cases involving motor-bus drivers.[5] However, this was just the tip of a much bigger iceberg of cases that never reached the Old Bailey or were heard at other courts. In three years, 1910–12, there were 37 fatal accidents involving motor buses, just within the narrow confines of the City of London. The maximum estimated speed of the buses was 7–8 mph (in 12 cases); 5 cases were 'very slow'; and, in 6 cases, the speed was 4–5 mph. But of particular interest here are cases where the deceased showed signs of 'hurry' – 'ran at a fast pace out of a side turning', 'suddenly ran across the road', 'ran across the carriage-way', 'alighting from an omnibus in motion', 'ran against the omnibus' – or indecision – 'hesitated', 'turned as if to go back', 'halted, turned, and tried to regain the foot-way'.[6] Even at barely more than walking pace, 'hurry' could prove fatal.

The earliest of the Old Bailey cases involved Thomas Chennell, an 18-year-old horse-bus driver, accused of causing the death of William Hancock near the Kensington turnpike gate in March 1833. William Clark told the Old Bailey he saw two omnibuses 'coming along the road very violently'. Another witness, Robert Bailey, implied that the two buses were in a race:

> I saw Mr. Kid's and Mr. Cloud's omnibuses – they were on the high road, on the near side; they had each two horses, and were going at full gallop, as hard as the horses could put their feet to the ground – Kid's omnibus, which the prisoner was driving, was about twenty yards behind Cloud's; they were on their proper side, and appeared to me to be racing one against the other . . . as the man got out of the way of Cloud's omnibus, the near horse of the prisoner's omnibus struck and knocked him down.

Clark reported shouting, 'You scoundrel, you have killed the man – this comes of your galloping and furious driving'. Chennell claimed he was not galloping. But Clark was sure: 'he was whipping the off-horse, and I am positive he was galloping; he was driving so fast it was impossible for him to pull up'. Hancock died in hospital a week after the accident. Chennell was sentenced to three months in prison.[7]

The speed associated with 'furious driving' varied according to location. In the City's narrow streets, speeds of 7–8 mph were regarded as excessive; on the New Road, witnesses were shocked by speeds they estimated to be around 12 mph. They judged speeds according to whether horses were 'trotting', 'cantering' or 'galloping' and whether drivers were using their whips. Some victims were children who had not learnt to be streetwise – a 6-year-old in High Street, Bloomsbury (St Giles'), in August 1833, a 5-year-old on City Road in February 1834, a 4-year-old on Goswell Street in October 1834, all girls; others were elderly folk – such as a 92-year-old in Deptford in February 1835 – who were too infirm to take avoiding action.[8] But there were also cases of panic, where victims were simply frozen to the spot as the speeding bus bore down on them, and cases where conductors, who normally stood on the rear step of their bus, were struck by the pole of another bus trying to overtake. Some articles in the press reported, not accidents, but cases where passengers panicked at the dangerous driving and excessive speed of drivers, who also refused to stop to allow anxious passengers to alight. When two buses were racing in Coleman Street in April 1830, 'A lady in one of the vehicles screamed aloud with fright, and begged to be let out'; when two buses on Old Kent Road raced to be the first to reach the Bricklayers' Arms (a public house where intending passengers often waited), two sisters, 'desirous of leaving the vehicle altogether, in consequence of the danger they were exposed to, owing to the furious manner in which they were driven . . . screamed aloud, but no notice was taken of their condition by either the defendant [driver] or the conductor of the omnibus'; and when two buses raced down Pentonville Road in June 1835, 'the passengers screamed out, expecting every moment that the vehicles would be dashed to pieces'. When the driver was at last persuaded to stop, 'all the passengers, 14 in number, jumped out, and proceeded home on foot'.[9]

At other times, impatient passengers accused their drivers of going too slowly:

> Two gentlemen were going in haste to Charing-cross, the omnibus which starts from the Bank was at the top of Ludgate-hill, and being told it was going immediately, and that no stop would be made except a minute or two in the Strand, they got in. Six minutes elapsed before the omnibus moved, it then dragged on slowly for a short distance and stopped again, when the two gentlemen got out, just below the Old Bailey, refusing to pay anything for being conveyed a distance not exceeding 150 yards in nine minutes.

The driver was fined 40s under the Hackney-Coach Act, which required cabs and buses to average at least 5 mph in the absence of congestion.[10] More generally, it was observed that buses went slowly while their conductors attempted to solicit more custom, but then accelerated as soon as they had acquired a full complement of passengers.[11]

Rivalry between different companies serving the same route was manifest not only in racing but also in 'nursing'. Two buses from the same company would corner the traffic: while the first ran at normal speed, the second would deliberately slow down and prevent its rival from overtaking; or a group of buses would box in a rival, preventing passengers from either boarding or alighting.[12]

The worst excesses of competitive 'racing' were gradually eliminated by the formation of associations of omnibus operators who agreed on timetables that allowed all their members fair access to potential passengers and, subsequently, their amalgamation into a few major combines, especially the London General Omnibus Company (LGOC), established in 1855; however, there were always outsiders, termed 'pirates', who interposed themselves into regular traffic.[13] There was a resurgence of pirate operators following the First World War. Virginia Woolf's description of a pirate bus on Victoria Street may be interpreted metaphorically, given the buccaneering adventure on which her protagonist, the teenage Elizabeth Dalloway, was embarking by riding on top of a bus, but it was also a literal account of how pirate buses behaved:

> Buses swooped, settled, were off – garish caravans, glistening with red and yellow varnish . . . .
>
> Suddenly Elizabeth stepped forward and most competently boarded the omnibus, in front of everybody. She took a seat on top. The impetuous creature – a pirate – started forward, sprang away; she had to hold the rail to steady herself, for a pirate it was, reckless, unscrupulous, bearing down ruthlessly, circumventing dangerously, boldly snatching a passenger, or ignoring a passenger, squeezing eel-like and arrogant in between, and then rushing insolently all sails spread up Whitehall.[14]

Until the 1920s, there were few speedometers or authoritative measurements of actual speed, merely calculations of average speed based on the time and distance from one end of a route to the other. A *Times* letter-writer in December 1929 urged the compulsory fitting of speedometers in buses, noting that he had 'only once seen a speedometer in a public road vehicle, and then it was out of order'. Six years later, a bus driver fined for driving at 36 mph on the Victoria Embankment claimed 'he did not realize that he was going so fast as he had no speedometer'.[15]

Average speeds in central London, including stops, were not much more than 5 mph in the 1880s, slightly higher for buses that began or ended their routes in suburban districts.[16] The Royal Commission on London Traffic reported in 1905 that horse-buses averaged 3.5–6 mph in 'crush hours', but 5.25–8 mph in 'slack hours', while new motor buses achieved 6.5–8.25 mph at peak-time and 8.25–11.5 mph off-peak.[17] The latter must have been very difficult to achieve legally, given that the speed limit for motor buses was only 12 mph (raised to 20 mph in the late 1920s, but only for buses equipped with pneumatic tyres). It was reckoned that the time saved by replacing horse-buses with motor buses averaged 25 per cent and, on some routes, was as high as 34 per cent, and that, for most journeys, buses were almost as fast as travelling by tube (where much of the time was spent waiting for and descending and ascending in lifts; escalators were rare until the interwar years).[18] The Select Committee on Cabs and Omnibuses (Metropolis) (1906) confirmed a speed limit of 12 mph for motor buses, though electric trams, which engineers considered more stable at speed, were allowed to operate at up to 16 mph.[19] A survey by London General in 1913 found that, on routes where there was competition from electric trams, average motor-bus speeds were higher (9–10 mph) than on routes with no competition (8.5 mph). On some routes, the average, including stops, was as high as 11.5 mph, again implying that the bus must have exceeded the 12 mph limit, especially as there were numerous places where the company imposed its own 8 mph restriction: when passing stationary tramcars, running downhill or crossing over bridges. The survey also measured the average time spent at each stop, ranging between 14 and 27 seconds on the fastest and slowest routes, with a median of 20 seconds per stop.[20] By the mid 1920s, surface transport (presumably including trams as well as buses) averaged 9.1 mph in 'normal areas' and 7.2 mph in 'congested districts', compared with an average speed of tube trains of 18 mph.[21]

In 1905, the motor-bus lobby had claimed that, 'as two Motor Buses can do more than three horse-buses, it is evident that the passenger vehicles in the streets can be reduced by at least one-third when the horse is replaced by the motor'. So they anticipated less congestion: 'The ride in an omnibus will no longer be a tedious trial of sheer endurance'.[22] In practice, and in the absence of many new or wider streets, the very success of the motor bus ('seat mileage' increased from 3,700 million in 1913 to 9,450 million in 1926) intensified congestion and reduced speeds: 'the bus creates congestion by its development'.[23] In 2014–15 (following further substantial increases in the average size of buses, but also the provision of extra doors to speed up loading and unloading), average bus speeds terminus-to-terminus (including 'dwell time' – time spent at stops loading and unloading passengers) ranged from 5.4 mph on the slowest route to over 20 mph on the fastest, with a mode of 11 mph. Routes passing through central London averaged roughly the same speed as in 1905.[24]

In the 1830s, along with the first horse-buses, there had also been 'steam carriages'. Walter Hancock built and operated the 'Enterprise', the 'Era' and the more ominous-sounding 'Autopsy' and 'Automaton', claiming an average speed of 12–13 mph for his service between Paddington and Moorfields. In 1836, trial runs of the 'Automaton' between City Road and Epping averaged 11.5 mph, and on Bow Road a top speed of 21 mph was recorded.[25] But horse-buses rarely attained such speeds. *The Times* asked in 1856: 'Shall we ever be able to rattle along our streets at anything like a reasonable pace? . . . We must be content if we can now and then get six miles an hour.'[26]

'Man about Town' reported on the trial of an electric bus in the West End in January 1897:

> No test of speed was of course possible, as the regulations bound the 'bus [*sic*] down to five miles an hour. But those who rode in it speak highly of the vehicle with regard to comfort. It was well ventilated, it was not overcrowded, and there was no jarring or vibration.[27]

Vibration was an aspect of speed that directs our attention to the experience, not only of riding by bus, but also living or being on the street close to passing buses. The 1906 Select Committee noted that:

> Private persons complain that their houses are damaged by the vibration, that the noise prevents them from working and sleeping, and that the smoke sickens them. Tradesmen complain that in certain cases the vibration makes it impossible for them to carry on their occupation.[28]

Chief Inspector Bassom of the Metropolitan Police described the situation in Albany Street (just east of Regents Park), where 'the structure of the houses was interfered with'. Noise and vibration were here attributed to motor buses returning to their garage in the early hours of the morning. When drivers were instructed to travel very slowly through the street at night, contrary to the natural tendency to drive faster because there was so little other traffic about, the problems declined: 'It seems, therefore, that a vehicle travelling at a less speed does not cause so much vibration as one travelling at a higher rate of speed'. However, vibration was attributable as much to 'heavy slow-moving mechanical traffic' – lorries and traction engines – as to speeding motor buses, and also to poorly maintained roads, 'full of holes'.[29] Unsurprisingly, witnesses from the motor trade disputed the negative effects of speeding buses: rubber tyres, more experienced drivers, better road surfaces and better-maintained buildings would solve the problems, and it was 'clipping the wings of the motor 'bus in a most unfair way to restrict its speed. If it is a mobile conveyance which can travel at a greater speed, why should not it?'[30]

Until after the First World War, most vehicles had hard steel tyres, and most roads, laid with horses in mind, had relatively soft surfaces. The ideal for fast motor vehicles was the opposite: hard surfaces and soft, yielding tyres; yet, although pneumatic tyres on buses were anticipated as early as 1894, it was not until the late 1920s that they became common.[31]

The imagination of hurry

The glorification of speed is often associated with Marinetti's 'Futurist Manifesto', issued in 1909, but, 4 years previously, George Swinton had suggested of a traffic board, recommended by the Royal Commission on London Traffic as a new authority to oversee 'London locomotion', that, 'when they come to die, graven on their hearts must be found the one word, "speed"'. More likely influenced by Futurism, Wilfrid Randell wrote in *The Academy* (1910):

> There is exhilaration in speed . . . there is a spiritual exaltation, doubtless, in charging through the city, twisting between clumsy horse-drawn drays, hunting for a stretch of clear road. The occasional rollicking spurt along a smooth vista . . . or, late at night, the sombre, echoing ravine of Victoria Street, is quite inspiring.[32]

Even among the earliest artistic representations of London bus travel in the 1840s, the bus was depicted as an icon of modernity characterized by exhilarating speed. James Pollard made his name as a creator of sporting prints of racehorses and horse-racing and applied the same artistic conventions in his depiction of speeding buses. Nineteenth-century photographs with long time exposures necessarily demanded a statuesque pose conveying the opposite of speed. By contrast, Pollard's *A Street Scene with Two Omnibuses* (1845; Figure 3.1) shows a 'Favorite', one of the most prominent early bus 'lines', passing rows of substantial five-storey townhouses, hotly pursued by a rival. The 'Favorite' is making a fast trot, but the horses pulling the second bus are close to a gallop, and their driver has his whip raised. Among the cluster of top-hatted gentlemen seated on

Figure 3.1 James Pollard (1792–1867), *A Street Scene with Two Omnibuses* (1845).

Source: Permission of Museum of London.

the top deck of the 'Favorite', beside and behind the driver, one is looking intently at the pursuers and grasps his own walking-stick as if excited by the prospect of a race. Other paintings by Pollard also convey a sense of 'hurry' through the body language of top-deck passengers. *The Upper Clapton Omnibus* (1852), viewed side-on, implies speed through the slight blurring of wheel spokes and a cloud of dust forming in their wake. The horse farther from us turns its head to fix its gaze on the artist, left behind as the bus passes.[33]

Although British artists disowned Futurism in the aftermath of the First World War, we can see its influence on two modernist images of the speeding bus. Cyril Power's *The Sunshine Roof* (1934) seats us inside, at the back of the bus behind the other passengers, as we lean into a bend as if we were racing at Brooklands. But it is Power's younger colleague, Claude Flight, who best portrayed the modern age of bus travel. His linocut of 1922, simply entitled *Speed*, depicts a succession of red double-decker buses, the foremost of which is passing so quickly out of the picture that we can only catch the letters 'S P E E' on the advertising panel along its side. The six-storey building in the background has a sinuosity that also suggests a view glimpsed at speed, and the street hints at Randell's 'echoing ravine of Victoria Street'. In the left foreground of the picture, pedestrians are anxiously hurrying across the road, shepherded by a policeman.[34] Their dilemma is the one first noted by Baudelaire attempting to cross the boulevard in his prose poem, 'Loss of a Halo', and subsequently elaborated by Marshall Berman – the increasing conflict between pedestrians seeking to cross the road safely and motorized traffic, now desensitized to speed by the smoothness of a tarmacked road.[35] Nor, in London in 1922, were there either traffic lights or pedestrian crossings to bring the traffic to a halt.

An earlier painting of motor buses – Charles Ginner's *Piccadilly Circus* (1912) – places more emphasis on the oppressive effect of constantly passing traffic, including, in this case, two motor buses and a motor taxi, on pedestrians hurrying past a flower-seller on one of Piccadilly Circus's 'refuges'. There is no sky: traffic, buildings and pavement fill the canvas.[36] The painting is a perfect match for Forster's near contemporaneous commentary of roads that 'smelt more strongly of petrol and were more difficult to cross' and streets where 'human beings heard each other speak with greater difficulty, breathed less of the air, and saw less of the sky'. We can imagine the 'clipped words, formless sentences, potted expressions of approval or disgust' being exchanged by the passers-by and passengers in Ginner's painting.[37]

Widening the perspective beyond that offered by Ginner, we can also conceive of the illuminated signs at Piccadilly Circus – advertising brands such as 'Bovril' and 'Schweppes' – as 'clipped words, formless sentences' designed to be glimpsed from the top deck of a passing bus. Likewise, C.R.W. Nevinson's *The Strand* (c.1913) shows a late-night crowd gathered around a pair of buses, bedecked with an exhortation to 'Keep Smiling', and surrounded by free-floating advertisements for 'Players Cigarettes' and 'Bovril': yet more clipped words.[38]

The bus, casually reduced from 'omnibus' (the name given to horse-buses in the 1820s and '30s, supposedly derived from the words 'Omnes Omnibus' on a shop owned by M. Omnes near the terminus of a horse-bus route in Nantes) to ''bus' and then cavalierly de-apostrophized, is itself a clipped word. The *Cornhill Magazine* (March 1890) devoted space to 'the 'bus': 'This short, smart, and useful word has perhaps contributed to the popularity of the vehicle itself'. It might be vulgar, reflective of lower-middle-class origins and signifying 'the independence of women, for girls and single ladies may travel safely under its sacred aegis', but phrases such as 'the last 'bus' had become 'familiar terms that will always be in vogue, and do useful service'. The following month, a lengthy correspondence in *The Times* was devoted to coining a word for travel by electric traction, to match the increasing use of 'to bus' and 'to cab', employed to describe travelling by bus or cab.[39] We should also note another feature of the twentieth-century London bus: the 'clippie', the conductor, latterly often female, who issued and 'clipped' your ticket.

Names and numbers

Pollard's buses carried no route numbers – merely a colour scheme that indicated their proprietor, such as 'the familiar "Royal Blue", the green and useful "Atlas", the red "Paddington", the yellow "Camden Town"'. Cruchley's *London in 1865: A Handbook for Strangers* listed destinations and colours, from '1. Bayswater: colour, *green*' to '28. Westminster: colour, *chocolate*'.[40] But there were more routes than colours, and so potential passengers also had to look out for line names, such as 'Favorite', 'Atlas', 'Caledonian', 'Perseverance', and for painted destination boards, usually affixed to the sides of buses. It was hard to see where a bus was going until it was passing by.

Frustrated passengers regularly requested that buses should carry simple destination boards affixed to drivers' footboards so that they would be visible as the bus approached. Writing to *The Times* in November 1850, 'A Merchant' suggested that, 'A plate should be affixed to the driver's footboard, with the destination of the vehicle conspicuously painted thereon'. Forty-seven years later, other *Times* correspondents used almost identical words: 'why should not the omnibus companies fasten a small upright board to the driver's footboard, with the destination plainly painted thereon'. But now they also wanted the board to be reversible or revolving: 'Can anything be more absurd than an omnibus racing towards the City . . . labelled in large letters "Hammersmith" or "Bayswater"?'[41] Other correspondents complained that the proliferation of advertisements on the front and back made it even harder to discern a bus's destination. As late as 1906, the Select Committee on Cabs and Omnibuses was recommending that neither front nor back should be used for advertisements on motor buses, but should have conspicuous (and, at night, illuminated) destination signs, while the sides should also show the route (i.e. the sequence of places served by the bus) and not the 'useless fancy names' currently displayed.[42]

Not everybody agreed. Writing from the Athenaeum, 'C.L.E.' (Charles Locke Eastlake, then Keeper of the National Gallery) thought that, 'To Londoners the very colour of the vehicle is a sufficient index of its destination'. Looking back from 1938, the poet and literary editor Sir John Squire (1884–1958) was certain that, in the past, 'the buses were more easily identifiable . . . . Now they are all red, all called General, and only distinguished by numbers, which not all of us can remember'.[43]

The LGOC had started to use index numbers in 1891 to denote different routes as part of an internal accounting system, and the company also began to paint large initial letters indicating route names on the fronts of their buses.[44] But this, the use of distinctive colours and the painting of destination names on side panels, still presumed that each bus operated permanently on only one route. The catalyst for change was the introduction of motor buses operated by new rival bus companies. In 1906, the London Motor Omnibus Company, which initially painted the name 'Vanguard' on its buses, began numbering its 'lines' from 'Vanguard 1' through to 'Vanguard 5'. The LGOC followed suit and began numbering its routes in 1908. By 1924, buses were *required* to carry route numbers. The numbering was revised in 1934, after the formation of London Transport, but many route numbers have shown remarkable resilience: the 9 served Kensington Gore in 1908 as it does today; the 24 has been running down Gower Street since 1910 (Figure 3.2).[45]

The introduction of route numbers and movable or reversible destination boards, back and front, benefitted both passengers and operators and made possible a speeding-up of bus travel. Operators had more flexibility to switch vehicles between routes. Passengers did not need to hail and stop buses, or chase after buses, that turned out to be going somewhere they did not want to go. In the pre-number era, Fred Jane, artistic editor of the *English Illustrated Magazine*, reported:

> seeing a man rush some hundred yards up Holborn after a bus. Near the Tottenham Court Road he caught it. 'Marble Arch', he gasped, as he stumbled into the conductor's arms. 'No, Piccadilly Circus; yours is the one you've been running away from', came the answer wrapped in a sardonic smile.[46]

However, route numbers were only useful if intending passengers knew the number of the bus they wanted, and this required either printed information attached to each bus stop (impossible while there were no official stops!) or information leaflets, such as guidebooks or printed maps. There were 'conveyance directories' and a 'London Omnibus Guide' in the mid nineteenth century, and a correspondent to *The Times* had suggested as early as 1879 that timetables should be posted in buses and sold by conductors, but bus maps displaying routes and numbers were first issued only in 1911. The LGOC's first map showed only 23 routes. Following the company's absorption into the Underground Group, advertising really took off. By spring 1914, more than 100 routes were marked on maps, which were updated monthly.[47]

Boarding and alighting

There were also improvements to the design of buses that facilitated a speeding-up in boarding and alighting safely, and some failures to copy best practice

Figure 3.2 A Union Jack motor bus (1906–14). The route number '9' and the destination 'Leyton' via 'Bank' are clearly displayed, but still in smaller letters than the advertisement for Seeger's Hair Dye.

Source: Permission of London Transport Museum.

elsewhere that slowed down the process, generating the frustration expressed in misplaced hurry. On the earliest horse-buses, entry and exit were through a door at the rear, policed by the conductor. Buses were free to stop on either side of the road, causing congestion if they crossed to the 'wrong', right-hand side to pick up or set down a passenger, or if they stopped in the centre of the road, not only preventing the flow of other traffic but also endangering and inconveniencing passengers, who had to cross busy and usually muddy roadways between kerb and bus. In 1846, 'A Small Tradesman' proposed that fines should be imposed for picking up or setting down passengers except by drawing up to the kerb.[48] But it was only in 1867, under the Metropolitan Streets Act, that buses were forbidden from pulling in to the 'wrong' side of the road and required to pick up from and set down at the nearside kerb. Despite the legislation, the *Pall Mall Gazette* was still noting in May 1869 that:

> It has always been a great delight to omnibus drivers and conductors to eject their passengers into the middle of the street. It was such fun to see an infirm old lady or gentleman bespattered with mud, endeavouring to avoid being knocked down by a butcher's cart on leaving the bus.

Now, the police had received strict orders to enforce the new rules.[49] Aside from reducing congestion and protecting passengers, one consequence of the change was a shift from a door in the back of the omnibus to a rear platform only on the near side. And a consequence of this was faster loading and unloading of passengers, as well as the ability of passengers to jump on or off buses between stops, a new encouragement to reckless hurry.

Another improvement – requested as early as 1851, but still being urged by *Times* letter-writers in 1879 – was the replacement of a vertical ladder by a spiral staircase as the means of accessing the top deck of buses (Figure 3.3).[50] The *Pall Mall Gazette* noted in 1885 that it was still only in 'some of the more modern omnibuses' that a winding staircase was provided, and a publicity item in the *Western Mail* in late 1886 lauded a local manufacturer whose buses with 'a graceful spiral staircase' were now in service in London.[51] Not only did this encourage women to ride 'upstairs', but it also made it safer (with occasional fatal exceptions) to climb or descend while the bus was in motion, meaning that the time spent while passengers boarded and alighted could be reduced, and passengers could satisfy their own desire to hurry by descending to the rear platform before their bus had reached their destination.

Meanwhile, in the absence of fixed stops, passengers needed to alert conductors and conductors needed to alert drivers to tell them when to stop and when it was safe to start. The earliest devices were check strings attached to drivers' arms: a tug on the left arm meant pull in to the left-hand side of the road; a tug on the right meant pull across the traffic to the right. The LGOC touted a pneumatic 'omnibus telegraph' activated by squeezing small elastic air vessels attached to the inside of the roof, but a common way of attracting a conductor's attention was simply to poke him in the back with your umbrella.[52]

Passengers paid their fares on alighting. Although this was accepted practice for travel by cab, it was perceived as a source of unnecessary delay by bus passengers, especially when those alighting disputed the fare or failed to present the correct money.[53] The obvious solutions – to sell tickets in advance, or for conductors to issue tickets as soon as passengers had boarded – seemed far from obvious to both management and conductors.

The source of most controversy concerned the designation of fixed stops. The earliest horse-buses were, in effect, big hansom cabs that followed a fixed route. They could be hailed anywhere, and passengers could ask to alight wherever it was safe for the bus to stop. Passengers expected a personal, door-to-door service. A *Punch* cartoon (1874) by Charles Keene showed a conversation between driver and conductor:

Figure 3.3 (Upper) LGOC knifeboard horse-bus, with door and ladder at rear, and conductor standing on the rear step (c.1856); (lower) 1890s 'Atlas' garden-seat horse-bus, with rear staircase, in Trafalgar Square (c.1890).

Source: (Upper) Permission of London Transport Museum; (lower) Public Domain, Wikicommons.

> Driver: (*impatient*) 'Now, Bill, what's it all about!'; Conductor: 'Ge'tleman wants to be put down at No. 20A in Claringdon Square, furst Portico on the Right after you pass the "Red Lion", private entrance round the Corner!'; Driver: 'O, certainly! Ask the Ge'tlemen if we shall Drive Up-stairs, an' set 'im down at 'is Bed-Room Door!'[54]

Fanciful as this may seem, it merely exaggerated the practice of some suburban routes on which early-morning services would tour outlying neighbourhoods, picking up regular passengers from their homes before setting off for the City.[55]

In practice, although there were no bus stops and therefore no shelters or waiting rooms to protect intending passengers from the elements, and no possibility of selling tickets in advance, which, in the absence of a queuing system, could have established priority for boarding the next bus (all practices said to be common in Paris), there were 'recognized stopping places' where at least those passengers familiar with the system would wait for passing buses. Contrary to the stereotype of twentieth-century Britons as inveterate queuers, nineteenth-century Londoners (especially, in the eyes of mostly male letter-writers, middle-class women) were represented as pushing and shoving others out of the way in their stampede to board the bus:

> At hot corners of the streets, such as Oxford-circus and Piccadilly-circus, we must fight like wild beasts or be unblessed. Now that travelling by omnibus is resorted to by all classes, from the proud Belgravian dame to the denizen of Whitechapel, and ladies think it not demeaning to ride outside as well as inside, what can poor male man [*sic*] do but quietly to allow omnibus after omnibus to pass without being able to secure a seat? If ladies push him aside and crowd round the too susceptible conductor, what can he do? His gallantry forbids him to complain or resist.[56]

There was even more self-interest when it came to alighting. Women were again judged the main culprits, insisting on stopping the bus exactly at the address they wished to visit, even if it had already stopped only a few yards earlier. The argument against such frequent stops was phrased in terms of speed, convenience to other passengers, and congestion causing delays to other traffic, but also in terms of cruelty to animals. Women were portrayed hypocritically as espousing animal welfare causes while inflicting suffering on omnibus horses, worn out by constant stopping and starting, especially in frosty weather and on steep hills.[57]

'Man about Town' wished the bus companies would adopt:

> the French system of defining certain places for embarking and disembarking their passengers. What a saving of horseflesh, time, and temper there would be if they would only do so! Women are terrible sinners in their want of thought for horses – and the other passengers. Only the other day I saw an omnibus stopped four times in less than a hundred yards by thoughtless, inconsiderate women who wanted to be put down at one particular spot and no other.[58]

The designation of 'omnibus stations' (with no stopping to pick up or set down anywhere else) was discussed by the City Court of Aldermen as early as 1863, but resisted by the LGOC. Stopping was prohibited in certain locations – in front of Mansion House from 1870, and in the immediate vicinity of busy junctions such as Hyde Park Corner and Piccadilly Circus in the late 1890s – but it was not until the London Cab Act (1907) that the Metropolitan Police Commissioner was authorized to fix stopping places. Even then, he had no powers to prevent buses from making additional stops anywhere outside the prohibited zones.[59]

There were counter-arguments – from those changing from one route to another at intersections, where they were now forced to walk long distances between safe stopping places, and from the 'nervous, timid and elderly' who would benefit from additional 'recognized stopping places'.[60] But how should such stopping places be indicated? The earliest proposals envisaged 'omnibus stations' equivalent to railway stations.[61] By the turn of the century, demands had been scaled down: *The Graphic* suggested painting lamp posts a distinguishing colour, whereas Miss Sophia Beale wrote to *The Times* proposing poles supporting boards on which the names of the vehicles which stopped there would be listed, a practice long common in German cities.[62] John Reed notes that the LGOC 'experimented with a few fixed stops' in 1913. By 1919, there were 59 bus stops with posts and signboards (compared with 17,000 in 2000). Fixed stops were established immediately outside 'exclusion zones' (the areas close to busy intersections where stopping was prohibited), and a brief experiment was conducted with fixed stops along the entire length of one west–east and one north–south route that each extended from a distant suburb to central London.[63] Stops were also erected outside some Underground stations, a product of the increasingly integrated public transport network following the absorption of the LGOC into the Underground group in 1912. Borough councils approved the locations of stop signs, which were to be of a standard design, including variants for compulsory and 'voluntary' (subsequently 'request') stops. However, it was not until the establishment of the London Passenger Transport Board that a comprehensive programme of erecting bus stops throughout the network was implemented, between 1934 and 1937.[64] The bus-stop queue became a symbol of discipline and efficiency, evidenced by modernist, originally vorticist, William Roberts' *The Bus-stop* (1924), an ink and watercolour sketch intended to become a poster, but also by Roberts' subsequent less orderly *Bus-stop* (1948–9) and the descent into anarchy in his *Rush Hour* (1971), focused on the fight to board the bus to Waterloo.[65]

Not everybody thought it necessary for a bus to stop in order to board or alight. A letter to the *Morning Post* in July 1896 asked why drivers could not simply slow down to let passengers on and off 'while the wheels are slowly turning'. The conductor could use different signals to tell the driver to 'stop' or 'slow'. If buses stopped only at 'stated places', it would be 'an intolerable nuisance to many travellers' who did not want to walk all the way to or from official stopping places and who were perfectly able to board or alight while the bus was in motion.[66] 'K.S.', 'a married woman, over forty', found 'no difficulty at all in descending when the horses are slowing down a little . . . anyone not absolutely decrepit can do it'.

'Seventy-eight' commended the 'healthy agility' of lady passengers burdened with luggage but still capable of 'springing into the moving 'bus, and afterwards out again', and boasted of 'jump[ing] in and out while at full speed – except in a very slippery and crowded road'.[67]

On the other hand, 'J.B.W.', aged 56, judged themselves 'still active' but objected to alighting from a bus in motion, and 'An Old Stager', 64, 'not quite so good at jumping as I used to be', had 'been twice thrown sprawling on the street'. In attempting to board a bus in Shaftesbury Avenue, they had missed their footing and were dragged 'fully fifty yards, holding on by the step rail'.[68]

'Man about Town' pleaded that buses should stop for the elderly:

> In company with a lady over sixty years of age one day I hailed a 'bus, the driver of which, to do him justice, slowed down to about half speed on seeing that the lady with me was not active, and she was hauled on to the 'bus (much to her surprise) as it passed at a useful rate. I myself had to get up speed and sprint after it to catch it again.
>
> Only a couple of days ago I saw a gentleman hail a 'bus (producing no effect), make a grab at the hand-rail as it flashed by, miss his footing, and be dragged – literally dragged – for about thirty yards till, as aforesaid, hauled up behind. It was in a string of 'buses, and had he let go he was almost sure to have been driven over. His life was actually in danger.[69]

Jumping on and off moving buses could harm more than just one's dignity. A 65-year-old woman was thrown into the road while alighting from a bus in Caledonian Road. The bus company claimed she had tried to alight before the bus had come to a halt, but the jury believed the woman's claim that the bus had restarted before she had safely alighted and awarded her damages of £120.[70] A more complicated case involved a Metropolitan Meat Market worker who ran after a bus in Hart Street, Bloomsbury, hailed it, climbed on board, but then fell from the top deck on to the roadway. A witness claimed he had boarded the bus while it was still moving. The jury agreed, but thought that the bus company was guilty of negligence in not stopping to allow him to board safely. The Lord Chief Justice, however, found that the plaintiff had taken the risk upon himself by boarding the bus while it was in motion. The public must insist that buses were brought to a standstill: 'So long as people were foolish enough to get on trains and omnibuses in motion, they must be taken to intend to take upon themselves the consequences of their own indiscreet act'.[71]

The Times reprinted *The Lancet*'s explanation of the laws of motion and the horrendous medical consequences of attempting to 'alight sideways' from the end of a slow-moving bus.[72] 'Chancery Lane' reminded drivers and conductors that they were legally obliged to stop when passengers wished to board or alight, provided it was safe to do so, regardless of the consequences for either the schedule or the health of the horses.[73] It took the introduction of motor buses, with faster acceleration and deceleration, to deter travellers from getting on or off moving vehicles. The LGOC mounted safety first campaigns in the 1910s explaining how to board and alight safely.[74] But the desire to save time and to alight from a moving bus or from one caught in traffic between stops has never quite deserted Londoners. Consider the acclamation that greeted the introduction of the New Routemaster ('Boris Bus') in 2012, reinstating the 'hop-on, hop-off' open rear platforms that had characterized most London buses prior to the 1980s. In practice, health and safety fears and economy cuts that have eliminated the 'customer assistants' needed to police the open platforms have led to the closing of rear doors to prevent passengers from alighting at traffic lights or between stops.[75] The right to hurry, even at one's own risk, has been denied.

Conclusion

The history of the London bus illustrates many of the components of an architecture of hurry: engineering technology; design (including placement of doors, stairs and gangways); flows of information (communications between and among passengers and crew; the advertising of routes, times and stopping places); the education of passengers in boarding, alighting, standing and sitting safely and efficiently; and the education of other road users in judging speeds and anticipating the movements of buses. Passengers and operators might wish for buses to speed up, but London's physical and social geography constantly worked to slow them down. Minor efficiencies could be introduced, usually at the expense of individual self-interest. Aspects of modernity associated with order, regulation and discipline countered technological improvements that, in theory, facilitated speeding-up. Real speed proved to be dangerous in streets with multiple and diverse uses and activities. Competition – between passengers scrambling for seats, buses competing for custom, and buses and other road users for priority in a congested road network – promoted the impersonality enshrined in Forster's 'clipped words'. In the absence of real speed, passengers, drivers and conductors could resort to hurry, cutting corners by taking risks, pushing to the front of the queue.

The language and body-language of hurry on streets and in buses were bound up with questions of gender and class and increasingly focused on the rights, risks and responsibilities attributable to pedestrians, passengers, drivers and conductors, and other road users. Fixed bus stops are a good idea, but only so long as they are fixed at points where *you* have always chosen to get on or off. Route numbers are useful if you are a stranger, but another form of impersonal, clipped words if you are making a familiar journey.

Objectively, journeys by bus in central London are little quicker today than they were a century and more ago. Technological improvements – the replacement of horses by motor power, more efficient internal design and means of communication between passengers and drivers, the purchase of tickets in advance from roadside machines – have been countered by the sheer density of traffic on a practically unimprovable road network. The frequency of 'stops' may be less than when passengers could cause buses to stop whenever and wherever they chose, but the ability to hop on and off between stops has been curtailed. On the other hand, passengers are more able to make choices to speed up their own journeys. Timetables and journey planners mean that you know when the next bus is due, and real-time information can tell you if the scheduled bus is on time. Victorian Londoners had no idea how long they might have to wait, and the frustration born of ignorance lay at the root of an anxiety of hurry that encouraged rash behaviour – running for a bus that might, or might not, be the right one, jumping on or off and dodging between moving vehicles. 'Racing' and 'nursing' have mostly been eliminated, though not outside London, where neo-liberal deregulation has replicated the worst aspects of Victorian liberalism.[76] The 'clipped words' of conductors, shouting, often unintelligibly, the names of stops, or of advertising slogans, or of casual conversation, have been replaced by the clipped words of individual passengers' mobiles and tablets, by visual displays indicating the next stop, and by the indifference and suspicion shown by most passengers towards their fellow travellers. We have different ways of indicating that we are in a hurry.

Notes

1 Michael Flanders and Donald Swann, 'A Transport of Delight' (original version 1956), www.donaldswann.co.uk/index.html (accessed 20 January 2017).
2 Tim Cresswell, *On The Move: Mobility in the Modern Western World* (New York: Routledge, 2006), 4; Mimi Sheller and John Urry, 'The City and the Car', *International Journal of Urban and Regional Research* 24 (2000): 741.
3 For general histories of London buses, see T.C. Barker and Michael Robbins, *A History of London Transport* (London: Allen & Unwin, 1963, 1974); A.L. Latchford and H. Pollins, *London General: The Story of the London Bus, 1856–1956* (London: London Transport, 1956); John R. Day, *The Story of the London Bus* (London: London Transport, 1973); David Lawrence, ed., *Omnibus: A Social History of the London Bus* (London: London Transport Museum, 2014).
4 *The Times*, 20 April 1830, 3; 29 January 1831, 3.
5 *Old Bailey Proceedings Online*, www.oldbaileyonline.org, version 7.2, 15 June 2015.
6 Report from the Select Committee on Motor Traffic (1913), Appendix B, 13.
7 *Old Bailey Proceedings Online*, April 1833, trial of Thomas Chennell (t18330411-184).
8 *The Times*, 12 September 1833, 3; 4 February 1834, 6; 29 October 1834, 3; 7 February 1835, 2.
9 *The Times*, 20 April 1830, 3; 23 August 1834, 3; 24 June 1835, 6.
10 *The Times*, 7 April 1832, 5.
11 'Omnibuses', *The Times*, 31 August 1868, 9; 'Omnibuses, and how to Use Them', *The Leisure Hour*, 19 November 1870, 744.
12 Day, *Story*, 22–3; see also H.C. Moore, *Omnibuses and Cabs: Their Origin and History* (London: Chapman & Hall, 1902), chapter XII; for specific examples, see *The Times*, 17 August 1863, 8; 18 September 1863, 3.
13 Moore, *Omnibuses and Cabs*, chapter XII; Day, *Story*, 62–4; Lawrence, *Omnibus*, 102–15.
14 Virginia Woolf, *Mrs Dalloway* (Oxford: Oxford University Press, 1992[1925]), 176–7.
15 *The Times*, 4 December 1929, 10; 11 September 1935, 9.
16 Latchford and Pollins, *London General*, 31.
17 As reported in 'The Coming of the Motor-Bus. IV', *The Times*, 29 November 1905, 7.
18 'The Coming of the Motor-Bus. V', *The Times*, 1 December 1905, 7.
19 Select Committee on Cabs and Omnibuses (Metropolis) (1906), xiii, 2014, 2449–52.
20 Select Committee on Motor Traffic (1913), xx, Appendix K.

21 Major-General Sir Frederick Sykes, 'London Traffic', *The English Review*, March 1927, 300.
22 'The Coming of the Motor-Bus. VI', *The Times*, 4 December 1905, 7.
23 Sykes, 'London Traffic', 300–1.
24 London Travel Watch, 'Bus Speeds' (2015) (accessed 20 January 2017), www.londontravelwatch.org.uk/documents/get_lob?id=4069&age=&field=file.
25 Day, *Story*, 8–9; H.C. Moore, 'The Transition of the Omnibus', *Good Words*, December 1905, 705–6; Moore, *Omnibuses and Cabs*, chapter IV.
26 *The Times*, 9 October 1856, 6.
27 'The Man About Town', *The County Gentleman*, 16 January 1897, 70.
28 Select Committee on Cabs and Omnibuses (Metropolis) (1906), xiv.
29 Select Committee (1906), 2272, xv.
30 Select Committee (1906), 2658, 2849.
31 'The Coming of the Motor-Bus. VI', *The Times*, 4 December 1905, 7; Fred T. Jane, 'The Romance of a London Omnibus', *English Illustrated Magazine*, April 1894, 695–6; Day, *Story*, 66–8.
32 George S.C. Swinton, 'The Traffic of London', *The Nineteenth Century and After*, September 1905, 402; Wilfrid L. Randell, 'Speed', *The Academy*, 2 July 1910, 16.
33 Mireille Galinou and John Hayes, *London in Paint* (London: Museum of London, 1996), 242–3, 525, 548; also http://artuk.org/discover/artworks/search/actor:pollard-james-17921867/page/2 (accessed 20 January 2017).
34 Clifford S. Ackley, ed., *British Prints from the Machine Age: Rhythms of Modern Life 1914–1939* (London: Thames and Hudson, 2008).
35 Marshall Berman, *All That Is Solid Melts Into Air* (London: Verso, 1983), 158–60, 164–7.
36 'Modernity and Metropolis', in Helena Bonett, Ysanne Holt and Jennifer Mundy (eds.), *The Camden Town Group in Context*, Tate Research Publication, May 2012 (accessed 21 January 2017), www.tate.org.uk/art/research-publications/camden-town-group/modernity-and-metropolis-r1105709.
37 E.M. Forster, *Howards End* (London: Penguin, 2000[1910]), 92–3.
38 'A Futurist's Conception of a London Street', *Manchester Guardian*, 5 March 1914, 10; Richard Ingleby, Jonathan Black, David Cohen and Gordon Cooke, *C.R.W. Nevinson: The Twentieth Century* (London: Merrell Holberton, 1999), 71.
39 'The 'Bus', *Cornhill Magazine*, March 1890, 298–305; 'A Verb for Electric Progression', *The Times*, 11 April 1890, 5; 14 April 1890, 6.
40 'The 'Bus', 302; 'Omnibuses', *Cruchley's London in 1865*, reproduced in Lee Jackson, 'The Dictionary of Victorian London' (accessed 22 January 2017), www.victorianlondon.org/index-2012.htm.
41 Letter from 'A Merchant', *The Times*, 5 November 1850, 6; Letters from William Vincent and Robert Johnson, *The Times*, 29 July 1897, 12.
42 *The Times*, 10 December 1895, 14; Select Committee (1906), xii.
43 *The Times*, 13 November 1890, 6; 'Modes and Manners Awheel: "Locomotion in Victorian London" by G.A. Sekon: an appreciation by Sir John Squire', *Illustrated London News*, 5 February 1938, 204.
44 Day, *Story*, 38.
45 Day, *Story*, 48; J. Graeme Bruce and C.H. Curtis, *The London Motor Bus: Its Origins and Development* (London: London Transport, 1973), 13; 'List of Bus Routes in London' (accessed 23 January 2017), https://en.wikipedia.org/wiki/List_of_bus_routes_in_London; 'London Transport Route Numbering' (accessed 23 January 2017), www.red-rf.com/lt_operations_-_general/lt_route_numbering.aspx.
46 Jane, 'Romance', 695.
47 Day, *Story*, 25; *The Times*, 26 April 1879, 12; Lawrence, *Omnibus*, 62.
48 *The Times*, 6 April 1846, 3.

49 Reprinted in *The Times*, 8 May 1869, 9.
50 *The Times*, 28 February 1851, 5; 26 April 1879, 12.
51 'A Wail from the Omnibus', *Pall Mall Gazette*, 24 February 1885; *Western Mail*, 27 October 1886.
52 Moore, *Omnibuses and Cabs*, chapters II, V; *The Times*, 5 November 1850, 6; 6 December 1856, 10; 'London Omnibuses', *Once a Week*, 9 April 1864, 428; Jane, 'Romance', 691; Day, *Story*, 12, 19; Latchford and Pollins, *London General*, 18.
53 *The Times*, 18 September 1851, 5; 'Omnibuses, and how to Use Them', 746.
54 Charles Keene, 'Precise', *Punch*, 5 December 1874, 242.
55 Day, *Story*, 38–9; Moore, *Omnibuses and Cabs*, chapter VI.
56 *The Times*, 14 June 1890, 6.
57 W.J. Gordon, *The Horse-World of London* (London: Religious Tract Society, 1893), chapter 1, 'The Omnibus Horse'. From a huge correspondence, see 'Occasional Notes', *Pall Mall Gazette*, 24 November 1879; 'Omnibus Horses', *Morning Post*, 17 July 1896, 3; 24 July 1896, 6.
58 'The Man About Town', *The County Gentleman*, 4 July 1896, 846.
59 'City Traffic', *The Times*, 24 October 1863, 7; 'Mansion-house', *The Times*, 20 June 1870, 12; 'Hippodamia', *The Times*, 11 August 1897, 6; 'The Police and Omnibus Stopping-Places', *The Times*, 12 December 1900, 3; 'London Traffic Problems', *The Times*, 13 June 1919, 10.
60 'Hippodamia', *Morning Post*, 2 January 1884, 3.
61 *The Times*, 31 December 1878, 11.
62 'The Bystander', *The Graphic*, 11 July 1896; 'Omnibus Stopping Places', *The Times*, 12 September 1900, 11.
63 John Reed, *London Buses: A Brief History* (Harrow: Capital Transport, 2007); Omnibus Society, London Historical Research Group Minutes of Meeting, 8 October 1998, 916: Bus Stops (London Transport Museum Library).
64 'London Omnibus "Stop Signs"', *The Times*, 23 May 1922, 7; Lawrence, *Omnibus*, 164.
65 'An English Cubist: William Roberts, 1895–1980', www.englishcubist.co.uk/ (accessed 24 January 2017).
66 *Morning Post*, 24 July 1896, 6.
67 'Getting Off Omnibuses', *Standard*, 18 January 1899, 9; 19 January 1899, 8.
68 *Morning Post*, 24 July 1896, 6; *Standard*, 18 January 1899, 9.
69 'The Man About Town', *The County Gentleman*, 8 October 1898, 1302.
70 'Bonner v. The London Road Car Company', *Morning Post*, 17 November 1898, 7.
71 'Lill v. The London General Omnibus Company', *The Times*, 26 January 1899, 9; 31 January 1899, 14.
72 'The Laws of Motion', *The Times*, 7 September 1872, 5.
73 'Getting Off Omnibuses', *Standard*, 19 January 1899, 8.
74 Lawrence, *Omnibus*, 67.
75 Dave Hill, 'How the New Routemaster Came Full Circle', www.theguardian.com/cities/2015/aug/03/new-routemaster-old-london-bus-boris-johnson (accessed 25 January 2017).
76 Ian Taylor and Lyn Sloman (2016), 'Building a World-Class Bus System for Britain', www.transportforqualityoflife.com/u/files/160120_Building_a_world-class_bus_system_for_Britain_FINAL1.pdf (accessed 20 January 2017).

4 The tales of two mobility infrastructures

The street and the underground railway of Buenos Aires, 1880s–1940s

Dhan Zunino Singh

Introduction

In December 1913, the first line of the Buenos Aires Underground (*Subte*) was opened, and Buenos Aires became the first city with an underground transit system in the Southern Hemisphere. It created a new space of circulation for the city and seemed to fulfil an extant promise of modern transport, namely: safety, comfort, and speed. Velocity, perhaps the ultimate expression of hurry, signified the 'annihilation' of space (distance), the reduction of travel time, and the elimination of traffic obstacles.[1] As symbols of the rational organization of space in relation to time, such features have made the underground railway a preferred solution to the related problems of urban congestion and suburbanization in modern metropolises. Because of this, during the inauguration ceremony of the first line, Mayor Joaquín Anchorena celebrated the new underground, as it epitomized Buenos Aires's coming of age as a modern metropolis. Like many others of his era, Anchorena associated modernization with the speeding-up of urban mobility. The Underground, he said, by reaching 'unknown velocities' and 'abolishing distances', will 'allow us not only to save time but to offer invaluable comfort to the anxious multitude that moves in the business field'.[2] Indeed, the increase in urban journeys facilitated by the underground railway prompted commentators to make parallels between velocity and the progress of the city and, in turn, of the whole nation – as claimed by the Central Society of Architects for the fourteenth anniversary of the first Subte line:

> Velocity is the foremost characteristic of progress and, hence, the permanent eagerness of the big cities is always to reduce distance [. . .] The more important and intense is the commercial movement of a country, the faster its transport service has to be.[3]

From a sceptical point of view, the writer Ezequiel Martínez Estrada noted that Buenos Aires had become a 'racetrack', and the surface space was not enough to contain an absurd velocity, therefore it needed to go underground. In so doing, underground space offered an 'exhaust valve for the excess of the yearning of velocity', channelling 'the surplus of energy' of the metropolis through pipes, cables, and subways.[4]

The Underground contributed to the spatial and technological transformation of Buenos Aires and occupied a central role within urban debates of the period 1880–1940. Plans sought mainly to change the street layout inherited from the colonial period (the checker-board) to improve aesthetics, hygiene, and circulation. Yet, the grid remained as the street layout of the city, and traffic problems persisted. In this context, the Underground was perceived as the best solution for public transport (moving the masses) and urban traffic (speeding up mobility). Nonetheless, it had to compete with the emergence of cars and buses and the importance that civil engineers gave to automobile infrastructure.

This chapter deals with the period of planning and construction of Buenos Aires's Subte (1880s–1940s), looking at spatial and symbolic aspects of urban and mobility spaces through the analysis of debates, plans, maps, photos, statistics, advertising, reports, and news. It argues that the early implementation of underground transit links necessarily to the growth of city traffic, especially as the latter involved, not only the flow – the speeding-up – of the new metropolis, but also the congested and dangerous streets of the colonial city. In this way, the Underground represents a crucial architecture of hurry in Buenos Aires, allowing for free-flowing mobility. Even so, the importance of the Underground decreased with the emergence of automobility and automobile infrastructure, a new circulation space for a new form of rapid transit. And through it we see that both the Underground and the automobile enabled the development of hurry by segregating diverse mobilities through the construction of dedicated conduits with standardized rhythms.

The argument unfolds thus: the first section demonstrates contemporaneous representations of deleterious street traffic, and its 'diagnosis' by urban planners. The main urban and transport plans, debates, and improvements are discussed next, in order to understand ideas about city circulation, the preference for going under the ground instead of above, and the later preference for the automobile over guided transport systems, particularly the tramway. Then, I deal with the ideas and values of modern transport that the Underground expressed, but also the ambivalent perceptions related to atavistic representations about subterranean space. The chapter ends with a discussion about the relational aspect of the surface and underground infrastructures of urban mobility.

Street views: old, congested, and dangerous

Between 1880 and 1940, Buenos Aires experienced rapid and spectacular 'metropolization'.[5] The city expanded in area (from 4,000 to 18,000 ha), and its population grew from 430,000 to 1.4 million between 1887 and 1914, owing to European immigration, reaching 2.4 million in 1936 because of a later rural–urban migration. Along with population growth and urban expansion, socio-spatial changes were associated with the experience of speeding-up led by the modernizing of transport technologies and infrastructure. In cultural terms, representations about metropolization were rather ambivalent, showing a permanent tension between ideas of progress and its material consequences. At the beginning of the process, the growth of Buenos Aires in terms of its size and population was a source of pride, although there was concern about how to control expansion and indicate the limit of the city.[6] By the 1920s, with the emergence of urbanism as a discipline and the city growing beyond jurisdictional limits, Buenos Aires was perceived as a sick and unwieldy metropolis. 'Macrocephaly' or 'gigantism', as the most relevant urbanist, Carlos della Paolera, called it, expressed the anxiety caused by the lack of a general 'plan' to guide or regulate rapid urban growth.[7] Urban debates filtered into public opinion, and the problem of size was epitomized by the figure of Goliath's head, the title of a book of influential essays about Buenos Aires by the writer Ezequiel Martínez Estrada.[8]

The city's fabric and evolution underlie perceptions about its traffic. Buenos Aires is a port city, with the 'centre' around the main square (Plaza de Mayo) beside the River Plate. In 1880, it became the capital city of the Republic, and new political boundaries were established to extend the territory of the old city towards the west and north. Reinforced by the construction of the new port (1887–98), commercial, administrative, financial, and even leisure and entertainment activities remained around the main square, as in colonial times. The first immigrants settled in old buildings, which became slums in the southern downtown. In the 1890s, Buenos Aires was still a walking city; the majority of the population lived within an urban area that extended about 2 km from the port. The process of suburbanization began by 1904: when the electrification of tramways was almost complete, when the municipality published a plan (the grid) that facilitated the urbanization of the capital city's new lands, and when the property market was boosted by the easy availability of affordable credit. The tram network extended to reach districts located between 6 km and 9 km from the centre. By the end of the 1920s, most of the capital city was fully occupied, and expansion continued beyond the city's political limits, connected to the centre by railway lines.

Modes of transport played a key role in urban expansion, allowing settlement on the periphery. Although railways were the first mechanized transport implemented in Buenos Aires (1857) and influenced settlement patterns through radial lines, the electric tramway was fundamental to suburbanization, moving people beyond the urban core (6 km or more from the main square) and offering a fast and affordable transport service. Not only did the network cover the urban core, but some lines extended into peripheral areas that had remained mostly rural since 1900.[9] In 1910, municipal engineers recounted how the tramway 'urbanised' the periphery:

> it has been seen how the line of La Capital Tramways, which went to Flores along completely deserted [. . .] streets, led in two years to the construction of buildings which are today on both sides of those streets and along many blocks.[10]

In the 1920s, with the introduction first of omnibuses and later *colectivos* (another bus-like form of shared transport), the auto-transport network covered many outlying districts. The first underground line (Line A), built in 1913 by the Anglo-Argentine Tramway Company (AATC), was part of a network that, for economic and political reasons, took three decades to be completed. In 1930, a second line (B), built by the national rail company Lacroze, was opened, and the next three lines (C, D, and E) were built between 1933 and 1944 by Spanish company CHADOPYF. The entire network was about 32 km. The four radial lines (east–west) reached districts about 6 km from Plaza de Mayo (Figure 4.1). From the beginning, the Subte carried annually 10 per cent of total passengers and many more passengers per kilometre than other modes.[11] By the late 1920s, surface railways were bringing commuters who lived beyond the capital city to the downtown. Private transport also grew and changed with the emergence of the automobile. For example, whereas in 1901 Buenos Aires had 5,200 bicycles and 1,200 cars, by 1930, there were around 52,000 cars and innumerable bicycles.[12] In short, Buenos Aires relied on varied modes of mechanized transport that had increased and expanded in two decades, reducing travel times and shrinking space.

Observing the images that circulated through the press, cinema, advertising, and so on, we see, not only the large daily volume of traffic between the suburbs and the city centre, but also the congestion in the central business district. This constituted one of the most striking features of Buenos Aires's metropolization.

Figure 4.1 Buenos Aires Subte network in 1950. The black dotted line indicates the limits of the capital city.

Source: created by the author.

As a popular magazine commented in 1932, urban traffic symbolized the city's vitality: the magazine thought that the city 'suffers from congestion' mainly because 'its arteries are insufficient to contain the huge volume of vitality that throws itself into them'.[13]

The image of traffic as vitality coexisted with the daily experience of blocked streets, delays, and collisions that turned the intense rhythm and volume of traffic into a peril of urban life. Paradoxically, traffic embodied uncontrolled and unplanned fluidity, even as it simultaneously generated blockages. Many of the problems associated with traffic were attributed to the city form inherited from colonial times – the narrowness of the streets, the checker-board layout, the location of the city centre, and the concentration of activities around the main square.[14] This historical influence persisted throughout the period of metropolization, regardless of the scale of traffic and the improvements carried out. In 1896, for example, a transport entrepreneur claimed that the layout of the 'old city', where 'the highest vitality of the organism of the great metropolis necessarily concentrates', was 'inadequate for circulation'.[15] In 1930, an engineer proffered a similar diagnosis: the 'narrow streets, reminiscences of a slow life, are today incapable of containing the volume of flow of a traffic that demands celerity and cannot develop it'.[16] Such a modernized representation of traffic progress reinforced tensions between backwardness and modernity. Modernization, hence, implied a radical change of urban fabric – the street layout, in particular – to cope with the rhythm of modern life.

Problems of street traffic were a recurrent topic in newspapers and magazines. Congestion, demands for strict controls on traffic, proper uses of a thoroughfare, relations between tramway staff and passengers, conditions of street paving, problems concerning timetables and delays, the perils of level crossings, the parking of private cars in central streets, and the lack of regulations about loading and unloading of goods were among the commonest issues. Frequent and recurrent collisions and fatal accidents reinforced the perception that traffic chaos inevitably made streets unsafe.

Visitors remarked on Buenos Aires's problems. Guests at the 1910 Republic Centenary, who widely admired the progress of the city, also turned critical when they discussed their experiences of traffic. Boulevards and promenades were perceived as beautiful because they were fashionable, clean, and bright and offered a broad prospect. The narrow streets of the centre, however, were busy, noisy, and dangerous for pedestrians. In the same year, Lloyd's Greater Britain Publishing Company included Argentina in its famous 'Twentieth Century Impressions' series. It observed the need for Buenos Aires's pedestrians to develop survival skills in an increasingly dangerous traffic milieu, one approaching 'tyranny'. 'How easily', the sarcastic observer suggested, 'the best-natured of pedestrians may be killed before he can complain'.[17]

The peril that street circulation represented for Buenos Aires's inhabitants was evident, for example, when the first underground line was opened in 1913. An editorial in the popular magazine *P.B.T.* explained that the success of the Subte derived from the 'terror' of the street: 'Carts, cars, bicycles, tramways and even the trolleys used by hawkers have made the surface of the city a public crushing place [*aplastadero público*]'.[18] Along with accidents, the prevailing image of chaos was the traffic jam. The important popular magazine, *Caras y Caretas*, in 1928 published a front-page drawing sardonically titled 'The Delights of Traffic'. It depicted in a picturesque way a typical scene of Buenos Aires traffic: crowds of people and vehicles in a futile hurry. In 'The Problems of Congestion', a cartoon published in 1930 in the cultural supplement *Plus Ultra*, the cause was ironically represented through the woman driver. The cartoon shows a group of taxis, an omnibus, and private cars stuck at a crossroads. The cause seems to be a luxury car driven by a woman who is applying make-up. The picture not only affirms increasing automobility in the 1930s, but also the emerging participation of women in public spaces – and a prevailing twentieth-century sexist attitude towards women drivers.

Even when some significant improvements were undertaken, there was still scepticism about the results they could achieve. Martínez Estrada suggested that widening of avenues, instead of relieving congestion, actually increased it: 'the paradoxical phenomenon is that the wider the streets are, the bigger traffic congestion is', as 'a new avenue is an attraction for the pedestrian and an opportunity to show off for the driver, especially if the car is new'.[19] The irony was that improved architectures of hurry contributed to what we now call 'traffic generation'.

The impression of traffic chaos was not only based on the experience of the street; it was also largely influenced by an ideal of circulation that, for all its proliferation in the press, was generated mainly by official voices – politicians, planners, and transport companies. Officials could agree with public opinion that traffic problems were inevitably a consequence of progress, as the Mayor stated in 1909: 'probably the phenomenon is common to all large cities of the world, because the conquests of modern progress have created new necessities and aspirations with the light of every new day'.[20] Urban traffic could be perceived positively, as a sign of progress and vitality, as long as such movement was orderly and regulated.

Body and medical metaphors of the city as an organism, and engineers and planners as physicians, are commonplace in urban planning history.[21]As Richard Sennett explains: 'the words veins and arteries applied to the city's streets sought to model traffic systems on the blood system of the body'. Hence, 'if motion through the city becomes blocked anywhere, the collective body suffers a crisis of circulation like the individual body suffers during a stroke when an artery becomes blocked'.[22] This body–medical language is well exemplified by the urbanist Carlos della Paolera, who understood that Buenos Aires suffered from 'stroke', and that the role of urbanist was that of 'doctor', to diagnose and heal the city. When the 9 de Julio Avenue was opened, della Paolera said that his project represented 'a real work for the decongestion of the city centre' and it also brought 'hygienic sunbathing'. To feel 'the sun's caress', he said, 'the Municipality's scalpel performs there a necessary amputation', and the roundabout where the Obelisk was placed was 'a bloodletting urgently carried out in the apoplectic Buenos Aires'.[23]

Urban reforms: the street layout and alternative spaces of circulation

Most of the urban plans for Buenos Aires were strictly street plans seeking to change the checker-board layout in order to 'modernize' the city.[24] Such modernization was based on three criteria: circulation, aesthetics, and hygiene. Urban designs were observed and borrowed from Paris, London, Vienna, and American cities – models of the modern metropolis shaped by planners, architects, and engineers. Whether it was professional Argentines travelling to Europe and the United States, or professional Europeans (Leon Jaussély, Weger Hegemann, Le Corbusier) journeying to Argentina, Buenos Aires became a laboratory for modernist planning ideas.

French urbanism – most particularly Haussmann's reforms – was influential in Buenos Aires, especially in the opening of boulevards and a street layout based on diagonal avenues and roundabouts, as exemplified by Mayo boulevard (1888–94) and represented by Joseph Bouvard's plan (1909). Only two short diagonal avenues were finally built one decade later, following this plan: Northern Diagonal and Southern Diagonal (Figure 4.2). Later plans such as the Organic Plan (1925) were more influenced by the American City Beautiful movement. Nonetheless, the main advisor was Jean-Claude Forestier from the French Association of Urban Planning. The Organic Plan proposed an important

Figure 4.2 A future Buenos Aires with the grid layout modified by diagonals, as illustrated in Villalobos's watercolour. The bird's-eye view highlights the city enhancement through the perspective created by the diagonal avenue.

Source: *Caras y Caretas*, 5 April 1913.

new component for Buenos Aires: decentralization. The new city had to be reorganized, 'with new civic centres in the barrio parks and new avenues connecting to the periphery', within the limits of the capital city.[25]

Although aesthetics, hygiene, and circulation remained central in urban plans and debates, the idea of Buenos Aires as a metropolitan region – as an extensive geographical and economic area – became more important in the 1920s. Carlos della Paolera – known as the father of Argentinian urbanism – studied in France and followed organicist planning principles in paying attention to the expansion of the city. For the new, local urbanism, led by della Paolera, form must follow function. In this sense, the large flow of passengers arriving in the downtown from the outskirts every day, shaping a pattern of circulation with a 'pendulum-like movement', made circulation one of the main concerns. But, unlike the traffic that earlier plans had dealt with – mostly carriages and tramcars (first horse-drawn and then electric) – by the end of the 1920s, the automobile was becoming the embodiment of future transport. Boosted by the American car industry as well as local motoring clubs, the growth of automobile culture required the creation of new roads.[26] Moreover, in the 1930s, important urban reforms were carried out following the national government's application of Keynesian policies favouring public works programmes. In the suburbs of Buenos Aires, sanitary and street improvements were made; many avenues were widened, and new ones opened. Roads were a visible symbol of public investment. In the city, the most important works were two old projects from the late nineteenth century which, in the 1930s, acquired a new scale and purpose: the 9 de Julio Avenue and General Paz expressway. Although both reflected the aesthetic and hygienic objectives of American-style parkways, the mobility they envisioned signified a change in the rhythm of the city: they were automobile infrastructures designed for metropolitan and regional flows.

The opening of the 9 de Julio Avenue started in 1936 with the creation of a roundabout (Plaza de la República), where the Obelisk was built at the junction of the northern diagonal, Corrientes Avenue and 9 de Julio. It was planned by della Paolera, who was chief of the Urbanization Plan Office. The idea of a north–south avenue in the downtown had been conceived in the 1890s. Act 8835 (1912) established a 33-m-wide street, but della Paolera proposed a 110-m-wide 'park avenue', which implied the demolition of entire city blocks. The current avenue took decades to build, but the first 500 m, opened in 1937, with a final width of 140 m, comprising five roads separated by lines of trees, exemplified the intention of creating an infrastructure for a large flow of cars, although still paying attention to aesthetics and hygiene. As shown by della Paolera's hygienic discourse, the new avenue signified a 'lung' and wide 'artery' for the city. The American 'parkway' model was adopted, as the lack of green spaces was one of della Paolera's main concerns. Engineers and architects, such as Ernesto Vautier who also proposed projects for the 9 de Julio, criticized him because he placed more emphasis on landscape than speed. Although the width of the avenue accommodated large traffic flows, there were crossroads every 100 m thanks to the grid layout of the city.

In the 1933 plan that della Paolera had proposed, and in his competitors' plans, the avenue was conceived as a sunken avenue (an open trench) to allow cars to move freely beneath the cross streets (Figure 4.3). However, the plan of placing the avenue below ground level was later modified, as Subte Line C would be built along the same route. As constructed, the avenue limited and segregated pedestrian mobility (with subways for pedestrians to cross the road) and provided underground car parking, seeking to resolve one of the main problems caused by cars downtown: parking in the middle of avenues.[27]

The second intervention, initiated in 1937, was the General Paz Avenue. Led by the National Highway Administration (*Dirección Nacional de Vialidad*), engineer Pascual Palazzo (who had proposed a plan for underground highways in the 1920s) and architect Ernesto Vautier laid out a reinforced-concrete 'ultra-fast bypass' encircling the city.[28] The bypass (24 km long and 100 m wide) was bigger than the original ring road (30 m wide) planned in 1887 as the limit of the capital city. It responded to the increasing numbers of cars (about 72,000 in 1937) and sought to distribute traffic flows from the surrounding towns and national roads towards the radial avenues that ran in an east–west direction within the city. The American parkway was a strong influence on this plan regarding the relationship between road and landscape, but aspects of the Italian *autostrada*, the German *autostrasse*, and even the Parisian boulevards were also considered.[29] Plans were adapted to different local and material needs, including ideas for creating a new landscape (a green belt) and improving, not only the speed, but also the smoothness of automobile travel, and considering diverse

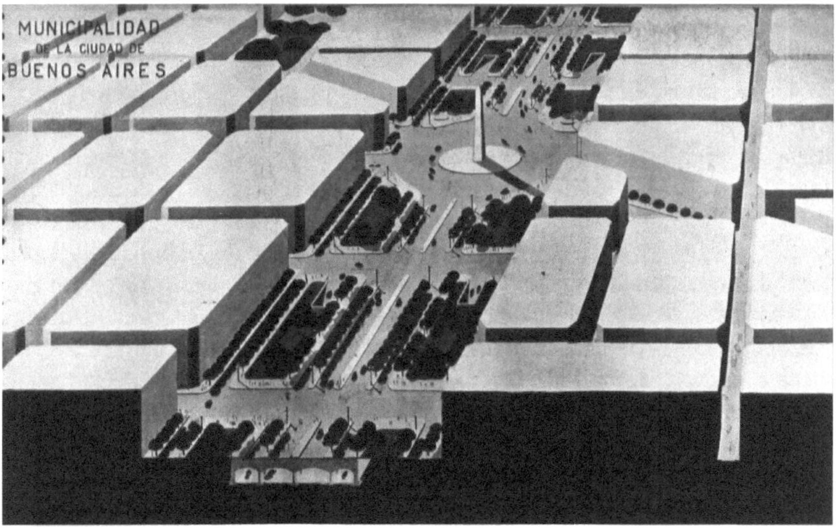

Figure 4.3 Plan for the 9 de Julio Avenue by Carlos della Polera, Buenos Aires Municipality, 1937.

Source: *Revista de Arquitectura*, August 1937.

uses: leisure, circulation (daily commuting), and commercial flows (buses and trucks). To accommodate speed, tunnels and bridges in different, but visually appealing, architectural styles were built to span the 24 streets and nine railway lines that crossed the ring road.[30]

In parallel to the street plans, from the 1880s there were projects for alternative ways of moving people and goods within the city. Elevated and underground railways were proposed by local and foreign entrepreneurs, railway and tramway companies, and even by the municipality, which in 1909 proposed a large underground tramway network (with radial and concentric routes) following the street layout planned by Bouvard and allowing for connectivity between future suburbs. In 1911, the main transport company in the city, the AATC, introduced its own plan for three underground lines (two running west–east and one north–south in downtown) to improve tramway circulation by connecting its surface network with tunnels using ramps, a model taken from Boston, Massachusetts.

As Buenos Aires streets were perceived as old, congested, and dangerous, and street reforms tended to be delayed, the electric underground railway was conceived as the best solution for traffic congestion and for a rapid connection between centre and suburbs. Masses of commuters could be transported without using the street. Congestion could be resolved, as the municipality specified that tramlines should be removed from streets beneath which underground lines ran. Tunnel transit, of course, allowed the circulation of trains without disruption to other vehicles and pedestrians, meaning that electric trains could run at higher speeds. Therefore, along with safety, the underground promised speed – a speed that electric tramways were prevented from achieving, partly because of other traffic, but also because municipal laws restricted speeds on downtown streets to about 12 km/h.[31]

Electrification was paramount, as the first underground railways in London had shown the advantages of rapid transit through tunnels but the disadvantages of steam traction. Electric underground transit in Budapest (1896), Boston (1897), Paris (1900), and New York (1904) confirmed the utility of electric underground trains, as electricity brought better lighting inside stations, cars, and tunnels, but, most importantly, it eliminated the bad atmosphere created by steam locomotives. But there were other reasons to go underground in Buenos Aires. One was aesthetic: elevated railways were ugly and spoiled the facades of buildings. It mattered that beautiful cities, such as Paris, refused the elevated railway option and, by 1900, New York's elevated railways were being fiercely criticized.[32] Another reason lay in the Argentinian elites' perception of the modern American city: the elevated railway was too redolent of American industrial cities, particularly New York, which offered a model elites tried to avoid.[33] In short, although several plans for elevated tramways were proposed for Buenos Aires at the end of the nineteenth century, by the turn of the twentieth century the electric underground was much more accepted as an alternative space of circulation. The underground retained its prestige status, and, at the inauguration of Line B (October 1930), politicians and transport businessmen made claims for a large underground network like a 'spider-web', 'a closely-woven web' of underground lines crossing Buenos Aires's subsoil in all directions.[34]

The type of metro implemented in Buenos Aires – a shallow underground tunnel (cut and cover) for electric trams – was shaped by a vigorous local debate in which political and economic interests were at stake. A dispute between the municipality and the national government about jurisdiction over the city subsoil, between 1905 and 1910, contended that passenger transport in the city must be served by tramways (under municipal legislation), and cargo transport by railways, subject to national legislation. This distinction also involved the placement of each tunnel: whereas underground tramways must be placed close to the surface, underground railways must run at a lower level. This was one of the main reasons for planning shallow tunnels for electric underground tramways for passenger service. Yet, cultural reasons related to representations about subterranean space were also influential on the conception of the underground as a shallow tunnel.

Underground culture

The underground system in Buenos Aires was perceived as a safer and faster form of transport than those modes that run on streets and, hence, more comfortable. The narrative of transport history tends to show transport technology as an 'evolutionary chain' in which new modes of transport 'naturally' replace older modes: from animal to steam, from horse-drawn to electric tramway, and so on. Such evolution is evident, for example, in the cover of an AATC pamphlet produced for the company's Golden Jubilee: the drawing represents the 'evolution' of the company and the city between 1876 and 1926, showing a horse-drawn tram on a dirt road beside an old, low building; next, the same street, paved, with an electric tram and a high modern building; and, at the bottom, inside a box, the underground tramway. The ideal of circulation, said the Organic Plan, quoting an AATC report, would not be 'achieved without certain sacrifices and some kinds of vehicles must succumb to the selective law of progress'.[35]

Expectations about speeding-up before and during the construction of the Subte, anticipating reductions in both journey time and waiting time, can be found, not only in advertising or technical reports, but also in cultural discourses. Consider engineer Stock's claim in 1930: the Underground was so rapid it allowed workers (particularly employees who worked in the city centre) to go home for lunch – in the suburbs. Journey time was the best indication of the Underground's advantage over street-surface modes of transport, as an advertisement from 1936 illustrates: different means of transport covering the same distance are measured by journey time (in minutes), showing that the underground railway was the best (20 minutes by underground against 31 by *colectivo*, 45 by omnibus, 60 by tramway).[36] Another advertisement (c.1940) exaggerated the speed of the underground, comparing it with an aeroplane. The poster showed an aircraft plummeting and a caption reading, 'Speed? Try the underground'.

Even before construction of the first line, a popular magazine, reporting an underground plan in 1901, imagined that, 'we will be able to run more than twenty blocks in the short time it takes to light a cigarette since the average speed of the underground railway will be 18.94 mph'.[37] On the eve of the inauguration of Line A, a prominent newspaper editorial envisaged that the train would run through the tunnel like a 'bullet' passing through 'the barrel of a rifle'.[38] A contemporary magazine claimed that 'innumerable trains will "fly" every 3 minutes'.[39]

Evidence for passengers' direct perception of speed has not been found. Even press reports about it are scarce. Nonetheless, as a chronicler recalled in 1930, the underground railway was a 'catacomb of vertigo' that, unlike the 'catacomb of mystic and fervent serenity', was devoted to velocity. The 'feeling of vertigo' was apparently based on the fact that speed was being experienced within a tunnel.

Those expectations were triggered by the maximum speed an underground train could reach – which varied between 50 and 65 km/h; however, the cruising speed was about 30 km/h, which was not very different from other means of transport. Nonetheless, the sense of speed in the underground railway was stimulated by the fact that one could see the walls of the tunnel rushing by only a short distance away, and exaggerated by the lights of the tunnels characteristic of the Subte (except on Line A). Another peculiarity of the Subte was that there were windows at the front and the back of every car, thus permitting passengers to see the tunnel from the driver's perspective or to have a view back down the line from the rear window of the last car.

The feeling of vertigo mentioned by the chronicler during the underground journey on Line B was shaped by the perception of movement in the tunnel:

> Out of curiosity, peering out of the back window of the last car, we could appreciate the speed of the train [. . .] A few metres after leaving the station, the train speeds up more and more. Peering out of the rear window, we have the sensation that above, below, and on both sides the roof, walls and rails flee backwards, chasing the station, whose lights will disappear on the first curve of the journey. On one side of the tunnel a series of small lamps, three staggered every three metres, seems, with the speed [of the train], to be a single cord of light attached to the wall.[40]

Finally, the visual impression of speed was also stimulated by the double-track tunnel, which enabled passengers to see trains coming from the opposite direction. At the opening of Line B, a newspaper showed an old man inside an underground car with a little girl seated beside him. He points with his finger at a train rushing by. The child looks at it, astonished.[41]

Yet, if the construction of the Subte signified a victory of modern engineering, at the same time, this artificial underground space was perceived as an oddity, characteristic of the apprehension that subterranean spaces triggered in many cities. There is a common trope within Western culture that links subterranean space to Hell, the Underworld, or the Abode of the Dead.[42] A combination of fear and fascination had been a typical reaction towards underground railways during the nineteenth century, in both America and Europe.[43] Those tropes were highlighted mostly, but not exclusively, when the first Subte line was inaugurated by humorous and satirical discourses in articles, editorials, and cartoons in illustrated and popular magazines. Underground was for troglodytes or burrowing animals; underground transit was a sign of backwardness rather than progress. Such sarcasm and humour seem to have been merely a jocular antidote to anxieties about the unfamiliar and the modern, rather than an actual attack on the idea of transforming the underground space of the city. Moreover, the same magazines, along with newspapers, celebrated the opening of the Subte with articles about, and especially photography of, the construction process, emphasizing creative destruction as a sign of modernity. The 100,000 people who attended Underground ceremonies on the opening day in 1913, and the even greater numbers at the inauguration of Lines B (1930) and C (1933), demonstrate the local interest in underground transport technology.[44]

The ambivalence towards going under the ground was also expressed in technical decisions. A tunnel situated as close as possible to the street surface was preferred by municipal engineers of the Public Works Department in 1905, on grounds of aesthetics, hygiene, and economy, as well as the jurisdictional controversy with the national state.[45] Not only would stations being close to the surface be well illuminated with electric light, but this would also allow sunlight to penetrate and fresh air to circulate. The platforms had to be easy and rapid to reach using stairs (with no need for expensive, time-consuming lifts). Such attributes became a singular characteristic of the Subte, particularly Line A, which was represented by the AATC as an improvement on other underground railways around the world.[46]

A concern about tunnel depth can be found in the construction of American subways, as a report by the chief engineer of the New York Subway, William Parsons, illustrates. Citing economic reasons to avoid deep tubular tunnels (like London's Tube) – such as the need for long escalators or deep lifts, or the cost of tunnelling instead of employing cut-and-cover methods – Parsons argued that a close-to-the-street, well-illuminated tunnel helped 'to destroy the popular antipathy to holes in the ground' and 'to remove the dark aspect' of the Underground.[47] The first American subway, in Boston, accomplished the ideal of a shallow tunnel or underground street, as claimed in the Boston Rapid Transit Commission's report in 1894: 'Such a subway will be in no sense a tunnel, but may be more correctly described as a basement street'.[48] The AATC, which found the Boston Subway (1897) to be an ideal model to maintain its tram business (because tramcars running on the streets went down through tunnels to access the central district), also highlighted the well-illuminated tunnel of Line A, because it gave 'the passenger the illusion of being in the street'.

Yet, this shallow-station common sense was difficult to implement in Buenos Aires's entire network, for practical reasons and political expediency. In the case of Line B, the National Congress approved a plan for an underground railway for passengers. The line started in the western suburbs as a shallow tunnel, then descended to a deeper level to pass under the Maldonado River, but, near downtown, the municipality required it to be about 14 m in depth to allow future underground tramways – like Line C – to run above. The American company that built Line B used cut and cover for the first section to construct a rectangular box tunnel in the western suburbs, but then the next section was built with a double-vault tunnel for the trains. Ventilation was to be produced by the movement of trains. The stations had a first level under the street with a hall and then a second level for the platforms. Escalators (the first in Latin America) were included as a novelty.

As the original scheme was interrupted by a conflict between the AATC and the city council about the cost of fares in the 1920s, new plans were submitted by private entrepreneurs. Although, the city council rejected a plan that proposed to build a London-type tube system passing under buildings, large sections of Lines C, D, and E, built by a Spanish company and approved by the municipality, were in relatively deep tunnels, necessitated by having to cross existing lines and by the impossibility of using cut-and-cover methods because of the disruption to street traffic they would have caused. Nonetheless, the construction of deep tunnels was an engineering feat signalling modernity, as was the insertion of the tunnel of Line C through a gap of only 5.6 m between the two current lines, and the construction of the junction of Lines B and D under the city's main monument (the Obelisk). These engineering achievements were disseminated through both technical journals and the company's own brochures, where technical sketches were re-deployed with a new aesthetic that emphasized speed through an underground infrastructure that normally remained hidden. The overlapping of tunnels, for example, resulted in a powerful image for advertising, not only the engineering feat, but also the way in which Buenos Aires was becoming a modern metropolis, like New York. In 1937, a CHADOPYF pamphlet showed the multilevel spaces of circulation in each city: although New York had five levels (three levels of tunnels, the street, and the elevated), Buenos Aires now had four (three tunnels and the street).

The Subte's designers thought about passengers' impressions of the stations. Whereas the first line adopted the look of the street, the deeper lines emphasized electric lighting in both stations and tunnels. Lines C, D, and E improved the architecture of the stations by decorating the walls of halls, passageways, and platforms with artistic tiles and murals of past rural and colonial landscapes. The carriages were also designed for comfort, and there was a concern of the municipality that the carriage companies provide 'modern' rolling stock. The smooth journey was good for health, said the engineer Jacobo Stock, in a discussion about Underground plans in 1930:

> Let's suppose for a moment the sacrifices that an employee or a worker has to make every day to move from the downtown to the suburbs to have lunch in his home. Exhausted not only by his work but also by the rocking of the tram [. . .] or omnibus he has to consume his food in record time [. . .] digestion is completed by the movement of the vehicles mentioned above (tram and omnibus) as he returns to work. [. . .] It is easy to deduce the terrible consequences for the physical and moral health of the individual.[49]

Finally, the Underground promised safety. Freedom from contact with other vehicles or pedestrians triggered the idea that the Underground was the safest mode of urban transport. In practice, fatal accidents that shaped negative representations of the Underground, such as the fire in the Paris Metro (1904), were rare in the Subte – fatalities were few over this period, and were mainly suicides.

From its inauguration, statistics supported the efficiency of the Underground in carrying more people than other modes of transport. The number of passengers increased steadily over the period, as rail companies advertised the convenience of underground travel. Politicians continued to support the construction of new lines. Yet, by the 1920s, experience showed that, in practice, the Underground had problems achieving both ideal arrival frequency (a train every two minutes) and comfortable convenience during the journey. Complaints about delays and trains crammed with people filled the press. The inception of women-only cars in 1928 implied the inconveniences that female passengers suffered while exposed to cars crowded with working men.[50]

Despite inconveniences, and unlike the number of tram users, the numbers travelling by Underground did not decrease: underground passenger journeys grew from 31.7 million to 68.4 million between 1918 and 1930, whereas tramway passengers decreased by 3.6 per cent.[51] The tramway, the city's once undisputed mass-transportation mode, was directly affected by the emergence of *taxi-colectivos* in 1928.[52] Moreover, there was increasing agreement that tramlines must be replaced by underground tramways, and that the street must be retained exclusively for cars and buses. In 1940, a report by the Transport Control Committee of Buenos Aires, chaired by the leading road engineer of Argentina, Justiniano Allende Posse, confirmed 'the public's preference' for the 'colectivo', which by then was a micro-bus carrying 11 passengers. According to the report, there was no future for the tramway, and the Underground, although efficient, was under-used and only supplemented other transport modes (railways and buses). Its main role now was to decongest the central city, rather than connect the city to the periphery.[53]

Conclusions

After the construction of Line E in 1944, there would be no major investment in the Subte network for decades. Between the 1950s and 1970s, plans for extensions were proposed, and a few improvements to the current lines were carried out, such as changing the layout of Line E to attract more passengers. In the 1980s, the *Premetro* (a surface metro line connected to Line E) was built, and, in the 2000s, the city council voted to extend the network, but during the last decade it only started the extension of existing lines and the construction of one new one (Line H). In contrast, plans and investment for automobile infrastructure have been privileged since the 1940s. Today, about 80 per cent of public transport journeys are by bus, and private car use represents around 40 per cent of daily mobility in the whole metropolitan area. The Subte still moves about 10 per cent of the total number of passengers. In this context, the bus rapid transit system, called 'metrobus' (separated lanes and stations for buses), has been implemented as a new solution for public transport problems. Yet, the underground railway as an efficient solution for Buenos Aires is still an idea supported by public opinion and technical discourse, even if the Subte is neither as comfortable nor as fast as it should be.

We can argue that this idea was shaped during the 'golden age' of underground railways in Buenos Aires, when several plans were proposed, supported by experts and politicians, and five lines were rapidly built. This age was in the early decades of the twentieth century and had a short life. Enthusiasm for the Underground was closely related to the negative perception of streets; it started to decrease when the automobile and its new infrastructure promised a new urban mobility experience.

Yet, the street and the Underground as spaces of circulation must be thought of as complementary rather than competitive. If the Underground was seen as fast, comfortable, and safe, it was also seen as a replacement for mass transport on the surface, removing tramlines from the street. As long as the Underground moves large numbers of people within the city without occupying the street, there is more space for other modes. The question here is which mode of transport would use the street when the trams were gone. Since the 1920s, the answer seems to be clear: the new, main protagonist of the street is the car. But, at the turn of the twentieth century, the answer was less obvious. For example, the image of the new street envisaged by urban plans showed a space marked by the rhythm of elegant carriages and the first automobiles, a continuous flow whose purpose was not merely acceleration. The imagined 'boulevard rhythm' was possible because the masses moved fast under the ground. If those spaces represented two distinctive rhythms, this difference seems to disappear with the increasing use of cars and the creation of expressways in the 1930s. Although congestion proved a common, daily experience on the surface, as well as underground, both infrastructures were symbols of speed, complementing one another for as long as they segregated different types of mobility.

Notes

1 John Tomlinson, *The Culture of Speed: The Coming of Immediacy* (London: Sage, 2007).
2 'La inauguración del tranvía subterráneo', *La Prensa*, 2 December 1913.
3 SCA, 'Las Grandes Obras (El Subterráneo)', *Revista de la Sociedad Central de Arquitectos* 112 (1917): 78.
4 Ezequiel Martínez Estrada, *La Cabeza de Goliat. Microscopía de Buenos Aires* (Buenos Aires: Centro Editor de América Latina, 1968[1940]), 22–24.
5 Horacio Torres, 'Evolución de los Procesos de Estructuración Espacial Urbana. El Caso de Buenos Aires', *Desarrollo Económico* 15 (1975): 281–306.
6 Adrián Gorelik, *La Grilla y el Parque. Espacio público y cultura urbana en Buenos Aires 1887–1936* (Buenos Aires: Universidad Nacional de Quilmes, 1998).
7 Carlos della Paolera, 'Urbanismo y Problemas Urbanos de Buenos Aires', *La Ingeniería* 660 (1929): 1–19.
8 Estrada, *La Cabeza de Goliat.*
9 James Scobie, *Buenos Aires: Plaza to Suburb, 1870–1910* (New York: Oxford University Press, 1974). Charles Sargent, *The Spatial Evolution of Greater Buenos Aires, Argentina, 1870–1930* (Tempe: Center for Latin American Studies, Arizona State University, 1974).

10 *Censo General de Población, Edificación, Comercio é Industrias de la Ciudad de Buenos Aires* (Buenos Aires: Municipalidad de Buenos Aires, 1910), 543.
11 Juan de Astirraga, 'Los Subterráneos de Buenos Aires', *Boletín del HCD 6*, 51–52 (1944): 60–7.
12 *Memorias de la Intendencia Municipal 1903* (Buenos Aires, 1904), 105. Source: Biblioteca de la Legislatura Ciudad Autónoma de Buenos Aires. Inventario 15368.
13 'Tráfico', *Revista Aconcagua* 8:2 (1932): 86–7.
14 Jacobo Stock, 'Los Subterráneos en Buenos Aires', *La Ingeniería* 664–668 (1930): 1–26. della Paolera, 'Urbanismo y Problemas Urbanos de Buenos Aires'.
15 Bernardo Caymari, *Propuesta para la Construcción en la Ciudad de Buenos Aires de un Tranvía Eléctrico Elevado Metropolitano* (Buenos Aires: Gunche, Wiebeck, Turtl, 1896), 4.
16 Stock, 'Los Subterráneos', 1.
17 Reginald Lloyd, *Twentieth Century Impressions of Argentina. Its History, People, Commerce, Industries, and Resources* (London: Lloyd's Greater Britain Publishing, 1911), 365.
18 'Charlas de PBT', *P.B.T.*, 13 December 1913, n.p.
19 Martínez Estrada, *La Cabeza De Goliat.*
20 *Mensajes, Informes y Decretos de la Intendencia* (Buenos Aires: Intendencia de Buenos Aires, 1909), 390. Source: Biblioteca de la Legislatura Ciudad Autónoma de Buenos Aires.
21 See John Olmsted, 'The Relation of the City Engineer to Public Parks', *Journal of the Association of Engineering Societies* 13 (1894): 594–5.
22 Richard Sennett, *Flesh and Stone: The Body and the City in Western Civilization* (London: Faber, 1994), 256.
23 Carlos della Paolera, *Buenos Aires y sus Problemas Urbanos* (Buenos Aires: Olkos, 1977[1937]), 51.
24 The most important plans were the 1904 *Plano Topográfico*, the Bouvard Plan (1909), and the Organic Plan (1925). Other urban projects, such as the Le Corbusier Plan for Buenos Aires (1937–8), also show the intention of reshaping the street layout.
25 Adrián Gorelik, 'A Metropolis in the Pampas: Buenos Aires 1890–1940', in *Cruelty and Utopia: Cities and Landscapes of Latin America*, ed. Jean-François Lejeune (New York: Princeton Architectural Press, 2005), 155.
26 Melina Piglia, *Autos, Rutas y Turismo: El Automóvil Club Argentino y el Estado* (Buenos Aires: Siglo Veintiuno Editores, 2014).
27 The projects competing with della Paolera's, such as Beretevirde's, Vautier's, or Guido's, considered having the avenue under ground level in order to have free space on the surface for monumental buildings (public and private) and parks. See Alicia Novick, 'Planes versus Proyectos: Algunos Problemas Constitutivos del Urbanismo Moderno', *Arquitextos* 057.01 (2005) (accessed 24 October 2016), www.vitruvius.com.br/revistas/read/arquitextos/05.057/497.
28 The entire team was composed of the road engineers León Laurent, Juan P. Tunessi, and Lauro O. Laura, and the architects Francisco Houloubek and Carlos Thays.
29 In the road journal *Caminos* 5:26 (1938), the architects Vautier and Holoubeck analysed American parkways as preliminary studies to General Paz Avenue.
30 Valeria Gruschetsky, 'Ingeniería Vial y Diseño Urbano en el Proyecto de la Avenida General Paz. Buenos Aires en los Años Treinta', paper presented at the Jornadas de Becarios y Tesistas, Universidad Nacional de Quilmes, Prov. de Buenos Aires, Argentina 2015.
31 Scobie, *Buenos Aires: Plaza to Suburb.*
32 Joseph Raskin, *The Routes Not Taken. A Trip through New York City's Unbuilt Subway System* (New York: Empire State Edition, 2014).
33 Dhan Zunino Singh, 'The Circulation and Reception of Mobility Technologies: The Construction of Buenos Aires's Underground Railways', in *Peripheral Flows: A*

Historical Perspective on Mobilities between Cores and Fringes, eds Simone Fari and Massimo Moraglio (Newcastle upon Tyne: Cambridge Scholars Press, 2016), 128–53.
34 'Hoy se Inaugura el Nuevo Subterráneo', *La Nación*, 18 October 1930, 9.
35 The AATC's report quoted in Comisión de Estética Edilicia, *Proyecto Urbano para la Urbanización del Municipio* (Buenos Aires: MCBA, 1925), 191.
36 CHADOPYF brochure (1936).
37 *Caras y Caretas*, 10 October 1901, n.p.
38 'El Subterráneo', *La Nación*, 1 December 1913.
39 *Caras y Caretas*, 29 November 1913, n.p.
40 *Critica*, 18 October 1930, 4.
41 *La Razón*, 18 October 1930, 6.
42 Dhan Zunino Singh, 'Towards a Cultural History of Underground Railways', *Mobility in History* 4 (2012): 106–12.
43 David Pike, *Subterranean Cities: The World beneath Paris and London, 1800–1945* (Ithaca, NY, and London: Cornell University Press, 2005); Michael Brooks, *Subway City: Riding the Trains, Reading New York* (New Brunswick, NJ, and London: Rutgers University Press, 1997).
44 Dhan Zunino Singh, 'Meaningful Mobilities: The Experience of Underground Travel in the Buenos Aires Subte (1913–1944)', *Journal of Transport History* 35 (2014): 97–113.
45 The municipality planned its own underground network in 1905, but failed to get companies interested in investing in this public work.
46 Zunino Singh, 'The Circulation and Reception of Mobility Technologies'.
47 Benson Bobrick, *Labyrinths of Iron: A History of the World's Subways* (New York: Henry Holt, 1994), 231.
48 Boston Transit Commission, *Statement of the Subway Commission* (Boston, 1894), 8–9.
49 Stock, 'Los Subterráneos', 3.
50 Zunino Singh, 'Meaningful Mobilities'.
51 'Tráfico Subterráneo en la ciudad de Buenos Aires', *Revista de Estadística Municipal* 42:4 (April 1930): 56–64.
52 In September 1928, 100 taxis (cars) started to provide public transport by carrying passengers (up to six) for a fixed fare following established routes (mostly competing with buses and tramways).
53 Control Committee of Buenos Aires Transport, *New Principles in Urban Transportation Economy* (Buenos Aires: Ministry of the Interior, 1941), 14.

5 Hurry-slow

Automobility in Beijing, or a resurrection of the Kingdom of Bicycles?

Glen Norcliffe and Boyang Gao

In *A Sentimental Journey through France and Italy*, Laurence Sterne's eighteenth-century protagonist, Mr. Yorick, revels in the pleasures of travelling slowly.[1] Sterne's verbosity results in long digressions, but that is the point: Sterne was not hurried because he felt important associations are made with the people one encounters en route, and not by swiftly ticking off a list of the celebrated sites of the Grand Tour. 'Sentimental', as used by Sterne, speaks for the leisurely stopovers along the way that foster convivial exchange. A century later, another travel writer, Elizabeth Pennell, travelling on a tandem tricycle with her illustrator husband, Joseph, wrote that, 'the oft-regretted delights of travelling in days of coach and post-chaise, destroyed on the coming of the railroad, were once more to be had by tricycle or bicycle'.[2] The title of the Pennells' book,[3] dedicated to Laurence Sterne, makes it clear that their journey was in the same spirit.[4] For Sterne and for the Pennells, travelling slowly resulted in an experience profoundly different from that of fast travel.

The Pennells' journeys by tricycle anticipated the great Western cycling boom of the 1890s, made possible by the advent of the safety bicycle. By 1900, the price of a bicycle, which had initially been high, had dropped dramatically, making it affordable to the average adult. At this point, China was still under the rule of a decaying Qing dynasty resisting change – until it was finally obliged to recognize the forces of modernity underway elsewhere. By then, the bicycle had entered the daily life of Europeans, whereas in China the delayed embrace of modernity led to cycling becoming an everyday event only in the Maoist era, 1949–76, a full half-century later than in Europe. Then, for a period in the mid to late twentieth century, China took to cycling en masse, developing a bicycle-friendly road infrastructure with little need for traffic lights because the bicycles simply flowed. From the 1960s, the bicycle was the people's customary mode of transport, thus earning China the title 'The Kingdom of Bicycles'.[5] Bicycle use at that time was more functional and utilitarian than it was on the Pennells' leisurely excursions, but cycles are not sealed-off bubbles: riders are exposed to humanity in ways that facilitate conversation and social interaction. For several decades, the bicycle and tricycle were part of quotidian life in China, used for ferrying young children to school, journeying to work, shopping, delivering all sorts of goods, transporting agricultural products to market, and carrying the aged and infirm in pedicabs.

And, on public holidays, bicycles were often used for recreation and social trips. Occasionally these were hurried trips, if the rider was late for class or work, or delivering an urgent item, but more generally cyclists go with the flow, which implies finding a comfortable cadence.

China's shift in late 1978 to the Open Doors agenda instigated by Deng Xiaoping saw the launch of programmes designed to catch up with other industrialized countries. This required new infrastructure, the rapid growth of coal-fuelled heavy industries and utilities, and diversification into higher-value-added industries, including an automobile industry with its elaborate supply chains. Social transformation accompanied these changes predicated on the principle of hurrying. China's twenty-first-century transformation into the world's leading manufacturing country has resulted in accelerated urbanization, massive migration from the countryside to cities, the rise of urban automobility, and extended periods of grim atmospheric pollution.

China's increasingly affluent society has embraced automobility, and many have abandoned cycling. In this, China presents a case of the developmental state, intent on rapidly transforming itself into an advanced industrialized country in per capita income, in technology, in housing, and in infrastructure. In transportation, China is in many respects a world leader. Measured in purchasing power parity, China's GDP has now surpassed that of the US.[6] China has the world's most extensive high-speed rail network, with more than 20,000 km in operation and close to 1 billion passengers in 2015. It has approximately 250 commercial airports; civil aviation carried 436 million passengers in 2015, second only to the US (with 798 million passengers).[7] Chinese cities have the most extensive mass-transit systems, with Beijing's network having 19 subway lines extending 700 km, with 319 stations. By 2013, 20 Chinese cities had an operating subway system, with several more under construction. All large cities have ring roads and are connected by expressways. Beijing, with a total population close to 20 million, now has seven ring roads (although the first is not strictly a ring road). The seventh regional ring road, some 940 km long (now known as G95 Capital Region Ring Expressway), was completed in December 2016. Nine expressways radiate from Beijing in every major direction, forming part of what Lefebvre calls the *architecture of enjoyment*; millions of its citizens find pleasure behind the wheel of an automobile, despite traffic congestion.[8] And, in 2012, China was the world's largest new car market, recording 18 million sales, against 14.5 million in the US.[9]

The adoption of modern culture and lifestyles has accompanied China's rapid industrialization and urban development in the major cities, but with a price in the form of problematic traffic jams and periods of air quality well below accepted international standards. The negative externalities resulting from the quest for speed have become sufficiently acute that city planners and administrators are taking steps to remedy them. Among the solutions is a revival of cycling, as planners seek alternatives to rising automobility and the environmental problems that motor vehicles and industry create.[10] And citizens themselves recognize that Beijing and other major cities have serious air-quality issues, urgently needing abatement. During periods of severe air pollution afflicting China's major cities, limits are placed on motor vehicle use and coal-burning industries, and power stations are sometimes shut down.

China entered the twentieth century for the most part following the age-old rhythms and practices of the Middle Kingdom. Society was still predominantly rural and hierarchically organized. Life was conducted at a measured pace, not yet perturbed by the architectures of modernity – and hurry – designed to speed up movement, work, and even play. One suspects that Laurence Sterne would have appreciated China's pre-modernity, little changed since the travels of Marco Polo. But the twentieth century saw it transform slowly, then open up to globalization in 1978 at an accelerating rate that culminated, by the twenty-first century, in a self-defeating quest for speed. The 'hurry-slow' outcome arises in two senses: impatient drivers anxious to floor the accelerator pedal of their Audis and BMWs soon come to a screeching halt, and the strict quotas now in place on purchasing a new vehicle are dramatically slowing the pace of vehicle acquisition. Recently, these circulatory embolisms have so severely hampered movement that serious consideration is now being given to a partial recovery of cycling, not as a reversion to the Kingdom of Bicycles but as a new, more Earth-friendly modulated lifestyle integrated with other forms of transport.

Rise of the Kingdom of Bicycles

Cycling in China was launched in the late 1860s by a report sent to the Imperial Court by members of a Chinese diplomatic mission visiting Paris.[11] They described many modern developments observed in France, including the appearance of the velocipede. But, even in the late 1890s when foreign diplomats, missionaries, and businessmen introduced a few bicycles to the major cities, they were regarded as a part of an alien culture:[12] wealthy Chinese travelled slowly and with dignity in sedan chairs and rickshaws. The first commercial imports were recorded in Shanghai in 1898, as the bicycle was recognized as a practical means of transportation with military potential.[13]

With the development of an efficient and reliable safety bicycle early in the twentieth century, young Chinese – particularly those returning from study overseas – took an interest in cycling. A range of modern Western commodities, from phonographs to bicycles, became acceptable consumer items for upper-class Chinese who dealt with the cultural dilemma of purchasing Western equipment by proposing that Chinese knowledge lay behind the technology. But the bicycle hardly penetrated the interior, where many citizens first saw bicycles only in the 1930s.[14]

The final collapse of the Qing dynasty in 1911 and the tentative modernization of major cities close to the coast (for instance by adopting the Western calendar) led to wider adoption of the cycle for leisure and for work. By 1930, Shanghai had around 20,000 bicycles, used mainly by letter carriers, police, and the military. They were also used to deliver small consignments of goods; indeed, during the Japanese occupation, considerable quantities of rice were smuggled from the countryside into cities on bicycle racks. In the early 1930s, the major cycle importers in Tianjin, Shanghai, and Shenyang began to assemble bicycles using imported components and then, through import substitution, shifted to making the whole bicycle. Eesfehani reports that, between 1937 and 1945, China produced around 10,000 bicycles annually.[15] By 1949, when the People's Republic was declared, China may have had half a million bicycles on the road, and the groundwork had been laid for the Kingdom of Bicycles.

The new Chinese government promoted the use of bicycles in five ways.[16] First, many smaller makers were merged into larger manufacturers. The resulting greater volume yielded some economies of scale, and prices fell. Second, bicycle makers had preferential access to rationed materials. Third, bicycle lanes (shared with other slower vehicles) became part of the urban fabric. In 1957, Beijing's first *sān bān shì* (three-lane carriageway), Sanlihe Road, was built following a Soviet model, with two 14-m motorized vehicular lanes and two 4-m bicycle lanes, with a green separation strip. Fourth, commuting workers had priority for a bicycle purchase ticket under the state rationing system, especially if they were blood donors, and could also obtain financial subsidies when purchasing a bicycle.[17] And fifth, during the pre-reform period, the work unit system (with residential blocks located close to places of work) shortened travel distance, thereby favouring walking and cycling.[18] The results were remarkable: by 1958, around a million bicycles a year were being manufactured, nearly all for the domestic market (compared with around 80 million today, of which two-thirds are exported).

From the 1950s to the 1970s, under Mao, bicycles were one of the four 'musts' (*sì-dà-jiàn*), along with a wristwatch, a radio, and a sewing machine (also known as 'three spins, one sound'). The Changcheng Bicycle Works, built in Tianjin in 1936 by two Japanese army veterans, began by mass-producing Anchor bicycles for the Chinese market. Converted to munitions production during the war, after the CPC takeover in 1949 this factory was re-converted to make Flying Pigeon (*Fēi Gē*) cycles, modelled on the 1932 British Raleigh roadster. Also founded in 1936 were Shanghai's Tongchang Chehang Bicycle Works, making *Yǒng Jiǔ* (Forever) bicycles, and the Daxing Bicycle Works in Shenyang, making *Bǎi Shān* (One Hundred Hills) bicycles.

By the 1960s and 1970s, large-scale production of bicycles and tricycles by the 'big four' makers, *Fēi Gē* and *Hóng Qí* (Red Flag) in Tianjin, and *Yǒng Jiǔ* and *Fèng Huáng* (Phoenix) in Shanghai, plus several other smaller makers, legitimized China's claim to be the Kingdom of Bicycles. These solidly built roadsters, with brand names intended to convey their durability (and, implicitly, the durability of the modern socialist nation promoting them), also had patriotic undertones. Urban infrastructure was built to reflect this priority, as bicycles became a major form of urban transport. On most city roads, dedicated rights-of-way (usually with a low barrier) separated thousands of slow-moving bicycles, rickshaws, and animal-drawn vehicles from motorised vehicles.[19] Traffic in the bike lanes moved at 10–15 kmph, and, unlike the motor vehicle lanes, they were rarely blocked by jams. This bicycle-friendly arrangement, with slower traffic segregated from motor vehicles, became institutionalized in China's cities during the 1950s, 1960s, and 1970s.

Sturdy bicycles were used to travel to work, to study, to shop, and to visit friends and relatives. A bicycle bestowed considerable kudos on its owner, whose children might be seated at the front or behind the cyclist. Most trips were relatively short, with longer trips usually made by bus. Also in widespread use were other forms of wheeled conveyance. Tricycle rickshaws carried the elderly and infirm to markets, to medical centres, and even to bus stops and local railway stations. Working tricycles were ubiquitous and sometimes incredibly overloaded, carrying furniture, water, farm produce, coal, construction materials, and all sorts of other freight. They moved slowly but could be manoeuvred along the narrow backstreets and *hútòng*,[20] between barriers on construction sites, and past parked motorized vehicles blocking other motorized vehicles. Operating as micro-enterprises, these working tricycles were the lifeblood of small retail markets, construction work, local fuel deliveries, and recycling. Pedal tricycles were also used by thousands of municipal workers cleaning streets and maintaining parks.[21]

Riders of these cycles moved slowly, although faster than cars do today.[22] They shivered in winter, sweated in summer, and were soaked by the rain, but they were mostly convivial.[23] Sentiments of the moment, greeting other cyclists, waving to friends and family when passing them, or shouting at someone blocking the way, were personal communications. Occasionally riders were in a hurry, but the nature of cycling is to find a comfortable cadence that maintains a rhythm. At times they resembled the slow safaris acclaimed by Laurence Sterne. There are still an estimated 430 million bikes in China today, some handed down through generations, but increasingly they are being replaced by e-bikes, and their use has declined dramatically.[24]

Open Doors, automobility, and de-bikification

When the Cultural Revolution ended in 1978, cars were rarely seen outside major cities, and those in evidence were mainly Chinese Red Flag saloons and Soviet Ladas driven by government officials. Private cars were rare: trips around the city were made by bus, by bicycle, or on foot. But, the opening of China to foreign investment, trade, and technology after December 1978, under Deng Xiaoping's Open Doors policy, set in motion a phase of rapid industrial growth, initially in special economic zones along the coast.[25] Importantly, this was not just an economic revolution but also a revolution in lifestyles, as China evolved into the world's leading manufacturing and trading nation. A rapidly growing economy and rising incomes fostered a younger millennial generation – the *bā líng hòu* (born in the 1980s) and the *jiǔ líng hòu* (born in the 1990s) – interested in consumer goods, global developments, travel, and new technologies denied to their parents during the Cultural Revolution. Private ownership of cars, apartments, and appliances, and the consumption of exotic foods, music, electronic devices, and their apps were all part of this cultural change. Fast food in the shape of international and domestic chains made an appearance.

Up to the mid 1980s, only a few thousand cars were produced annually in China, but then private car ownership was declared acceptable, and taxis appeared in major cities. A flood of Japanese cars followed, making China the second most important export market for Japanese makers after the US. The resulting trade deficit led to a moratorium on imports and efforts to boost local vehicle manufacture. But 'the launch of the National Automobile Industry Policy in 1994 had a detrimental impact on bicycle use and marked the beginning of a dramatic decline in bicycle use in Beijing'.[26] In 1996, it was estimated that there were 9.2 million bicycles in Beijing for the 12.5 million residents (about 2.5 bicycles per household).[27] But the use of bicycles was in steep decline, as cyclists who had purchased cars switched to using automobiles for most trips, and others, fearful of the growing vehicular traffic, switched to public transport. Following the launch in 2000 of China's Western Development Plan, major cities in the mid west of China joined this new phase of industrialization. For instance, the Chang'an Automobile group headquartered in Chongqing (which formerly made military equipment) now has partnerships with several Western car makers.

Developments since 1990 included the opening of branch motor vehicle assembly plants by a number of European, American, and East Asian car makers seeking to gain a slice of China's domestic car market, normally by creating a joint venture with a state-owned corporation.[28] Thus the largest brand by sales (Volkswagen) is made in a joint venture with FAW, a state-owned manufacturer headquartered in Changchun that also sells cars with its own brand names. Several domestic car makers are also fighting for market share in the world's fastest growing car market. Within 20 years, private motor vehicles were jamming traffic in all major Chinese cities, and bicycles no longer ruled the streets. 'In 1986, bicycles made up 63 per cent of transport on the roads. Last year (2014), they accounted for just 15 per cent.'[29]

By the 1990s, motorized vehicles had become a problem, not only on the streets, but also in housing complexes designed with parking spaces for bicycles but not for cars. In 1990, Beijing introduced a proof of parking certificate, required to purchase a car. Alas, many of the certificates were forgeries, the system proved hard to police, and it was eventually abandoned. Shanghai's adoption of a lottery to reduce the growth of traffic in 1994 proved fairly successful. Beijing did not implement its own vehicle quota system until 2010 and, instead of Shanghai's revenue-generating licence auctioning, adopted a random lottery system with no revenue (although it was more equitable).[30]

Feng and Li report that, 'since China joined the World Trade Organization (WTO) in 2001, its automobile industry has expanded significantly. The output has risen from 2 million vehicles in 2000 to 18 million in 2010'.[31] In fact, China is now the world's largest car maker, producing nearly one quarter of the world's output. Older bicycles, on the other hand, have been abandoned by the thousand and lie in neglected piles outside apartment buildings and elsewhere. The tensions resulting from this switch to automobility will be examined in the context of Beijing.

The rise of automobility in Beijing has been documented, but the rise and decline of bicycle use has not been recorded very accurately. The modal split of journeys to work from 1986 to 2010 is shown in Figure 5.1.[32] It shows: a steady growth of subway use from 1 per cent to 12 per cent; bus use holding steady around 26–28 per cent; bicycle use dropping rapidly from 63 per cent to 17 per cent;

and private cars and taxis rising from 5 per cent to 40 per cent. As widely reported, cycling dropped dramatically by nearly three-quarters; bus transport was unchanged; following the expansion of the subway system, rail trips rose twelvefold; and motor vehicles (cars plus taxis) rose eightfold. Meanwhile, the time taken for journeys to work has grown, although the data are somewhat contradictory, with average daily journey times between 52 and 97 minutes recorded, but there are consistent data indicating that Beijing workers have the longest average commute in China at 19.2 km.[33]

The number of vehicles registered in Beijing (see Table 5.1) more than doubled from 2.1 million in 2005 to 5 million in 2012 – or 63 private cars for every 100 households.[34] But the proportion of trips in vehicles rose only about a quarter between 2005 and 2010, partly because of attempts to reduce transport demand discussed below, and partly because many households have two cars to avoid one of the rules of transport demand management. Since 2005, a national public-transit-priority strategy has been promoted. Many cities invested heavily in building public transit infrastructure and announced policies to encourage people to use buses and subways.

The Beijing Olympics and the first slowing

The 2008 Olympics put Beijing in the world's eye as never before. Well before the actual games, the Bird's Nest Olympic Stadium was receiving favourable coverage worldwide. But the reports came with a coda: thick pollution (CO_2 and airborne particulates) made the intricate structure hard to see. Beijing's citizens had been living with this increase in pollution for some time, and the wearing

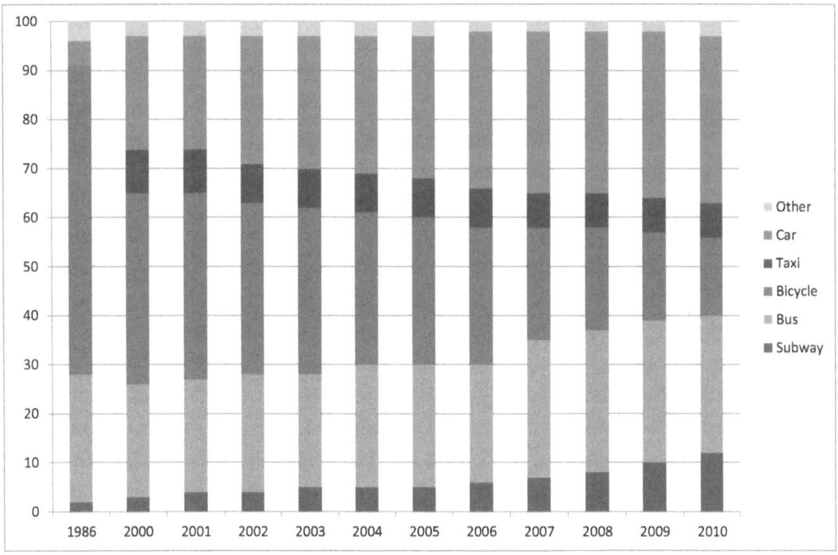

Figure 5.1 Modal split of travel in Beijing, 1986–2010.

Source: Beijing Transportation Research Centre, 2011.

Table 5.1 Population and numbers of vehicles in Beijing, 1953–2016[1]

Year	*Population of Beijing municipality (millions)*	*Number of vehicles*	*Number of vehicles per 1,000 persons*	*Bicycle use HH = household*
1953	2.7	n.a.	n.a.	n.a.
1964	7.6	n.a.	n.a.	753,000 bikes 48.5 bikes/100HH
1979	9.0	c.100,000	10	2.6m. c.70% of all trips
1986	10.3	267,000	26	62.8% of all trips
1990	10.8	389,000	36	c.60% of all trips
2000	13.6	1.507m	81	38.7% of all trips
2004	14.9	2.296m	111	32% of all trips
2005	15.4	2.583m	168	31% of all trips
2010	19.6	4.809m	245	16.7% of all trips 13 million bikes in use (some rarely)
2012	20.6	5.0m	243	14% of all trips
2016	21.7	5.6m	258	2.5m electric bikes

Source: compiled by the authors.

Note: [1]Wang, *A Shrinking Path for Bicycles*; China Statistics Yearbook; 60 Years of Beijing from 1940 to 1990; Beijing Statistical Bureau (1999); www.cbsnews.com/news/there-are-9-million-stolen-bicycles-in-beijing/; Tom Phillips, 'Bike-Sharing Revolution Aims to Put China back on Two Wheels', *The Guardian*, 28 December 2016 (accessed 15 October 2015), www.theguardian.com/world/2016/dec/28/bike-sharing-revolution-aims-to-put-china-back-on-two-wheels.

of face masks had become commonplace. Urban planners in Beijing and other major cities were well aware of this problem, and discussions were underway to curb pollution, but the Olympics crystallized these rising concerns. Promising strategies included accelerating the up-grade of technologies by shifting industrial structure to cleaner, higher-value-added activities, reducing the fossil carbon component in the energy mix, and developing Clean Development Mechanism projects (as defined in the Kyoto Protocol of 2007).[35] During the Olympics, the solution was an almost total ban on the use of private cars by Beijing's residents, but since then vehicular pollution has continued to grow, and the 5.6 million vehicles in Beijing in 2016 were estimated to have added 500,000 million tons of atmospheric pollution.[36]

Faced with this crisis, Beijing's government has pursued several solutions. First, a number of heavy industries and utilities (such as coal-burning power stations) located close to the central city were moved out to the urban fringe.[37] This removed much (diesel) truck traffic from central Beijing.[38] Second, changes in power generation were initiated, including a switch to nuclear power and cleaner fuels to reduce the carbon imprint. Between 2000 and 2008, China doubled its hydroelectric capacity, increased its wind-generating capacity 30-fold, and its nuclear capacity fourfold, although coal still accounts for about 70 per cent of generating capacity.[39] Third, the expansion of subway and commuter rail systems was accelerated so as to shift the modal split in favour of rail transport. And fourth, transport demand management (especially of vehicles) became a priority. Steps were taken to reduce vehicular traffic by offering cheaper fares on the subway system, and by limiting vehicle use with restrictions according to licence plate numbers – plates with odd and even numbers could be used on different days of the week.[40] This policy was initially effective: it reduced the number of cars on Beijing's roads by 700,000 and vehicle exhaust emissions by 40 per cent. This rule meant that a vehicle could only be used half of the time, and so vehicle taxes were reduced – a concession that cost the municipality 1.3 billion yuan. But, in the longer term, those who could afford it bought a second vehicle with a licence plate showing the opposite number, with the result that the number of vehicles began to rise, even as their use declined. Within 3 years the effects of this road rationing were reversed.

Countering automobility

Transport demand management slowed down growth of car ownership in Beijing, particularly after 2011. Owing to quotas, the threshold of 5 million registered cars was hit in early 2012 – nearly a year later than expected. With a quota of 240,000 set in 2011, the number of new cars was about one-third of the 2010 number (720,000).[41] Road space rationing was taken further in 2012, when a lottery system was adopted for the purchase of small passenger cars: eligibility for the monthly lottery required applicants to have *hùkǒu* status in Beijing, to have paid city taxes for more than 5 years, to not already own a passenger car, and to own a driving licence.[42] On days with red-alert smog levels, further restrictions were imposed on motor vehicle use, although electric cars were exempt. By November 2013, only 18,000 new vehicle licences were awarded in the monthly draw – and there were 1.74 million applicants. Between 2014 and 2017, the number of new licences awarded was reduced from 240,000 a year to 150,000. The push for electric cars was taken further in May 2016, when 60,000 of the 150,000 annual new licence plates were reserved for electric cars. The city has pledged to provide the infrastructure needed for this shift to electric cars, with 20,000 roadside vehicle chargers planned by 2020 (China had 49,000 public charging stations operating in 2015, compared with 32,000 in the US). Most of these electric cars are small, light two-seaters suited to urban travel and intended for use in bicycle lanes (they are not designed for long-distance travel).

There has also been change on the bicycle front, as sales of e-bikes have grown rapidly since 1996, with an estimated 2.5 million electric bicycles and tricycles now operating in Beijing.[43] The number of e-bikes in China rose from 40,000 in 1998 to 10 million in 2005.[44] Many of Beijing's cyclists have taken advantage of battery technologies requiring less physical effort, especially as mass production has lowered unit production costs, and lithium technology has increased battery efficiency. Also, rising real incomes make e-bikes affordable for many residents. A related factor in this resurgence of e-bikes is the explosive growth of online retail sales in China, which are estimated to have totalled 3.9 trillion yuan in 2015 (US$580 billion). Efficient deliveries by couriers of messages and packets purchased online are growing rapidly, as has the delivery of precooked meals, especially to working couples. Cycles are crucial to this e-commerce in two ways: first, the logistic costs are kept low (many of the delivery persons are rural migrants working for low wages); and second, purchasers want cheap, quick deliveries, which e-bikes achieve because they are scarcely slowed down by traffic congestion. The growth of e-bikes has, however, triggered a clash between motorists and cyclists over priority on the road: cars frequently drive in, park in, or block the lanes reserved for cyclists, and cyclists frequently ignore traffic lights and other road signs, sometimes travel in the lanes reserved for motor vehicles, travel in the wrong direction, and jostle pedestrians on the sidewalk. Motorists have demanded that controls – even a complete ban – be placed on delivery tricycles in some parts of the city.[45] Given, however, the recent shift to promoting electric cars and cycling in Beijing and the growing popularity of e-commerce, it seems likely that cyclists will survive 'a smouldering class war', pitting the rising middle class against the blue-collar worker, although with stricter enforcement and tighter regulation.[46] At present, limits to e-bike weight (40.8 kg) and speed (21 kmph) are not enforced, so that many silently exceed speed limits; indeed machines with high-powered batteries are capable of speeds up to 48 kmph. The weight and speed of e-tricycles are also not much controlled. With bicycles and e-tricycles now banned from Chang'an Avenue (Beijing's main downtown street, used for major parades), tighter regulation is evidently coming, but a total ban would compromise travel to work, to study, to visit friends and family, and to shop, and delay essential deliveries for many.

There is also increasing realization, as Zacharias and Zhang stress, that working tricycles provide the bottom rung of one of the most efficient hierarchical recycling systems in the world.[47] Tricycles are able to wheel around *hútòngs*, alleyways, and apartment buildings with narrow entrances designed to block cars, and follow mobile shops to gather waste at the neighbourhood level. These materials are then sorted at a local collection point and sent in large consignments to major recycling facilities.

The emphasis on modal splits in travel surveys overlooks the fact that trips can be multimodal and include bicycles.[48] Train stations in the Netherlands are surrounded by vast bicycle parks. Tokyo permits folding bikes on commuter trains, so that commuters can cycle from home to a station and, on leaving the train, cycle to their workplace. In Beijing, there are many possibilities for multimodal trips involving the bicycle, in one or more stages, that have yet to be pursued. For instance, there is a need for safe bicycle parking spaces in most of Beijing's buildings and transport hubs.

A new dawn

'In 2010, a woman on a popular Chinese television programme sparked heated debate on materialism nationwide when she said she would rather cry in a BMW than smile on a bike.'[49] But shifts in attitudes to transport, to air pollution, and the cultural values that they reflect are clearly coming. On 5 March 2014, Premier Li Keqiang declared a formal *War on Pollution* at the annual People's Congress (China's parliament).[50] He described pollution as 'nature's red-light warning against the model of inefficient and blind development', indicating that China must not only shut down coal-fired furnaces, but shift to a different kind of development, requiring difficult structural adjustments.[51] A Pew poll from October 2016 shows that 70 per cent of Chinese are worried about the country's choking air pollution, with smog that afflicted northern China for much of December 2016 into January 2017 a possible turning point.

> On the first day of 2017 in Beijing, concentrations of tiny particles that penetrate deep into the lungs climbed as high as 24 times levels recommended by the World Health Organization. More than 100 flights were cancelled and all intercity buses were halted at the capital's airport . . . Across northern China 24 cities issued red alerts on Friday (30th Dec) and Saturday (31st), while orange alerts persisted in 21 cities through the New Year holiday . . . Decades of economic development have made acrid air a common occurrence in nearly all major Chinese cities, with government-owned coal burning power stations and heating plants and steel manufacturing concentrated in northern provinces the main source of pollution.[52]

Planners and politicians share these concerns and have launched a series of initiatives.[53] First, ambitious electrification efforts are underway, including expanding the subway system with a ridership that continues to grow with trips to study, to work, to airports, to mainline stations, and to the central business district to shop or have a night out. The new lower-density suburbs of Beijing are car-friendly, but the switch to electric cars is rapidly changing the vehicle mix across the city.[54] And e-bike sales indicate a rapid switch to power-assisted cycling. Meanwhile, working tricycles, which were originally human-powered but modified to use small two-stroke engines around 20 years ago, are now mostly battery-powered.

The recent rise of electric vehicles (EVs) demonstrates how a developmental state can move ahead of the market by anticipating major technical changes. Vanderklippe reports that several small firms have begun to manufacture small two-seater cars that resemble batteries with wheels. Enthusiasts see Beijing and Shenzhen replacing Detroit and Mannheim as the leading centres of innovation in automobile production. Western car makers are moving in the same direction, but China's efforts are 'led in part by a government that has used every tool at its disposal to kick start an industry it sees as a future economic pillar'.[55] It is estimated that, between 2013 and 2015, various levels of government in China spent US$9.5 billion in subsidies supporting the infant industry, in addition to investments by a band of e-commerce billionaires looking for a new field in which to participate. BYD in Xi'an is in partnership with Daimler-Benz, making luxury EVs: they sold 62,000 EVs in 2015. In September 2016, the Chinese government ratified the Paris climate agreement to mitigate greenhouse gas emissions and has since actively promoted these new technologies.[56] But so far, private sales of EVs have been weak, with 80 per cent being institutional purchases. Under construction in Beijing's Daxing district is China's largest research facility for new-energy vehicles – mostly electric cars.[57]

Second, Beijing's transportation is clearly on the cusp of major technical and cultural changes. The future promises to be cleaner, quieter, and healthier. And since technology is socially and geographically constructed, culture will be important to this transition. If lifestyles do not change, Beijing's residents will remain in a hurry through replacing the internal combustion engine with a battery to create a world of EVs on congested ring roads. But members of the younger generation who have travelled widely, seen other lifestyles, consume new things vicariously on the web, and enjoy films and exotic food are becoming nostalgic about an era they did not experience – the age of the Kingdom of Bicycles. They differ from the West's Y generation: instead of a consumer boom, their parents experienced the intensification of heavy industry, the collectivization of agriculture, and the Cultural Revolution. After 1979, they mainly grew up as only children under the one-child policy. And, according to Beijing's Municipal Bureau of Human Resources and Social Security and the Bureau of Statistics, in 2013, they enjoyed modest prosperity, with the average annual salary of employees in Beijing (GDP per capita) being 69,521 yuan (US$11,130).[58] We are also witnessing a cultural shift, with a new generation enjoying the fun of recreational cycling, including informal alley-cat races,[59] weekend family outings, and MTB trail rides.

In 2012, Beijing implemented a bicycle-sharing programme, with 40,000 bicycles available for rent near subway and bus stations. The bike-share system took a while to take off, but, by 2016, another 10,000 bikes were added. Tom Phillips envisages a big growth in bike-sharing:

> From Shanghai to Sichuan, schemes are being rolled out to slash congestion, cut air pollution – and spin a profit. Even through Beijing's nicotine-tinged smog you can make out the multi-coloured frames, gliding through the pea soup towards a greener future. In recent months an unmissable fleet of fluorescent orange, canary yellow and ocean blue bicycles has hit the streets of urban China as part of a hi-tech bike-sharing boom that entrepreneurs hope will make them rich while simultaneously transforming the country's traffic-clogged cities . . . [A start-up company] Ofo, so named because of the word's resemblance to a bicycle, has put about 250,000 of its bright yellow bikes to work since late 2015 [under the Campus Bicycle Sharing Project], of which around 40,000–50,000 are in the capital. The . . . company was founded by five students looking to improve transport options on university

> campuses. [It] has attracted about 3 million users in cities such as Beijing, Shanghai, Xiamen and Guangzhou. Its bicycles make about 1.5m trips each day between them.[60]

This bike-sharing programme had serious teething problems, with many of the rental bikes being stolen and left in heaps, probably the result of expanding the system too quickly without allowing for a learning curve. Besides, such programmes are only a partial solution for one category of citizen. Recognizing this, a consortium consisting of the Dutch sustainability consultancy Ecofys, the China Academy of Transportation Science, and the Dutch design and project management firm Royal Haskoning DHV Engineering was formed to come up with a comprehensive plan for an overall sustainable transport system for Beijing. Their February 2016 report, *Beijing Bicycle Strategy and Policy*, which sought to reverse the decline in bicycle use, was presented to the Beijing municipal government and the Asian Development Bank, which funded the project. Proposals included: extending the bike-sharing programme; high-quality bicycle routes and parking facilities; placing more trip destinations within bikeable distances; and integrating cycling with a high-quality public transport system (i.e. planning for multimodal trips). They state that: 'Our plan identifies the need for an institutional setting that recognises cycling as an integral part of urban transport planning, with high-level officials promoting cycling with enthusiasm.'[61]

The recent rise of automobility in Beijing is clearly unsustainable: the hurry-slow domain of automobiles is rapidly shifting to the slow side. And public transport is only a partial answer, as users do not have the freedom to choose a route, stop and start on a whim, or change plan en route. Part of the solution lies in recovering the best parts of the Bicycle Kingdom, but now with electric-assist for those who need it; with bike-sharing for convenience; with separate, safe bicycle lanes; with improved infrastructure, including intermodal shifts; and stricter enforcement of road-safety rules. Bicycles also offer a new arena for slow life as a recreational machine away from the cusp of modernity, one rooted, not in global traditions, but in a phase of China's unique recent past, and as a social and convivial machine that is responsive to human sensibilities.

Conclusion

The value of the cultural turn in economic geography, certainly for the story told here, is illustrated by Wang Xiaoshuai's film *Beijing Bicycle*, a work in social realism. Two very different 17-year-olds living in Beijing struggle over possession of one desirable, stolen mountain bike used for deliveries. Li suggests that:

> The bicycle, while necessarily invoking nostalgia and being reminiscent of a bygone era, becomes a contested site where the relationship between the individual and society, globalization and tradition, upward mobility and social stratification, desire and dystopia, and belief and disillusion destabilizes and fluctuates.[62]

The deliveries could easily have been done on a Flying Pigeon, but the employer wanted to project a trendy image with the mountain bike, a sign of social affluence. The implication is that a subtle cultural shift is making the bicycle a symbol for an environmentally aware younger generation.

Laurence Sterne and the Pennells would undoubtedly approve of a bicycle used for 'coolness, fashion and freedom', implying a convivial machine used in daily life. This is happening in Beijing, as the private Mobike bike-share system (launched on 1 September 2016) rapidly expands on the streets (the target is 2,000 bicycles). With a GPS chip in them, a mobile phone can find the nearest available Mobike and make the payment: after a 299 yuan deposit, the rental cost is only 1 yuan (15 US cents) for 30 minutes. Orange-wheeled Mobikes now have several rivals, including the campus-based yellow Ofo bikes, Bluegogo, and Beijing's public bike-sharing system. Indeed, on 8 September 2017, the BBC reported that, with nearly 2.4 million shared bicycles in Beijing, new deliveries of bikes to the city's 15 bike-sharing schemes would be suspended to curb the mayhem of abandoned bicycles and a lack of bicycle parking spaces. In November 2017, Bluegogo (the third largest bike-sharing company) closed in financial difficulty after several marketing errors, including a failed attempt to set up in San Francisco. It was not the first bike-sharing company to collapse, but was the largest, as the industry began to consolidate and spread globally to many other cities.

This all points to a new cycling culture that avoids the frustrating hurry-slow characteristics of automobility: bicycles generally flow. The sharing economy is sharing bicycles on the streets, around transport hubs, and on campuses. Booming e-commerce is efficiently delivering small packets and hot meals to apartment buildings on tricycles and bicycles. Teenagers find it cool to hang out on mountain bikes. Battery-powered e-bikes are taking much of the effort out of cycling. Folding bikes are slipped into the trunks of cars, safe bike parking facilities are appearing at stations and elsewhere, while an environmentally conscious public looks for ways to improve air quality. Politicians and planners accept that automobility has reached its limits. Investors are jumping in: 'Chinese investors, including the tech giants Didi Chuxing and Tencent, are throwing their weight behind the bike-sharing start-ups, pumping tens of millions of pounds into their operations since the autumn'.[63] And there are also signs of nostalgia for the days of the silent steed. Given the central role of the bicycle in China's recent history and the intractable problems associated with automobility, a cycling renaissance seems imminent. And, if this happens, Laurence Sterne and Elizabeth Pennell would surely be delighted to join a leisurely weekend outing of the Beijing Flying Pigeon Club, meandering around the back roads and parks of the city, avoiding the frustrations of automobile-bound citizens stalled in traffic jams on overcrowded city streets.

Notes

1 Lawrence Sterne, *A Sentimental Journey Through France and Italy* (Reprinted in New York by Golden Cockerel Press (1928[1768]).

2 Dave Buchanan, 'Pilgrims on Wheels: The Pennells, F.W. Bockett, and Literary Cycle Travels', in *Culture on Two Wheels: The Bicycle in Literature and Film*, eds Jeremy Withers and Daniel Shea (Lincoln, NE: University of Nebraska Press, 2016), 26.

3 Joseph Pennell and Elizabeth Robins Pennell, *Our Sentimental Journey through France and Italy* (London: T Fisher Unwin, 1893).
4 Dave Buchanan, *A Canterbury Pilgrimage/An Italian Pilgrimage by Joseph Pennell and Elizabeth Robins Pennell*, ed. with Introduction (Edmonton, AB: University of Alberta Press, 2015).
5 The origin of this term – *zì xíng chē wáng guó* in Pinyin – has not been traced, but it probably first appeared in a journal in the 1960s.
6 Michael Dunford and Weidong Liu, eds, *The Geographical Transformation of China* (Abingdon, UK: Routledge, 2017).
7 'Air Transport, Passengers Carried', *World Bank Database* (accessed 23 June 2017), http://data.worldbank.org/indicator/IS.AIR.PSGR.
8 Henri Lefebvre, *Towards an Architecture of Enjoyment*, ed. Łukasz Stanek (Minneapolis: University of Minnesota Press, 2014).
9 Tania Branigan, 'China and Cars: A Love Story', *The Guardian*, 14 December 2012 (accessed 15 September 2015), www.theguardian.com/world/2012/dec/14/china-worlds-biggest-new-car-market.
10 Mimi Sheller and John Urry, 'The City and the Car', *International Journal of Urban and Regional Research* 24 (2000): 737–57.
11 Amir Moghaddaas Eesfehani, 'The Bicycle's Long Way in China: The Appropriation of Cycling as a Foreign Cultural Technique 1860–1940', *Cycle History 13: Proceedings of the 13th International Cycle History Conference*, eds Nick Clayton and Andrew Ritchie (San Francisco: Cycle Publishing, 2003), 94–102.
12 J. Min, 'Early History of Bicycle in China' 中国自行车早期历史. *Yanhuangchunqiu*, 2 (2003), http://news.163.com/09/0624/15/5CJ81VOS00011247.html.
13 Qiuning Wang, *A Shrinking Path for Bicycles: A Historical Review of Bicycle Use in Beijing* (unpublished MA thesis, Department of Community and Regional Planning, University of British Columbia, 2012).
14 Eesfehani, 'The Bicycle's Long Way in China'.
15 Ibid.
16 Lixia Mei and Jici Wang, 'The Changing Geography of Chinese Bicycle Industry: The Case of Tianjin Bicycle Cluster and Its Evolutional Trajectory', paper presented at the Annual Meeting, Association of American Geographers, Boston, MA, 22–27 April 2008; Eesfehani, 'The Bicycle's Long Way in China'.
17 Nathan Jones, 'Life on Two Wheels: The Bicycle Kingdom', *CCTV* (2006) (accessed 23 June 2017), www.bicyclekingdom.com/bicycle/Bicyclekingdom1.htm.
18 J. Allaire, 'China and Bicycle: The End of the Story?', *Structure* 7 (2007), (accessed 10 September 2016), http://webu2.upmf-grenoble.fr/iepe/textes/JA_poster-Velocity-June2007.pdf.
19 John Zacharias, 'Bicycle in Shanghai: Movement Patterns, Cyclist Attitudes and the Impact of Traffic Separation', *Transport Reviews* 22 (2002): 309–22.
20 Hútòng are traditional narrow streets found in northern Chinese cities, especially Beijing, forming residential communities served (historically) by a water well.
21 See Chapter 10, 'Neoliberal Mobility and Its Discontents: Working Tricycles in China's Cities', in Glen Norcliffe, *Critical Geographies of Cycling: History, Political Economy and Culture* (London: Routledge, 2016).
22 Gwynn Guilford 'A Big Reason Beijing is Polluted: The Average Car goes 7.5 Miles Per Hour' (12.1 kmph), *Quartz* 3 (January 2014) (accessed 21 November 2016), https://qz.com/163178/. This compares with 15.5 mph for vehicles in New York. Another estimate is 9 mph, slower than a bicycle (Sintana Vergara, 'From One Billion Cars to One Billion Bicycles', *Sustainable Cities*, 28 June 2012 [accessed 21 November 2016], http://blogs.worldbank.org/sustainablecities/from-one-billion-cars-to-a-billion-bicycles).
23 Ivan Illich, *Tools for Conviviality* (New York: Harper & Row, 1973).
24 Fabiowzgogo, 'China Bicycle', *China Travel.com* (accessed 23 June 2017), www.chinatravel.com/facts/china-bicycle.htm.

25 George Lin, Developing China: Land, Politics and Social Conditions (London: Routledge, 2009).
26 Debra Bruno, 'The De-bikification of Beijing', *The Atlantic Citylab*, 9 April 2012 (accessed 28 September 2016), www.citylab.com/commute/2012/04/de-bikification-beijing/1681/; Wang, *A Shrinking Path for Bicycles*, ii.
27 National Bureau of Statistics, *Beijing Statistics Yearbook, 1996* (accessed 28 September 2106), www.google.ca/?gws_rd=ssl#q=beijing+statistical+yearbook+1996.
28 Weidong Liu and Peter Dicken, 'Transnational Corporations and "Obligated Embeddedness": Foreign Direct Investment in China's Automobile Industry', *Environment and Planning A* 38 (2006): 1229–47; Weidong Liu and Henry Yeung 'China's Dynamic Industrial Sector: The Automobile Industry', *Eurasian Geography and Economics* 49 (2008): 523–48.
29 Anne Renzenbrink 'Bid in Beijing to Bring Back "Kingdom of Bikes"', *South China Morning Post*, 27 July 2015 (accessed 15 October 2015), Document SCMP000020150726eb7r0003p.
30 Jun Yang, Ying Liu, Ping Qin and Antung A. Liu, 'Review of Beijing's Vehicle Registration Lottery: Short-term Effects on Vehicle Growth and Fuel Consumption', *Energy Policy* 75 (2014): 157–66.
31 Suwei Feng and Qiang Li, 'Car Ownership Control in Chinese Mega Cities: Shanghai, Beijing and Guangzhou', *Journeys* September (2011): 40–49 (accessed 28 September 2016), www.lta.gov.sg/ltaacademy/doc/13Sep040-Feng_CarOwnershipControl.pdf.
32 These data were documented by the Beijing Transportation Research Centre and illustrated by Wang, *Shrinking Path for Bicycles*.
33 Stephen Chen, 'Beijing Workers Have Longest Daily Commute in China at 52 Minutes Each Way', *South China Morning Post*, 27 January 2015 (accessed 15 July 2016), www.scmp.com/news/china/article/1692839/beijingers-lead-chinas-pack-longest-daily-commute; Du Juan, 'Beijing Likes Shared Bicycles – However . . .', *China Daily*, 2 May 2017 (accessed 23 June 2017), http://africa.chinadaily.com.cn/china/2017-05/02/content_29159877.htm.
34 Renzenbrink, 'Bid in Beijing to Bring Back "Kingdom of Bikes"'.
35 Hongguang Liu and Weidong Liu, 'Decomposition of Energy-Induced CO_2 Emissions in Industry of China', *Progress in Geography* 2 (2009): 285–92.
36 Tom Phillips, 'Bike-Sharing Revolution Aims to Put China Back on Two Wheels', *The Guardian*, 28 December 2016 (accessed 15 October 2015), www.theguardian.com/world/2016/dec/28/bike-sharing-revolution-aims-to-put-china-back-on-two-wheels.
37 Boyang Gao, Weidong Liu, and Michael Dunford, 'State Land Policy, Land Markets and Geographies of Manufacturing: The Case of Beijing, China', *Land Use Policy* 36 (2014): 1–12.
38 Trucks delivering coal to Beijing's power stations from Inner Mongolia on the Beijing–Tibet expressway undergoing road works created the world's worst traffic jam in 2010 – a 60-mile back-up that lasted 12 days.
39 Robert Dreyfuss, 'The Nation: China: A Kingdom of Bicycles No Longer', National Public Radio (accessed 25 November 2009), www.npr.org/templates/story/story.php?storyId=120811453.
40 Nick Mulvenney, 'Beijing to Launch Olympic Odd–Even Car Ban', Reuters, 20 June 2008 (accessed 15 November 2015), http://uk.reuters.com/article/beijing-car-ban-idUKNOA03972720080620.
41 Suwei Feng and Qiang Li, 'Car Ownership Control in Chinese Mega Cities'.
42 *Hùkǒu* is a form of residential regulation with origins in China's imperial age. A family is granted a permit to live in a specific place, with rights to social services in that place. Assigning rural versus urban *hùkǒu* status is particularly important to urban and economic development.

43 Anonymous, 'Beijing's Electric Bikes, The Wheels of e-Commerce Face Traffic Backlash', *Today*, 1 June 2016 (accessed 15 November 2016), www.todayonline.com/chinaindia/china/beijings-electric-bikes-wheels-e-commerce-face-traffic-backlash; Jonathan Weinert, Chaktan Ma, and Christopher Cherry, 'The Transition to Electric Bikes in China: History and Key Reasons for Rapid Growth', *Transportation* 34 (2007): 301–18.
44 Lin Lin, 'Bicycles in China: The Lost of the Kingdom of Bicycles' (accessed 5 April 2017), www.google.ca/?gws_rd=ssl#q=faculty.washington.edu/abassok/bikeurb/resources/presentations/lin.pdf.
45 John Zacharias and Bingjie Zhang, 'Local Distribution and Collection for Environmental and Social Sustainability – Tricycles in Central Beijing', *Journal of Transport Geography* 49 (2015): 9–15.
46 Anonymous, 'Beijing's Electric Bikes'.
47 John Zacharias and Jian Ming Zhang, 'Estimating the Shift from Bicycle to Metro in Tianjin', *International Development Planning Review* 30 (2008): 93–111; Norcliffe, *Critical Geographies of Cycling*.
48 Haixiao Pan, Qing Shen, and Song Xue, 'Intermodal Transfer Between Bicycles and Rail Transit in Shanghai, China', *Transportation Research Record* 2144 (2010): 181–8.
49 Renzenbrink, 'Bid in Beijing to Bring Back "Kingdom of Bikes"'.
50 Tania Branigan, 'Chinese Premier Declares War on Pollution in Economic Overhaul', *The Guardian*, 5 March 2014 (accessed 23 June 2017), www.theguardian.com/world/2014/mar/05/china-pollution-economic-reform-growth-target.
51 Tania Branigan, 'Chinese Premier Declares War on Pollution in Economic Overhaul'.
52 Benjamin Haas, 'China Smog: Millions Start New Year Shrouded by Health Alerts and Travel Chaos', *The Guardian*, 2 January 2017 (accessed 23 June 2017), www.theguardian.com/world/2017/jan/02/china-smog-millions-start-new-year-shrouded-by-health-alerts-and-travel-chaos. *The Guardian*.
53 Tania Branigan, 'Chinese Premier Declares War on Pollution in Economic Overhaul'.
54 Pengjun Zhao, 'Sustainable Urban Expansion and Transportation in a Growing Megacity: Consequences of Urban Sprawl for Mobility on the Urban Fringe of Beijing', *Habitat International* 34 (2010): 236–43.
55 Nathan Vanderklippe, 'Electric Cars Drive China's Superpower Zeal', *Globe & Mail* (Toronto), 10 September 2016: B9.
56 'Greening the Dragon: China's Search for a Sustainable Future', *Green Futures Magazine* September (2006), Department for Environment, Food and Rural Affairs (Accessed 15 November 2015), www.forumforthefuture.org/sites/default/files/images/GreenFutures/Greening_the_dragon.
57 www.ebeijing.gov.cn/BeijingInformation/BeijingNewsUpdate/t1455559.htm.
58 On 26 February 2016, the BBC announced that Beijing had overtaken New York as the city with the most billionaires, with 100 to New York's 95.
59 'Beijing Hosts Alley Cat Bicycle Race', China.Org.CN (accessed 5 April 2017), www.china.org.cn/video/2012-03/23/content_24967527.htm.
60 Phillips, 'Bike-Sharing Revolution Aims to Put China Back on Two Wheels'. The classic example of this is National Taiwan University's large campus in Taipei, where students travel mostly on bicycles. At the centre of the university is a large bicycle repair shop.
61 Wim van der Wijk, *Beijing Bicycle Strategy and Policy: The Way Back to 'Bicycle Capital of the World'* (Royal Haskoning DHV Engineering, Ecofys, China Academy of Transport Science, and Asian Development Bank, 2016).
62 Jinhua Li, 'Beijing Bicycle: Desire, Identity and the Wheels', in *Culture on Two Wheels: The Bicycle in Literature and Film*, eds Jeremy Withers and Daniel Shewa (Lincoln, NE: University of Nebraska Press, 2016), 281–2.
63 Phillips, 'Bike-Sharing Revolution Aims to Put China Back on Two Wheels'.

Part II

Local and global infrastructures

6 An architecture of sluggishness

Organic infrastructure and anti-mobility in Toronto, 1870–1910

Phillip Gordon Mackintosh

> To one standing at the corner of King and Yonge streets when the employees of business houses and workshops are going to their homes at the close of the day, it is obvious that Toronto is a growing city. He observes it in the closer throng on the walks, in the greater scramble for the cars, in the tenser look of human faces, and in the hurry, hurry, hurry of human feet.
>
> [Toronto] *Globe* (1908)[1]

Moderns dare not dispute the *Globe*'s early twentieth-century observation of a city organizing around 'hurry, hurry, hurry'. People, business, transportation, and technology kept modern cities month upon month, 'stepping livelier', as novelist E.M. Forster noted of London.[2] Thus, scholars of technology and culture demonstrate how modern time–space compression has been socially interruptive but ineluctably continuous – in large part because of the urbanization of capital and the production of fixed configurations of space (such as transportation and communications networks).[3] It is now given that moderns are 'constituted' through such urban technologies and infrastructure. These, whether in developed or developing nations, mediate and necessitate urban life.[4]

The nineteenth century offered urban moderns one transportation advancement after another, from the draisine (Karl Drais's running-powered bicycle) to the automobile, each demanding a surface infrastructure to accommodate its deployment and production (this is the principle behind *traffic generation*). Hence, the Victorian impetus to construct, not only canals, railway tracks, and engineered roads, for example, but also the modernized ports, railway stations, road houses, and hotels and boarding houses (see Olson and Poutanen's chapter in this volume) that accommodated and expanded production of cargo ships, railway cars, bicycles, automobiles, and people that used them. All such mobilities necessarily impelled the relatively simultaneous embedding of hydroelectricity infrastructure, logging and mining industries, smelting and manufacturing, mechanical and chemical engineering, and petroleum production, and all their logical consequences in urban and urban hinterland landscapes. And, because all such technologies and infrastructures fade into the 'naturalized background' for most moderns, little wonder they think of modernization and its attributes as inevitable.[5]

Despite the modern common sense of urban infrastructure, under careful scrutiny we find cracks in the certainties that encase the urban geographic and material order of things and that affirm Paul Edwards's contention that 'most technology is not salient for most people, most of the time'.[6] One such fissure appears in the municipal quest for street surface infrastructure. In Toronto, in the late eighteen and early nineteen hundreds, the construction of surface infrastructure was constrained by the city's local improvement by-law, which gave property owners control of the street surfaces and pavements abutting their properties. Apparently, property owners' mean disposable incomes, and not 'liberal dreams' of modernized technology, determined the quality of mobility in Toronto.[7] Instead of privileging an expensive architecture of hurry, Toronto property owners preferred an architecture of affordability, an architecture of 'good enough' – an architecture of sluggishness grounded in the cheap organic materials – wood, stone, and gravel – acting as pavements on roads and sidewalks.[8]

Organic infrastructure

Toronto's surface infrastructure, like that of most modern industrial cities circa 1900, was largely 'organic'. This means it was comprised of wood (usually cedar roadways and spruce plank sidewalks), stone (typically macadam, described below, but also flagstone sidewalks and granite block pavements), gravel (from the city's own gravel pits), and plain old dirt (accumulating in the dozens of miles of unpaved streets and roads).[9] In 1892, Toronto owned 416 miles of plank sidewalks and 116 miles of cedar block roadways – both perishable.[10] Toronto's organic street surfaces, when not wooden, were stone and gravel, or not paved at all. By 1900, these surfaces, all having little to no real utility – or lifespan – comprised 50 per cent of the city's roadways: macadam (47.81 miles), gravel (5.34 miles), and unpaved (77.18 miles). Combined with a (then) declining percentage of cedar block roadways (33 per cent, or 87 miles) as other bituminous surfaces appeared – principally asphalt, but also tarmacadam and bitulithic – organic street surfaces nonetheless covered 83 per cent of Toronto's 259 miles of roadway.[11] Such a prevalence of organic street surfaces suggests a motivation for the emergence of 'architectures of hurry': the sluggishness of organic infrastructure before the era of chemically enhanced asphalt in the early twentieth century.

Cedar block paving, the predominant pavement before 1900, captured the imagination of city councils and city engineers throughout the late-Victorian West, and no council or engineers' office more forcefully than Toronto's. Toronto laid and relaid hundreds of miles of cedar blocks between 1880 and 1907. By 1885, Toronto had laid 48 miles of cedar block pavements and, by 1891, almost 117 miles – in a city of well over 200 miles of street but only 127 miles paved.[12] Cedar pavements so preoccupied municipal thinking that Aldermen Foster and Dunn insisted, in 1894, that the city stop laying cedar blocks, urging their colleagues to 'consider the advisability' of paving with cedar when other classes of pavement were available.[13] Torontonians, as we will see, disagreed, settling on a pavement they believed amenable to their thrifty purposes. Near the end of the cedar block era (c.1900), the city engineer still reported that 85.78 of the 181.85 miles of paved streets were cedar; the rest were stone, brick, and gravel, with 30.81 miles of asphalt.[14] When city engineer Charles Rust announced 'only one street was [newly] paved with cedar blocks' in 1907, it turned out to be the final year that new cedar block pavements were included in the annual engineering report.[15] His statement indicated a dramatic change from the mileages laid from the 1880s, when the *Globe* claimed the municipal paving problem could be 'solved' through the assiduous employment of cedar blocks.[16]

Initially, cedar blocks offered promise. Other organic pavements, specifically macadam and gravel, produced copious dust.[17] Despite offering durability and firmness when dry (bicyclists, it was asserted, liked dry macadam), gravel and stone surfaces abjectly failed in wet weather, especially given the quaternary substrate upon which Toronto was founded.[18] Macadam was a water-bound pavement constructed of layers of larger to smaller stones, crushed to a powder that rain turned to a binder as the road dried, and it drew the sardonic attention of Charles Baudelaire: wet macadam was a deleterious mire in which to lose the halo of one's modern soul.[19] In dry weather, macadam cloaked everything in a fine dust, especially in Toronto's windy, Lake Ontario clime. Local newspapers roiled with complaints about the pall of dust suppressing the city's vibrant commercialism, a dust to which macadam, gravel, and even cedar blocks contributed fine particulates. A 'veritable fog', the dust cloaked downtown shops and shoppers.[20] Dust was especially irksome on streets that had been paved with asphalt *specifically to reduce dust*. It occurred too late to Torontonians that dust-making streets bounded all the newly asphalted ones.[21]

Cedar, alternatively, retained moisture and could hamper dust. It muted the heavy animal traffic that thundered through the city day and night. In Toronto's formidable winters, it even provided the impoverished with fuel for their meager fires (they pilfered blocks from failing roads).[22] Importantly, cedar met road-building rules of thumb: despite growing technocratic predilections for bituminous surfaces, the best materials were (1) locally sourced; (2) affordable, in a municipal infrastructure environment dominated by local improvement by-laws; and (3) readily accessible, making repairs and maintenance unchallenging and cost-effective.[23] Ontario's eastern white cedar forests solved all three issues.

The problem, despite cedar's popularity with property owners, was that cedar roads eschewed practicality. Lasting only about 5 years, cedar blocks on Toronto's roadways became a ratepayer favourite because of their inexpensiveness, not their durability. Cedar pavements collapsed under repeated mechanical and climatic assault: steeled wheels and hooves, rain, snow, ice, and freeze/thaw cycles. A failing cedar pavement mimicked a garden plot, looking (and smelling) convincingly like compost, as does Brunswick Avenue in Figure 6.1.[24] This created a dilemma for property owners: cedar blocks' infuriatingly short lifespan could leave ratepayers paying for a pavement long after its expiry.[25] Yet, the blocks' low first-cost (or initial outlay) resoundingly recommended them over the acknowledged high first-cost of better materials, such as asphalt.

Figure 6.1 A worn-out cedar block pavement on Brunswick Avenue north of Harbord Street, November 1899.

Source: Permission of City of Toronto Archives, City Engineer's Department, Fonds 200, Series 376, File 2, Item 89.

And cedar blocks, and organic infrastructure generally, *were* cheap. Circa 1900, asphalt cost between $2.40 and $2.80 per square yard. Bricks, of which Toronto had plenty, were laid for $1.75–$2.00. Prices for wood, macadam, and gravel were considerably lower. Macadam could be purchased at $.90–$1.50 per square yard, but cedar's $0.60 per square yard was simply irresistible.[26] Gravel was even cheaper, but most property owners knew it was hopelessly inferior as a surface in Toronto – this likely explains why, even in a local improvements environment, gravel appears least often on the city engineer's pavements plans.

If cedar was ubiquitous (memoirists remark on its smell in Toronto), use of macadam and gravel was hardly modest.[27] The city laid 37.3 miles of macadam and 8.5 miles of gravel between 1890 and 1907.[28] So, despite the pavement reform being championed by Toronto's liberals and liberal institutions, most Torontonians stoutly resisted.[29] Indeed, some thought a good macadam roadway maintained thoroughly was unparalleled; so too a well-attended gravel road.[30] In such a milieu, Toronto's local improvements petitioners between 1890 and 1907 paid for 95 miles of new organic road surfaces for the city's increasingly mobilizing streets.[31]

Sluggishness

Torontonians' attraction to sluggishness-inducing pavements arose from a simple recognition that many did not use roads themselves. Torontonians whose homes abutted poorly paved streets walked on – but seldom moved in wheeled vehicles over – their own and neighbouring streets. Hence, the *Globe* posed one of the most relevant questions of the era: 'for whose benefit is the roadway constructed?'[32] For property owners required to pay for urban infrastructure under local improvement by-laws (and, by extension, their renters), the clear answer was *other people*. Paved roadways existed for carters and/or the companies they worked for; carriages and hacks, their owners or their fares; bicyclists looking for speedier and more convenient routes to work and leisure; automobiles and their chauffeurs hoping to challenge old uses of the street on bituminous pavements; city councillors employing modernization as bait to catch new businesses; city engineers whose technocratic vocation and calling consisted of urban improvement; and urban reformers, liberals who believed a city was only as moral as its worst moments of environmental decrepitude.[33] Well-paved street surfaces were decidedly not in the interest of home-owners and renters – most of them walkers and transit riders. They knew that bad pavements signified a street 'free from vehicular traffic' and thus 'safe for small children' who used the street as a playground in the era before park and playground building in Toronto.[34]

Sluggishness mattered to property-owning parents in a world of incipient hurry, because organic infrastructure guaranteed impeded vehicular mobility. In the case of speeding bicycles, it was well known in Toronto that 'scorching' bicyclists routinely hurt children (and adults) playing on the streets. In one instance inter alia, a 'dastardly cyclist' severely injured a little girl (breaking her shoulder and lacerating her skull), then sped off, 'with more regard for his own convenience and safety than for the ordinary demands of humanity'.[35] Toronto's modernizing pavements encouraged speed, introducing a relatively new mobilities circumstance: the *hit and run* collision. By the late 1890s, it was too common 'for cyclists to knock people down and then ride away'.[36] Then there were the automobiles, a small but feisty population of hurriers in the early 1900s. Largely chauffeur-driven, these new vehicles persistently challenged streetcars, passing on either side, whether the streetcars had stopped to pick up or let off passengers. It was 'a daily experience in Toronto to sit in a street car and see an automobile come panting alongside, draw ahead, and pass the car'.[37] On residential streets that had paid for asphalt paving, it was 'common to see a machine dash along at tremendous speed'. Trepidatious parents knew '[i]t would be impossible for the driver to do anything to prevent an accident should a child happen to run across the road as this whirlwind was passing'.[38] It seems organic infrastructure was also an anti-modern response to modernist speeders on streets where children played.

Well-paved streets, then, existed to benefit those who were required, or wished, to navigate them unhindered. And these regular users regarded the city's wooden pavements, especially on residential streets, as inferior and hazardous. For example carters, iconic denizens of the Victorian street, delivered their goods in open carts pulled by animals often over cedar block pavements that resisted passage. The *Star* noted that, when a milk cart approached one of Toronto's myriad dilapidated cedar streets, the 'milkman' preferred to park the wagon and deliver the milk on foot, rather than risk churning a load of milk to butter. Toronto's 'hack men' (cabbies) were always on the lookout for pavements 'fit to drive on', and they did not include cedar blocks.[39] Bicyclists jumped in a panic from their bikes 'to escape being dashed to pieces', when suddenly finding themselves riding on such a street (we will see below that local improvements meant that a single street, from one block to the next, could be paved with differing materials). Children living on a cedar block street would pull up the blocks, sometimes to use them to build 'castles' in the roadway, which older children would knock down and put back, presumably incorrectly.[40] The drivers and horse owners of Toronto 'thoroughly condemned' the condition of cedar block roads, which inflicted such 'jolts and jars' and exposed horses and carts to such a 'terrible series of pitfalls' that the cedar roads were deemed not only a 'dangerous and discredited nuisance', but cost the city in 'damages' suits (discussed below).[41]

Toronto's gravel-paved roads were notoriously impassable, in part because of the quaternary substrate supporting them. Toronto's age-old epithet 'muddy York' was literally grounded in the city's lacustrine glacial geomorphology; the sand and silty clay of the prehistoric Lake Iroquois (on whose ancient soils Toronto sits) provided a wholly inadequate physical geography for gravel pavement.[42] The result was absurd quantities of mud passing as streets and roadways (Figure 6.2). The City of Toronto Archives has collected numerous images of morass roadways bogging carters, carriages (and hearses), automobiles, and people at the turn of the twentieth century.[43] And, of course, mud halted hurry. The Toronto Street Railway Company, for example, complained to the city council in 1902 that the mud accumulating on the macadam roadway on Bloor Street, between Yonge Street and Avenue Road, was splashing into the streetcar motors, causing 'considerable damage and delay'.[44] The seasonally muddy suburban streets of West Toronto created difficulties for 'heavy motor cars and vehicles', which were '"stuck" daily in the more outlying streets', as Toronto automobilized.[45]

But even the so-called superior pavement, asphalt, offered unique challenges to traffic on wheels. Victorian asphalt was not the near-cement of our experience and had only about a 10-year lifespan, or less depending on local conditions. Most of it was petroleum-based, making it susceptible to heat, water, and freezing. Summer heat melted it, dust-suppression with mobile sprinklers degraded it (it was, astonishingly, not waterproof), and winter weathered it. A freshly laid asphalt pavement on Spadina Avenue succumbed to the first hot summer day after its construction, to the dismay of the property owners who bought it.[46] Disgusted by asphalt's susceptibility to water during summer water sprinkling, Alderman Lamb exclaimed to the council, '[f]or every thousand dollars worth of water you drench this pavement with, you lose a thousand dollars in deterioration'.[47] Winter frost cracked Jarvis Street's expensive new asphalt surface, the same on Bay and Simcoe Streets.[48] Oil-based asphalt could even catch fire.[49]

Figure 6.2 Mud roadway on Ashdale Avenue, 1908. Notice the concrete sidewalk. Apparently, pedestrian mobility mattered more in this neighbourhood.

Source: permission of City of Toronto Archives, Fonds 1244, f1244, it0024.

Then there was the experience of travelling on bad asphalt. The *Star* likened bicycling on Adelaide Street to canoeing in rough weather.[50] Bicyclists on Queen Street East (Yonge Street to the Don River) compared the ride to a teeter-totter, unless the rider was in a hurry, when the bumps came a little too frequently for comfort.[51] Streetcar track allowances had become de facto bicycle lanes in the 1890s, but the weight and unsteadiness of the streetcars disintegrated the allowances. 'Bicycle face' was the expression of terror a bicyclist assumed while traversing these asphalted 'devil strips' in heavy traffic.[52] The *Globe* suggested that, barring any sufficient solution to the paving problem, 'bicycle face in its acute form will become the most marked characteristic of the Toronto' bicyclist.[53] All these pavement shortcomings suggest Toronto's roads waxed sluggish.

Toronto's interaction with plank and flag sidewalks was no better, and, likely, worse, because most people walked. City Engineer Rust reported 100,000 riders a day on the Toronto Street Railway by 1900. The 36,061,867 individual rides for the year are convincing evidence of a walking impulse, as the riders walked to their streetcar stops.[54] These same walkers were frequently the property owners choosing organic pavements, who also preferred cheap plank sidewalks to more expensive concrete. It seems that, even when Torontonians used an infrastructure, they could tolerate a substandard one, if they were required to pay for it.

The 1890s stand as the heyday of plank sidewalks. Hundreds of miles were laid, despite grave warnings from officials and newspapers regarding their inferiority as an urban walking surface and the outright dangers they posed when failing. Plank sidewalks lay perpendicularly across the scantling (frame), which raised the planks from 4 to 6 inches above the ground. Broken planks created a significant tripping hazard, exacerbated by low light levels in early gas- and electric-lighted streets. The *Globe* growled about the continuous damages the city paid 'for injuries sustained while stumbling over or through' plank walks. Torontonians had the power to petition for better sidewalks, but they refused; in 1893, the city engineer received 196 petitions for plank sidewalks and two for concrete.[55] And yet the same petitioners allowed their plank sidewalks 'to fall into decay'; hence the damages. Such lackadaisical citizens needed, the *Globe* surmised, 'sidewalks by coercion', implemented 'whenever citizens will not petition for a proper sidewalk under the local improvements plan'. The city would try this in the 1900s, with respect to road surfaces, as discussed below.[56]

Although property owners preferred cheap plank walks despite their dangers, it was also true that they were not responsible for the quality and durability of a constructed sidewalk. An 1881 investigation into contracted sidewalk construction revealed 'that neither in quality of lumber nor in the manner of laying' did the year's sidewalks 'strictly comply with specifications'. The builders used 'coarse, common lumber' and much of it 'green' (not cured) that warped and split in the sun, making sidewalks 'unsightly and very defective' – and they were the first to rot and fail. These planks were not uniform thicknesses or widths, but laid as they were pulled off the woodpiles. Neither did the builders take care to place the 'waney' (uneven edge of a board) of the sleepers (constituting the scantling) away from the plank, thus inhibiting proper securing of the plank. Clearly, good construction aided durability, and we can imagine property owners' anger in paying for sidewalks – sometimes over 10 years – that failed the same year of their laying.

Hence the many damages claims against the city. In the era of 'wild-west' capitalism, poorly built sidewalks made the pedestrian's mobility difficult – likely the same walker who paid for them. One Margaret McKelcan sued the city for $9,000, after falling on a defective sidewalk on Yonge Street, just after dark. According to the suit, McKelcan broke her leg after tripping over a 4-inch rise in the sidewalk as she walked from Gould Street on to Yonge Street.[57] A Mrs Ford sued the city for $10,000 after stepping through a gap in a plank walk on Seaton Street and falling 'with great violence'.[58] Even an alderman severely injured himself, falling on worn out planks.[59] By 1909, the Civic Claims Commission was engaged in 'a lot of work' paying out numerous damages to sidewalk and infrastructure users in general. This included various payments, such as to a Mrs Woodley for her daughter's mishap on a 'defective walk'; to the Salvation Army for damages to a wagon on a 'defective roadway'; and to a Mr R.G. Kearney for damages to his horse, injured on a 'defective roadway'.[60]

The politics of sluggishness: local improvements

If reformers and road-users abhorred organic street surfaces, property owners had a different opinion. Repeatedly, and over decades, they chose organic surfaces, and any other less expensive pavement they could afford, to pass before the front doors of their homes and businesses. Moreover, it was comprehensively understood by all who lobbied for street surfaces in Toronto that property owners would 'get as good a pavement as they are willing to pay for, and no better'.[61] This unwillingness to consume what we might call over-engineered ('high-tech') pavements, such as asphalt, complicates popular assumptions of the inevitability of modernized, technological mobilities. Inevitable they may have been, but not without committed resistance – anti-modernism being an ironic characteristic of modernism.[62] In late Victorian and Edwardian Toronto, the facilitation of mobilities connected to the pocket books of property owners highly reluctant to spend.

Under Toronto's local improvements by-law, property owners obtained the right to petition for infrastructure materials abutting their properties. When the city engineer exercised his authority to scrutinize a city street for infrastructure problems (worn out sidewalks and roadways, privy pits and cesspools, or no infrastructure at all), he could initiate an infrastructure improvement, without property owner endorsement. This much-detested power of the city engineer was called the 'initiative principle'.[63] However, a street having been identified and an infrastructure improvement legally initiated, affected property owners could then petition for the type of material they preferred to purchase. They did this through a debenture (a transferable, unsecured loan repaid usually over 10 years) procured on their behalf by the city. Property owners' reluctance to buy new architectures of hurry, under local improvements legislation, created a significant obstacle for a council and engineer's office seeking to modernize hundreds of miles of unnavigable street surfaces.

Thus, Toronto's local improvement 'By-law respecting Local Improvements and Special Assessments therefor' legislated council compulsion of property owners to improve their neighbourhood streets on behalf the city – with the legal caveat that the council must grant property owners the democratic right to choose the quality of the street surface owners could afford.[64] By-law 4298 accommodated property owners' choices through the right of 'sufficiently signed' petitions. This meant that any street surface petition presented to the city engineer must contain signatures from two-thirds of all property owners whose combined property equalled at least one-half of the total value of all private property on the identified street.[65] Such a requirement exerted excessive strain on everyone involved: the city clerk, who had to accept and reject petitions; the city engineer, who received petitions that contradicted his expert recommendations; interested neighbourhood petitioners, who disagreed with the choices of their neighbours whose signatures they must procure; self-promoting pavement contractors, who hawked their own materials in petitions they peddled, often *before* neighbourhood petitioners had organized.[66]

The pavement petitions themselves have disappeared, and yet contemporaneous newspapers confirm, not only their prolific use in Toronto, but also their democratic efficacy in citizen disputes with City Hall. By the 1890s, newspapers brimmed with engineers' recommendations 'on the initiative' and public notice of the city clerk John Blevins's (1884–1900) receipt of sufficiently signed petitions – and insufficiently signed petitions, which we can imagine was an indirect indication of petitioner unrest on the street for which it was submitted (neighbourhood petition discord is beyond the scope of this chapter, but the antagonism associated with petition-signing at least suggests the type of paving material to be used on any given Toronto street cannot be assumed).[67] Year upon year, newspapers posted notices of local improvement recommendations on the initiative that included this standard sentence:

> Take notice that the Municipal Council of the Corporation of the City of Toronto intends to carry out the Local Improvement works set out in the schedule hereunder, and to assess the final cost thereof upon the property abutting thereon to be benefitted thereby.

The notice would detail the work to be undertaken (roadways, sidewalks, sewers) and, if a pavement, the recommended material. It would then conclude: 'Persons desiring to petition the said Council against undertaking any of the proposed works must do so on or before' a given date, within 30 days.

Thus, budget-conscious Torontonians spied the public notices for local improvements news about their streets and submitted sufficiently signed petitions, *usually against the recommendations of the city engineer*. The consequence was bad pavements in need of repairs that property owners resisted undertaking. Worn-out cedar roads, such as in Figure 6.2, prompted the *Globe*'s sarcasm: 'Toronto's pavements are evidently intended to be golf links for beginners'.[68] The *Star* complained that the 'woes of the city engineer' (Charles Rust, 1899–1912) connected directly to pavement petitioners, in this instance on Front Street:

> The local improvement law is the cause of Mr. Rust's trouble. He cannot pave the streets in the neighborhood of the Wm. Davies Pork Packing Factory, in Front street east, because the property-owners petition against the work every time he recommends it.[69]

This inclination of property owners to reject over-engineered pavements – expensive and patented materials such as Trinidad Lake Pitch Asphalt – presented a considerable obstacle to Toronto's attempt to facilitate mobility.[70] The annual reports of the city engineer show just how effectively petitions staved off pavement continuity in the city. From the 1890s to the 1910s, the *Annual Report of the City Engineer* included a pavements plan or map showing 'the different classes of pavement' in the city at the conclusion of every year. The plans, colourful depictions of the inability of the City to effect a single, continuous class of material throughout the city, illustrate the power of the local improvements petition.[71]

The *Plan of the City of Toronto* included in the annual report of 1898 reveals a city beguiled by choice, its key a rainbow of paving materials: asphalt (blue), brick (red), scoria block (peach), cedar block (yellow) cedar block out of repair (yellow stripe), gravel (brown), macadam (green), unpaved (white).[72] The central business district is predominantly green and yellow stripe. Main streets – Yonge, King, Queen, Adelaide, and Wellington – have stretches of blue, interspersed with red, green, and yellow stripe. Church Street is largely yellow and yellow stripe, with red at intersections with King, Queen, and Front/Wellington. University, the avenue famous for its promenading from the mid century, is green from Queen Street to Queen's Park. Jarvis is blue; the street of tony Victorian piles and their wealthy owners was asphalted (not without discord) in 1889.

To fathom the efficacy of the pavements petition in facilitating property owner choice (commercial and residential), and to understand why the city council, the city engineer, and the newspapers repeatedly called for an end to the democratic process, we can look at two streets in Toronto in 1898: Queen Street, from Roncesvalles Avenue in the west to the Don [River] Bridge in the east; and Euclid Avenue, between Queen Street West and Follis Avenue, above Bloor Street (Figure 6.3). The former largely commercial, the latter ostensibly residential, both offer instruction. Queen Street West, approximately 5 miles between Roncesvalles and Yonge, passed through the central business district (CBD) and was an ex-urban and suburban street.[73] Euclid Avenue, alternatively, is a roughly 2-mile-long inner suburban street (then and now), but of the sort described by Richard Harris and Robert Lewis, and Robert Fogelson: simultaneous suburban residential and business expansion.[74] Yet, irrespective of the business or domestic activities occurring on either street, property owners chose 'under'-engineered pavements.

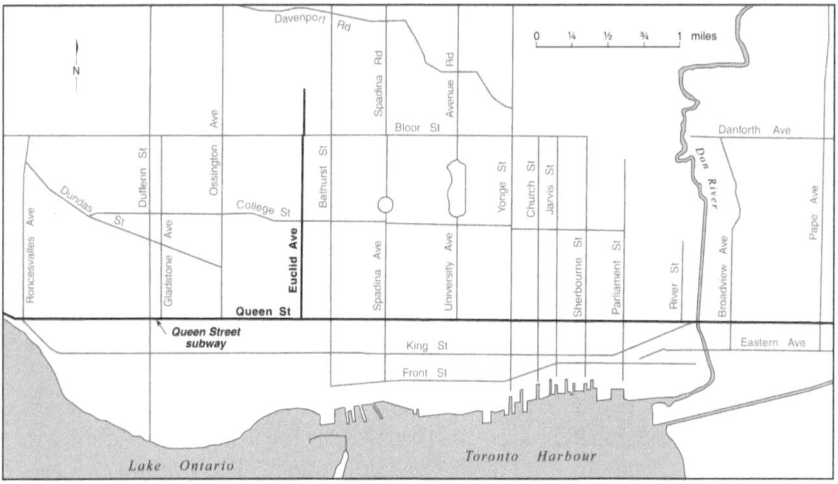

Figure 6.3 Queen Street and Euclid Avenue, Toronto.

Source: drawn by Loris Gasparotto (Brock University) and the author.

As the 1898 pavements plan illustrates, Queen Street East was asphalted (blue) from Yonge Street almost to the Don Bridge, property owners in that section perhaps able to see practicality in the expense. At River Street, Queen turned to cedar for a block and then to macadam. In the CBD, Queen Street West was worn-out cedar between Yonge and John Street, where the intersection was macadamized. West past John Street, Queen was asphalted to Bathurst Street, with a brick intersection at Spadina, which was worn-out block on both sides. Beyond Bathurst, worn-out block runs to Bellwoods Avenue, where new cedar block continued west to the Queen Street railway underpass, which was brick (paid for by the city). From there, Queen continued west as worn-out cedar to Roncesvalles, while intersecting along its way with north–south suburban streets paved with gravel, macadam, worn-out block, two asphalted streets, and one brick. All but one intersection remained worn-out block.

In the 1898 plan, the intersection of Queen Street West and Euclid Avenue shows a mix of worn-out cedar block and macadam. As Euclid moves north from Queen, it stays macadam for a block to Robinson Street, where it turns to old cedar block, which continues for a city block to Arthur Street, at which point a new cedar roadway begins.

This new cedar runs to College Street, including the intersection, and then turns into asphalt for a block up to and including the intersection at Ulster Avenue. From here, old cedar carries on past the four shorter blocks to Bloor Street and beyond for the final three shorter blocks to Follis Avenue.

Both streets' plentiful commercial and residential property owners plainly rejected expensive paving recommendations by the city engineer. In the case of Queen Street, City Engineer Keating, in 1896, 'condemned' as worn out Queen West from Yonge to Bathurst – Keating did renew the blocks in the track allowance from Bathurst to Gladstone Avenue.[75] Two years later, the 1898 annual report still depicted that stretch of Queen West as worn-out cedar. But, when, in the fall of 1898, Keating recommended cedar block on Queen West from Niagara Street to Gladstone, property owners agreed, and the 1898 pavements plan shows a new cedar block pavement.[76] This occurred repeatedly in the newspapers and pavement plans records. For example in 1896, Engineer Keating recommended asphalt for the stretch of Dundas Street between Queen and Arthur Street.[77] Shortly after, City Clerk Blevins received a sufficiently signed petition against asphalt.[78] The 1898 pavements plan shows a new brick roadway. Interestingly, the city engineer succeeded in asphalting a crucial section of Queen West: Yonge to the railway subway (at Gladstone Avenue), asphalt now covering about two-thirds of the entire length of the street west of the Don River by 1900.[79]

This regular countermanding of the city engineer's recommendations irked those trying to manufacture hurry in Toronto. By 1905, the council flirted with the idea of abolishing the local improvements petition.[80] The *Toronto Star* griped about the democratic nature of the pavements petition, which, although laudable, was a considerable obstacle to the city's modernization:

> It has been the practice to allow ratepayers to exercise [the right to choose] in presenting their petition, but it could not have been the intention when the law was framed, nor can it be considered good public policy, that any such right should be exercised under the conditions that now exist.

The city engineer knew more about paving matters than 'property owners on any street can know'. They should simply defer to his 'expert' opinion.[81] Thus, despite a legal mandate to validate all sufficiently signed petitions, City Engineer Rust and the city council began dismissing petitions. In the case of a recommendation for asphalt on Bloor Street (from Yonge Street to Avenue Road), locals petitioned overwhelmingly (60 of 66 property owners) for macadam. Engineer Rust 'strongly recommended the use of asphalt on so prominent a street', and the works committee agreed.[82] The Bloor Street petitioners' lawsuit failed, in part because the city moved, in 1906, to ban petitions. In November, the council forwarded legislation to grant it the power to 'force improvements' to the Ontario legislative assembly. Yet, the province declined, wary of 'taking power out of the hands of the electors'.[83] Toronto's property owners would continue for another decade to choose an architecture of 'good enough' for their streets, despite a formidable modern discourse of hurry.

Conclusion

Whether on residential or commercial streets, as the brief investigation of Queen Street and Euclid Avenue revealed, Toronto's residential *and* commercial property owners deliberately hampered modernized mobility for at least 25 years. This doubtless had nothing to do with mobilities per se. Torontonians simply refused to pay for expensive pavements they believed had little relevance in their mostly pedestrian lives – or they could not or would not afford them. Indeed, in the 1897 mayoralty campaign, Alderman John Shaw told his colleagues it was time for the council to recognize this circumstance and to seize from property owners the control of pavements, by abolishing the local improvements system. As he put it:

> There are pavements in all parts of the city which are disgrace to Toronto. They are dangerous and uncomfortable to travel on, as well as expensive to keep clean. Property owners will not renew them on the local improvement plan. Even if they would it is an unjust, costly system that has done more to injure Toronto than anyone has any conception of, and cannot be done away with any too soon.[84]

He was stumping for votes. In August, after Mayor Robert Fleming resigned to take a position as assessments commissioner, Shaw assumed the mayor's mantle – yet the local improvements system lay intact for years to come. But Shaw was not wrong in his interpretation of the local improvements system (ironically, the alternative of financing paving from the city's own funds promoted 'ward grabbing' corruption).[85] The local improvements by-law alone made Toronto's street surfaces a perennial nuisance for everyone, including the property owners themselves (curiously, by the 1910s, automobiles were 'denounced as being destructive' to roadways and needed to be taxed accordingly).[86]

Torontonians' tolerance of poor street surfaces confounds our assumptions about modernity and modernization. Jackson Lears, in an ageing but wonderful study of modernism, and arguing for the presence of a parallel and dialectical anti-modern impulse in the (North) American experience of modernism, suggests that beneficiaries of modernism, the moderns themselves, felt that they were also its 'victims'.[87] Local improvements in Toronto loosely align with Lears's observation. Toronto property owners who felt besieged by a relentless and expensive modernism found a way to resist it through the pavements petition. In this way, Toronto's street surfaces between 1870 and 1910 maintained a sluggishness that both dogged and impelled the city's quest for modern architectures of hurry.

Notes

1 'See How Toronto Has Prospered', *Globe*, 21 November 1908, 1.
2 E.M. Forster, *Howards End* (London: Penguin, 2000), 93.
3 Select titles include: Marshall Berman, *All That Is Solid Melts into Air* (New York: Verso, 1983); Stephen Kern, *The Culture of Time and Space, 1880–1918* (Cambridge, MA: Harvard University Press, 1983); David Harvey, *The Urbanization of Capital: Studies in the History and Theory of Capitalist Urbanization* (Baltimore: Johns Hopkins University Press, 1985), 35–45; Stanley Schultz, *Constructing Urban Culture: American Cities and City Planning, 1800–1920* (Philadelphia: Temple University Press, 1989); Clay McShane, *Down the Asphalt Path: The Automobile and the American City* (New York: Columbia University Press, 1994); Thomas Misa, Philip Brey, and Andrew Feenberg, eds, *Modernity and Technology* (Cambridge and London: MIT Press, 2003).
4 Thomas Misa, 'The Compelling Tangle of Modernity and Technology', in Misa, Brey, and Feenberg, *Modernity and Technology*, 4.
5 Paul Edwards, 'Infrastructure and Modernity: Force, Time and Social Organization in the History of Sociotechnical Systems', in Misa, Brey, and Feenberg, *Modernity and Technology*, 185.
6 Ibid.
7 James Lemon, *Liberal Dreams and Nature's Limits: Great Cities of North America Since 1600* (Toronto: Oxford University Press, 1996).
8 I use the term 'pavement' in the North American sense, meaning any paved street surface, wood, stone, asphalt, or otherwise.
9 Toronto had its own gravel pits and quarries. See the 'Preliminary Geological Map P138—Pleistocene Geography of the Scarborough Area', which identifies among others the old gravel pit near York Station, south of Danforth Avenue on Main Street (accessed 23 March 2017), www.geologyontario.mndmf.gov.on.ca/mndmfiles/pub/data/imaging/PR1962-01/PR1962-01.pdf. The city also experimented with other forms of wooden block (City Engineer of Toronto, *Annual Report of the City Engineer of Toronto for 1900* (Toronto: The Carswell Company, 1901), 3.
10 'Wooden Sidewalks By-Law', *Globe*, 19 December 1892, 4.
11 City Engineer of Toronto, *Annual Report of the City Engineer of Toronto for 1900* (Toronto: The Carswell Company, 1901), vii, 13.
12 City Engineer of Toronto, *Report of the City Engineer, Toronto, 1890* (Toronto: J.Y. Reid, 1891), 27, 9. See, also, 'Toronto Progress in 1886', *Globe*, 1 January 1887, 1.

13 'For Better Pavements', *Daily Star*, 25 September 1894, 4.
14 City Engineer of Toronto, *Annual Report of the City Engineer of Toronto for 1899* (Toronto: The Carswell Company, 1900), vii.
15 City Engineer of Toronto, *Annual Report of the City Engineer of Toronto for 1906* (Toronto: The Carswell Company, 1907), 12; City Engineer of Toronto, *Annual Report of the City Engineer of Toronto for 1907* (Toronto: The Carswell Company, 1908), 17.
16 'Detroit Pavements', *Globe*, 26 April 1880, 3.
17 Brick pavements were moderately popular in Toronto, but existed somewhere between organic and modern. Like organic pavements, brick produced much (fine red) dust on the streets it surfaced. Unlike organic pavements, brick provided better resilience to industrial traffic, and exhausted animals and carts. It generated noise and was susceptible to freeze/thaw cycles. Yet, because Toronto had a brick works, a brick pavement was easily repaired. See, 'Brick vs. Asphalt', *Daily Star*, 4 December 1900, 4. Engineer Keating liked brick, because worn brick pavements made fine foundations for asphalt (City Engineer of Toronto, *Annual Report of the City Engineer of Toronto for 1896* [Toronto: The Carswell Company, 1897], 6).
18 'Best Road for Cyclers', *Daily Star*, 27 February 1897, 9; 'Toronto Roads', *Globe*, 17 August 1875, 4. On Toronto's quaternary soils paving dilemma, see Phillip Gordon Mackintosh, *Newspaper City: Toronto's Street Surfaces and the Liberal Press, 1860–1935* (Toronto and Buffalo: UTP, 2017), 98–100.
19 Charles Baudelaire, 'Perte d'Auréole', in *Le Spleen de Paris: Ou les Cinquante Petit Poemes en Prose de Charles Baudelaire* (Paris: Chez Emile-Paul, 1869), 152. See Marshall Berman's discussion of Baudelaire's poem (Berman, *All That Is Solid Melts into Air*, 115).
20 'Cleaning of the City', *Globe*, 1 September 1904, 6.
21 'Will Sprinkle Asphalt Roads', *Daily Star*, 2 June 1906, 23.
22 'Theft of Old Blocks', *Daily Star*, 29 October 1902, 3.
23 George Tillson, *Street Pavements and Paving Materials – A Manual of City Pavements: The Methods and Materials of Their Construction* (New York: John Wiley, 1900) 136.
24 On the smell of Toronto's streets, see Chapter 5, 'Farmlike City', in Mackintosh, *Newspaper City*.
25 City Engineer of Toronto, *Annual Report of the City Engineer of Toronto for 1897* (Toronto: The Carswell Company, 1898), 9–10.
26 Two other organic materials, granite and scoria (basaltic lava) blocks, were used but very little – 0.68 miles in 1900 (City Engineer, *Annual Report of the City Engineer of Toronto for 1900*. Toronto: The Carswell Company, 1901), vii; 'Cannot Abolish Block Pavements', *Daily Star*, 29 November 1900, 3.
27 Gordon Hill Grahame, *Short Days Ago* (Toronto: Macmillan, 1972), 3, 15; Algernon Blackwood, *Episodes before Thirty* (London, New York, Toronto, and Melbourne: Cassell, 1923), 18.
28 City Engineer, 'Table No. 2: Mileage of Different Classes of Pavements, Roadways and Sidewalks Laid from 1890 to 1911', *Annual Report of the City Engineer of Toronto for 1911* (Toronto: The Carswell Company, 1912), 167–8.
29 Mackintosh, *Newspaper City*.
30 'A Good Old Pavement', *Daily Star*, 22 June 1895, 4; 'Macadam the Best', *Daily Star*, 21 April 1897, 4; 'Old Pavements the Best', *Daily Star*, 13 June 1899, 4.
31 City Engineer, *Annual Report of the City Engineer of Toronto for 1907* (Toronto: The Carswell Company, 1908), vii.
32 'Cedar Blocks', *Globe*, 10 February 1897, 5.
33 On these points, select literature includes: Clay McShane, 'Transforming the Use of Urban Space: A Look at the Revolution in Street Pavements, 1880–1924', *Journal of Urban History* 5 (1979): 279–307; Christine Boyer, *Dreaming the Rational City: The Myth of American City Planning* (Cambridge, MA: MIT Press, 1983); Richard Foglesong, *Planning the Capitalist City: The Colonial Era to the 1920s* (Princeton: Princeton University Press, 1986); Schultz, *Constructing Urban Culture*; Neil Larry

Shumsky, ed., *The Physical City: Public Space and the Infrastructure* (New York and London: Garland, 1997); Glen Norcliffe, *The Ride to Modernity: The Bicycle in Canada, 1869–1900* (Toronto: University of Toronto Press, 2001); Phillip Gordon Mackintosh, '"The Development of Higher Urban Life" and the Geographic Imagination: Beauty, Art, and Moral Environmentalism in Toronto, 1900–1920', *Journal of Historical Geography* 31 (2005): 688–722.

34 'Adventure Street', *Daily Star*, 29 May 1900, 6.
35 *Mail and Empire*, 10 March 1895, 6.
36 *Mail and Empire*, 10 May 1895, 6.
37 'The Speed of Automobiles', *Daily Star*, 21 June 1905, 6.
38 'The Automobile Accident', *Daily Star*, 17 June 1905, 6.
39 Reporters of The Toronto News, *Toronto by Gaslight: Nighthawks of a Great City* (Toronto: Edmund E. Sheppard, 1884), 6.
40 'Adventure Street', *Daily Star*, 29 May 1900, 6.
41 'Cedar Blocks', *Globe*, 10 February 1897, 5.
42 See S. Morley Wickett, 'Municipal Government of Toronto', in *Municipal Government in Canada*, ed. S. Morley Wickett (Toronto: Librarian of the University of Toronto, 1907), n. 1: 37.
43 Although such archived images attract our suspicion, their contents remain instructive. See Ian McKay, *The Quest of the Folk: Antimodernism and Cultural Selection in Twentieth Century Nova Scotia* (Montreal and Kingston: McGill–Queen's University Press, 1994), xvii, xviii.
44 'Mud in the Motors', *Daily Star*, 10 December 1902, 2.
45 'Spring Brings Smells to West Toronto', *Daily Star*, 30 March 1914, 7.
46 'Asphalt Is Soft, Mr. Rust Likes It', *Daily Star*, 7 July 1903, 1.
47 'Water, Asphalt and Wheels', *Daily Star*, 12 April 1899, 7.
48 'Asphalt Pavements Cracked', *Globe*, 16 March 1892, 8.
49 'From City Hall', *Globe*, 29 August 1896, 17.
50 'Wood Must Go', *Daily Star*, 25 May 1900, 6.
51 'Like a sea voyage', *Daily Star*, 4 July 1900, 2.
52 'Allowances' refers to that section of the roadway supporting the streetcar tracks, and the strip of pavement running between the two opposite running tracks. The allowance was typically paved with brick or asphalt, with granite setts bracing the rails. Bicyclists called the strip between the two tracks the 'devil strip', because it was hellish to get caught between two opposite-running streetcars.
53 'Dangerous Pavements', *Globe*, 10 May 1898, 6. On wheelwomen, see Phillip Gordon Mackintosh and Glen Norcliffe, 'Flâneurie on Bicycles: Acquiescence to Women in Public in the 1890s', *The Canadian Geographer* 50 (2006): 17–37.
54 City Engineer of Toronto, *Annual Report of the City Engineer of Toronto for 1900* (Toronto: The Carswell Company, 1901), xi.
55 City Engineer of Toronto, *Annual Report of the City Engineer of Toronto for 1893* (Toronto: J.Y. Reid, 1894), 11.
56 'The City Council', *Globe*, 23 August 1881, 9.
57 'Spring Assizes', *Globe*, 31 March 1882, 5.
58 'The Winter Assizes', *Globe*, 3 March 1885, 6.
59 'City News', *Globe*, 10 May 1869, 1.
60 'Claims Against City Kept Out of Court', *Daily Star*, 3 October 1909, 12.
61 'Wooden Sidewalks By-Law', *Globe*, 19 December 1892, 4.
62 Select titles include Berman, *All That Is Solid Melts into Air*; T.J. Jackson Lears, *No Place of Grace: Antimodernism and the Transformation of American Culture, 1880–1920* (Chicago and London: University of Chicago Press, 1994); Ian McKay, *The Quest of the Folk*; Lynda Jessup, ed., *Antimodernism and the Artistic Experience: Policing the Boundaries of Modernity* (Toronto, Buffalo, and London: University of Toronto Press, 2001).

63 City of Toronto, *By-laws of the City of Toronto, of General Application and also Showing Those Passed since 13th January, 1890, to 22nd February, 1904, Inclusive* (Toronto: William Briggs, 1904), 65–6.
64 Ibid., 61–72, 66. See specifically 4298.9 and 4298.10.
65 By-law 4298.7.1 (Ibid., 65).
66 For a more detailed discussion, see Chapter 4, 'Discordant City', in Phillip Gordon Mackintosh, *Newspaper City*.
67 See, for example, the petition and counter-petition for and against asphalt on Avenue Road ('Avenue Road in Need of Pavement', *Daily Star*, 10 October 1903, 6).
68 'Notes and Comments', *Globe*, 26 May 1906, 8.
69 'Woes of the City Engineer', *Daily Star*, 26 March 1903, 1.
70 See 'Trinidad Asphalt', *Daily Star*, 28 June 1901, 4.
71 Logistics prevent the publishing of a sample plan, but a quick search under 'Annual Report of the City Engineer of Toronto', at archive.org will provide numerous reports containing pavements maps.
72 Pavement Plan – City Engineer of Toronto, *Annual Report of the City Engineer of Toronto for 1898* (Toronto: The Carswell Company, 1899), 4.
73 Queen also continues 4 miles east from the Don River to the Beaches, through east Toronto. As East Toronto was developing in 1898, I have chosen to ignore it.
74 Richard Harris and Robert Lewis, 'The Geography of North American Cities and Suburbs: A New Synthesis', *Journal of Urban History* 27 (2001): 262–92; Robert Fogelson, *Bourgeois Nightmares: Suburbia, 1870–1930* (New Haven and London: Yale University Press, 2005). The 1900 City Directory indicates businesses on Euclid Avenue included bakeries, blacksmith, dairies, lumber yard, shoemaker, beverage maker, butchers, business college, confectioner, builders and contractors, dressmaker and embroiderer, druggist, delivery service, grocers, hairdresser, music teachers, reed maker, photographer, physicians, publisher, vinegar maker, churches, and fraternal lodges (*The Toronto City Directory Volume XXV* [Toronto: Might City Directory, 1900]).
75 'Bad Pavements: Streets that the City Engineer Has Condemned', *Globe*, 1 June 1896, 5; City Engineer of Toronto, *Annual Report of the City Engineer of Toronto for 1896* (Toronto: The Carswell Company, 1897), 28. Toronto's agreement with the Toronto Street Railway Company allowed the city engineer to make recommendations for the track allowances (Mackintosh, *Newspaper City*, 126–7).
76 'The Queen Street Pavement', *Daily Star*, 11 October 1898, 1.
77 We now call this Ossington Street. 'Big Batch of Pavements', *Globe*, 12 September 1896, 21.
78 'Preventing a Pavement', *Daily Star*, 24 November 1896, 1.
79 Pavement Plan – City Engineer of Toronto, *Annual Report of the City Engineer of Toronto for 1899* (Toronto: The Carswell Company, 1900), 15; 'Cost of Asphalt', *Daily Star*, 7 April 1898, 7. This report notes a new pavement on Euclid between College and Bloor: bitulithic, a tar and gravel road surface.
80 'City May Abolish the Petition System', *Daily Star*, 6 July 1905, 1.
81 'Make Use of the Expert', *Daily Star*, 3 June 1905, 6.
82 'Asphalt Bloor Street', *Globe*, 17 June 1905, 24.
83 'City's Bill for the Assembly', *Daily Star*, 22 November 1906, 9; 'Will Not Allow City to Force Improvements', *Daily Star*, 19 March 1907, 1.
84 'Local Improvements', *Globe*, 13 July 1897, 1.
85 'Wooden Sidewalks By-Law', *Globe*, 19 December 1892, 4.
86 'Tax Automobiles to Keep Roads in Good Repair', *Daily Star*, 27 February 1913, 9.
87 Lears, *No Place of Grace*, xv.

7 Keeping pedestrians in their place

Technologies of segregation on the streets of East London

David Rooney

Introduction

On Thursday 28 May 1936, the UK's minister of transport, Leslie Hore-Belisha, fixed into place a steel guard rail alongside the East India Dock Road, a busy traffic thoroughfare in Poplar, east London. It was part of a 1.5-mile installation constructed that year at the request of the Metropolitan Police traffic commissioner, Alker Tripp, and its purpose was to prevent pedestrians from stepping into the roadway except at fixed pedestrian crossings. The barriers were just over 3 feet high and installed 12 inches from the edge of the kerb, enabling vehicles to drive close to the edge of the carriageway without fear of striking pedestrians, thus, it was said, making better use of road space. Gaps were provided at bus and tram stops, side roads and garages, and businesses requiring goods loading were provided with hinged sections under their lock and key for use during deliveries.[1] The entire installation was completed by September 1936 and was the country's first ever large-scale installation of guard-rail technology (see Figure 7.1).

Yet the guard rails were only the start of Alker Tripp's plans. Ultimately, he wanted laws preventing pedestrians from entering the roadway except at designated crossings, and for those crossings to be controlled by lights and barriers, just like at railway level crossings. Tripp's pedestrian control scheme was thus a profound departure from the laissez-faire traffic policing culture of his predecessors. It involved the physical partition of the streets and a radical set of prohibitions. Tripp believed that 'promiscuous pedestrian crossing will ultimately have to be prohibited',[2] and the word he consistently used to describe his approach was 'segregation'. Through consideration of the wider context of this term, Tripp's pedestrian segregation can be seen to demonstrate links to racial segregation in the USA at this time, with a culture of categorization and exclusion common to both situations. Thus, in this chapter, we will explore how pedestrians on pavements were, in Tripp's intolerant and technocratic characterization, racialized, gendered, sexualized, aged and classed. These pedestrians formed a diverse, populous East End group whose supposed uncontrolled and disordered presence in the roadway was slowing those 'driving' through the streets – the relatively well-to-do and largely white, male car drivers whose vehicles travelled most quickly. Thus, in

Figure 7.1 Guard rails and pedestrian crossing on East India Dock Road, 1936.

Source: William Whiffin; Permission of Tower Hamlets Local History Library & Archives: 'Photograph Library: Street Scenes: East India Dock Road', P02766.

Tripp's view of automobilizing London, Poplar's sidewalk users were breaking down the 'proper' order of the road.

Tripp joined the Metropolitan Police as a Home Office civil servant in 1902. In 1920, he became chairman of the police recruiting board and, in 1932, was promoted to become assistant commissioner with responsibility for traffic, a post he held until his retirement in 1947 (he died in 1954). He is often cited in the urban planning literature as the inspiration for ideas such as neighbourhood precincts, limited-access motor roads and vertical separation. He appears in that literature as a marginal figure in the network of people and ideas, not quite a planner, but strongly influential as a traffic practitioner. But, in a different literature, that of policing, Tripp occupies a more central position. Thus, so that we can understand his approach, Tripp's role as an actor in the culture of policing will be examined here, including the ways his police experience influenced his views on the pedestrian's place on London's roads, both from a safety point of view and owing to more complex ideas developed through his experiences of American policing at a time of particular social and racial tension. This *widening* of the network, and the consequent centring of a marginal figure, draws a broader assemblage of concepts and beliefs into the story of London's traffic, helping us interrogate concepts such as pedestrian segregation that can easily be taken for granted.

Traffic on the East India Dock Road

For Alker Tripp, pedestrian safety and vehicle circulation were two sides of the same problem, which makes his guard-rail installation an 'architecture of hurry' as well as one of safety. Because pedestrians in the roadway did slow traffic and contribute to accidents, the solution to both in all modern cities lay in keeping pedestrians out of the roadway – in novel ways. His East India Dock Road scheme, Tripp observed, was 'aimed at pedestrian safety equally with (or even more than) vehicular circulation'.[3] The East India Dock Road was chosen for the guard-rail experiment as it had a high pedestrian accident record.[4] It was also – as contemporary photographs, including Figure 7.1 show – wide and straight and carried a mixed traffic, including motor cars, lorries, horse-drawn vehicles, steam engines, cyclists and pedestrians, as well as buses and trams. Trams travelled along the central crown of the road; buses tended to do so too, in order to avoid the numerous cars, vans and carts that pulled in along the kerb to load and unload. There were few formal restrictions as we might imagine them today; this was the laissez-faire road landscape that Tripp sought to bring under tighter control.

To explore the effect of the East India Dock Road guard rails, we may look at three main sources of commentary: the press, motor and pedestrian lobbies, and local business organizations. Much of the press reaction to the scheme took Tripp's line, which was that road traffic and foot traffic should be separated. This view assumed the road was a traffic artery. *The Manchester Guardian*, for instance, predicted:

> We are probably approaching a time when no pedestrian will be allowed on the roadway at all (at least in towns) except at stated points under special protection. If motor transport continues to expand at its present rate the segregation of urban wheeled and foot traffic must some day be complete.[5]

Once the Poplar rails were up, the same newspaper suggested that, 'the principle of putting something more solid than common sense between the motorist and the pedestrian is bound to grow in strength'. It went on to state that, 'Whether one will consider it a division of the sheep from the goats or the lambs from the slaughterer will depend, no doubt, on which side of the rail one finds oneself'.[6] This might be seen as a view driven by the motor lobby, and the car interests certainly approved of guard rails as a means to shift blame for accidents from motorists to pedestrians.[7] The Automobile Association, for instance, lobbied for guard rails on the basis of their successful use in mainland Europe, which had a tendency 'to deal with the pedestrian as a traffic unit rather than a separate problem'.[8]

To find a countervailing view, one needs to look at two groupings who might have felt aggrieved at this interference in the conduct of the streets. The first is the pedestrian, and the response of the Pedestrians' Association (now Living Streets), founded by Viscount Cecil of Chelwood in 1929, vacillated between qualified support and trenchant criticism. The association had already expressed unease in 1934 about the idea of non-junction pedestrian crossings (known then as 'crossing lanes'), fearing that plans to make their use compulsory would be 'a grave interference with the liberty of the public', commenting that:

> a man whose office or shop is midway between two crossing lanes would have to walk a quarter of a mile to make a call on the opposite side of the road and to return, or a householder would have to walk the same distance to post a letter in a pillar box on the other side of the road.[9]

Age, class and gender categories were mobilized by the association in its attack on the plans, as follows:

> If the crossings are widely adopted, it is right that pedestrians, and particularly children and old people, should be urged to use them as much as possible, and they will have the strongest inducement, that of self-preservation, to do so – but it is unreasonable to ask busy business men to go considerable distances out of their way . . . The prohibition would slow up pedestrian traffic enormously and cause congestion.[10]

The result of the association's representations to Hore-Belisha on legislative prohibition was positive: he abandoned the idea. But when, the following year, the East India Dock Road guard-rail experiment was announced, the association swiftly denounced it as a material alternative to legislative prohibition, stating that:

> it represents a blow at the existing rights of foot passengers, and imposes on them an enormous inconvenience. It is one more step in the policy of handing over the roads to the motorist and depriving all other road users of their rights.[11]

Yet the association went on to produce a report for local authorities agreeing that 'anything that deters persons, especially the young, from stepping thoughtlessly from the pavement on to the roadway will tend to render less frequent one common form of accident', so long as the goal was to promote safety, not speed. Its biggest concern was the danger to bus passengers boarding or alighting outside the guard rails, but it was also keen to stress that there must be 'no legal limitation on a pedestrian's right to cross the road at any point'. A further cause for concern was, perhaps surprisingly, the convenience of the road-user:

> Has he to stop opposite the house or shop at which he wishes to call and to climb through the railings, or to shout and gesticulate until he can attract the attention of the frontager to come with a key and remove the railings?

The association pressed for the rails to be removable by road-users every few yards and advised an evaluation of the experiment after 12 months.[12]

By 1938, the association's position had hardened into a belief that the continuous guard rail idea was part of 'the movement, encouraged by the motoring organisations, for limiting the pedestrian's right to the use of the road'. It recognized that 'modern conditions call for a certain measure of regulation of the movements of all road users', including the use of short sections of railing at particularly dangerous junctions or crossings, but that, 'the general invasion of pedestrians' rights to the use of the Highway under the guise of safety measures as an alternative to the restriction of speed is a policy which the Committee has stoutly resisted and will continue to resist'.[13] A year later, it concluded that the experiment was dead, with no further installations planned, as 'the rails proved very unpopular, for they not only caused great inconvenience, but in some respects added to the dangers'.[14]

A similarly equivocal reaction came from the second grouping for whom the street was a place as much as an artery, namely the local business community. The East India Dock Road was largely commercial.[15] For these people, the street was an intensely local place, control over which was not to be given up lightly. In 1936, as the rails were being erected, the Poplar Chamber of Commerce passed a resolution urging:

> that in the interests of business houses and shops along the routes, and the avoidance of congestion by delivery vans at fixed points that prior to the commencement of such work, the Borough Councils be instructed to consult with local business organisations, and the firms concerned, with a view to their convenience being met as far as possible.

It observed that, 'customers, rather than climbing through the rails preferred to go further for their purchases instead of dealing with their regular tradesmen'.[16] Yet even this response was muted, and was more a chance to criticize the local council than a sustained campaign against prohibition. It certainly was not a criticism directed at either the Metropolitan Police or the Ministry of Transport.

It can therefore be concluded that the practical effect of Tripp's guard rails was modest, partly as they were only intended to be the first phase of his wider project to protect pedestrian crossings with lights and place prohibitions on pedestrians from crossing at any other time or place. Since the guard rails had been erected, Tripp claimed in 1938, 'the effect has been to localise the casualties at the crossings. The protection of those crossings [with traffic lights] is obviously the next move'. He concluded that the police were 'powerless without the aid of the Ministry of Transport and the Local Authorities, and, if they show so little response, there is not much hope of really getting the casualties down'. The solution, Tripp had long argued, was for the police to take control of traffic across London away from local authorities, placing London 'in the same position as the Police in other large cities, such as New York, Chicago, Berlin, Paris, etc.'.[17] But it was not to be. The outbreak of the Second World War put all plans on hold, with the experiment only half built.

Prohibition and the police state

The rapid and widespread adoption of motor cars after the First World War profoundly reshaped the relationship between the police and the public, as Clive Emsley has described. 'The development of motor vehicles travelling at much greater speeds than previous road traffic', he says, 'constituted a problem of a new dimension. By the early 1920s the use of the law to control motor vehicles was jamming the magistrates' courts and creating friction, hitherto unknown, between the police and the middle classes.'[18] William Plowden's account of the motor car and politics further highlights changes in the public–police relationship with increasing regulation in the 1920s and 1930s.[19] And Keith Laybourn and David Taylor have provided a detailed account of the effects of motorization on policing in the interwar period, describing governments that were 'convulsed with the fear of communism and consumed with concern about motorised transport and the "road holocaust"'. They note that:

> the seemingly narrow issue of the regulation of traffic raised wider questions of individual freedom and, much to the concern of the police, brought ordinary, working-class constables into contact (and conflict) with members of the middle classes, who previously had held a positive, if somewhat patronising, perception of the British bobby.[20]

Alker Tripp was clear about the role the motor car had played in reshaping the relationship between the police and the public. 'Motor-car law', he said in 1928, 'has brought a new stratum of the public into frequent relations with the police. The area of contact, and of potential friction in consequence, increases daily'. He went on to note that:

> the individual constable enters on task already handicapped by having to some extent lost his reputation as a monument of stolid and tolerant common-sense and a barrier against interference with individual liberty. He is too often regarded as a busybody, a stickler for trifles . . . The police must vindicate anew their quality in the public eye.[21]

One of Tripp's first major tasks as traffic commissioner was to go on a fact-finding visit to the USA and Canada to study control methods. His trip took place over 20 days in October 1934 and included visits to New York, Boston, Detroit, Chicago, Toronto and Montreal. Upon his return to London, Tripp prepared a substantial report for his police colleagues and the Ministry of Transport detailing his findings. The tension between accidents and free-flowing traffic was expressed clearly in the report, which was split into two main parts – accidents and circulation. But the report conveyed a wider set of issues that shed light on the London traffic scene post-1934.

For Tripp, the difference between America and England was the difference between prevention and cure. 'One is conscious', he began, 'of what appears to be a constant solicitude on the part of the House of Commons for the motorists' interest – a solicitude which arises largely from the desire to protect personal liberty'. He went on:

> the Americans seem in one important particular to have been clearer-sighted than ourselves: they have consistently relied upon prohibitions intended to prevent bad habits of driving (e.g. by speed limits, stop-streets, etc.) instead of merely leaving dangerous driving to be dealt with after it has occurred.[22]

Tripp called for immediate action:

> Such action lies along the lines of (i) physical safeguards – traffic lights, duplicate carriageways, railings . . . and (ii) a code of exact law (based on the advice of traffic experts to the exclusion of interested parties) which is firmly, rigidly and exactly enforced by a more specialised police organisation with the help of courts which are more alive to the public danger and less concerned with the liberty of action of the motorist.[23]

By 1932, when Tripp had been appointed traffic commissioner, society had changed dramatically compared with a generation before. The rise of the motor-car city had coincided with the arrival of the political Labour movement and the development of an increasingly interventionist state. Martin Pugh has surveyed a major social and political shift at the start of the twentieth century and describes what he terms 'a sustained extension of the role of the state between 1906 and 1914'.[24] Fears of racial deterioration following the Boer War (in part) led to demands for more state intervention in everyday life, and the requirements of the First World War dramatically increased state control of industry, food and wages. This was reversed after the war ended, but it signalled a 'period of confusion' in the interwar years in which collectivism vied with capitalism. Under British Prime Minister Stanley Baldwin's leadership, from 1924 to 1929, social and economic policies were implemented in order to prevent a class war. By 1931, with depression, unemployment, strikes and fear of revolution and the power of the masses, there was a clear concern about liberal laissez-faire governance and an increasing voice calling for state control.[25] Thus, this culture of state intervention in more and more aspects of everyday life, along with fears of the masses, can be seen to relate to the shifts in police control of London's streets outlined in this chapter.

Tripp's promotion in 1932 had come just one year after a change at the top of the Metropolitan Police, with the retirement of the 'gentle' commissioner, Julian Byng, and the arrival of the 'autocratic, intolerant' Hugh Trenchard, as David Ascoli has described them.[26] Tripp himself brought a new style. His predecessor, Frank Elliott, had been described as a 'kindly' and 'much-liked' old policeman.[27] Tripp, by contrast, was 'forceful and cogent in conference and a born fighter, he was impatient of compromise and delay and he both gave and took hard knocks'.[28] In many ways, Tripp's 1934 visit to America was symbolic of a harder-edged approach to match the changing times. We are looking at a shift in the characterization of congestion in the early 1930s paralleling a shift in wider relationships between police and public, both in American cities and London. Tripp's rhetoric was about public safety, not individual liberty, as had been paramount in a previous generation. His was more an American (interventionist, controlling) police model, replacing an older British one of consent and freedom. And, in his report of the 1934 trip to America, we can discern disturbing influences that were to shape his approach.

Alongside Tripp's statistical reporting of accident rates and so on, he offered verbatim accounts of confrontations between local police and the citizenry that throw a frank light on the problems of control in 1930s America. In fact, the report's potentially incendiary nature was recognized by Hugh Trenchard, who, asking for it to be made confidential, observed, 'I do not want to see extracts from the report published in every newspaper. I think it would have a very bad effect if some of the sentences in it were taken away from their context'.[29] Tripp's visit to America came just after the end of the Prohibition era. A country-wide ban on alcohol had been in operation from 1920 to 1933 and had led to a breakdown of respect for law-enforcement bodies, among other effects. Cities such as Chicago had become the focus of organized bootlegging gangs operating violent black markets, with Al Capone becoming an iconic figure in American criminal lore, and it is difficult to overstate the negative effects (though in many cases unintended) on American urban culture of this movement.[30] Tripp was in Chicago just months after Prohibition was repealed, and made the following observations about the breakdown of civil society in the city and others like it:

> it must be remembered that in these American cities (Chicago especially) there is a mixed population, much of which is drawn from the dregs of Europe. Of the murders in Chicago about 30% are negroes killing negroes, and, as the Police put it, 'when a negro gets a drop of liquor into him he kills. He will kill for anything. He will kill for an electric light globe, a cigar, or really for nothing at all.' Another 10% are Italians killing Italians, and the majority are gangsters killing gangsters, or criminals, criminals. The risk to life run by the ordinary white man who behaves himself and manages to keep clean of rackets is much lower than might be inferred from the statistics.[31]

Apart from forcefully articulating the era's typically racialized discourse, what this translated to was a 'bold and rather extreme policy', in Tripp's words, of active policing.[32] He described night-time police cruises in which people deemed suspicious were stopped, searched and, if still suspicious, arrested and taken to the police station for questioning. His trip was ostensibly to learn about traffic control, but, as he later reported, it was these night-time experiences that gave him a thrill. 'I confess that I found it most entertaining', he recalled. 'There was a spice of excitement and adventure about it which was not to be denied.'[33]

There was also endemic corruption of both the police and the court system as a result of Prohibition. Police corruption took the form of large-scale political interference in the service, individually corrupt officers, corrupt practice in compacts with criminals such as Capone, and the overlooking of offences in order to obtain political advantage. The last type was particularly relevant to the traffic service, where widespread traffic offence ticket-fixing took place.

The tension here was about individual liberty versus public order, and it is clear that Tripp had some sympathy with the American approach. He began by remarking as follows:

> one of the main reasons for the results produced in America is that the Police (armed with appropriate laws) contrive to make the motoring public 'toe the line' (so to speak) in a way that we never have succeeded in doing. Speed is *lower* and *more uniform* and vehicles keep line better. That effect is not produced by exhortation of Press campaign; it is produced partly by mechanical control, but especially by intensive law enforcement.[34]

This was not about lower individual speeds leading to better journey times across the population of vehicles; it was more fundamental (and less nuanced) than that: cops want order. He expanded on his theme, stating that the police in all countries had been 'put to the test by the sudden menace' of motor traffic in the first 30 years of the twentieth century, and some had reacted better than others:

> American institutions and methods seem to have reacted more freely and better than our own, which have proved too inflexible and conservative. We appear to have shown less adaptability of mind; and as a result we have by contrast failed to a greater degree. To say this is not of course to decry the British character or to exalt the American by contrast. Undue quickness of reaction may easily lead to instability; the fiasco of Prohibition is a case in point.[35]

Yet the idea of American people submitting to curbs on personal liberty for the greater public good seems as surprising as expecting the British people to do so. Clearly, Prohibition is an important concept in understanding Tripp's policing culture. But there was more to it than that. By looking more closely at the particular conditions in America in the 1920s and 1930s, we can find a more complex story of why, it seems, Chicagoans, New Yorkers, Londoners and Detroiters reacted to police traffic controls in the ways they did. In particular, and in light of the racist culture of policing witnessed by Tripp, we will look at the role of segregation in US cultures of governance.

Segregation, cosmopolitanism and 'cordoning off'

Following his return from America in 1934, Tripp made the following observation about exclusion in the context of traffic in US towns:

> Regulation of traffic is all in favour of the private car; commercial vehicles are rigorously excluded from many of the main traffic arteries in towns, while in the country the parkways, which are specially constructed motor roads, exhibit notices 'No buses, no trucks, no commercial traffic'.[36]

Tripp became a prolific writer on matters relating to traffic and the public. Throughout his work, there is a strong rhetoric of exclusion – as part of a culture of categorization – and it was in the categorization of pedestrians that the language of separation was clearest. The term Tripp consistently used for the separation of pedestrians from vehicles was 'segregation', giving a new meaning to a term that had been used by his predecessors, from the early 1920s, to mean separating horse-drawn traffic from motor vehicles, or freight and public-service vehicles from private cars.[37] In Tripp's influential 1938 book *Road Traffic and Its Control*, he described two forms of pedestrian separation, namely place-segregation and time-segregation, as follows:

> In the case of new roads required for the needs of wheeled traffic, pedestrians can be excluded altogether. In the case of existing roads, footpaths provide the means of segregation, so far as pedestrians proceeding longitudinally are concerned – more especially if guard-rails are added. For pedestrians crossing the road, however, a system of place-segregation (by bridge or subway) is very costly, and the less straightforward plan of time-segregation must generally be invoked.

In this case, pedestrians and motor vehicles needed to share the same physical road space, but not at the same time, and crossings protected by traffic signals provided the segregating technology. However, care needed to be taken, cautioned Tripp, to balance the needs of pedestrians to cross the road with the need for 'vehicular fluidity', in order to prevent the latter's 'virtual immobility'. And he concluded with the warning that pedestrians would never be fully safe as long as they were 'free to neglect such means of safety as are provided for their benefit'.[38]

In Tripp's East India Dock Road project, he seemed to identify three issues: bad driving, bad walking and bad policing. As was discussed earlier, he failed to resolve the latter – his attempts to take control of all traffic matters – owing to the complexities of London's politics and the intervention of the Second World War. But why did his guard rails seek to control pedestrians, leaving drivers a clear run? Part of the answer must lie in his own prejudices – about who had the right to use the road. He had been presenting the notion of pedestrian segregation since his appointment as traffic commissioner. In 1933, in front of the Institute of Transport, he made a bold set of claims about the unfairness of the road scene towards motorists:

> Pedestrians demand the right to use any part of the footway or carriageway at their discretion. The footway is forbidden ground for vehicles; but the carriageway is not forbidden ground for pedestrians . . . Vehicles are herded into droves on each side of the road, and are ordered about by police officers and traffic signals and made to use one way streets and so forth; pedestrians are completely free agents.[39]

The solution was that, 'steps should be taken wherever possible to segregate vehicular and pedestrian movement, and that in the case of arterial routes, the segregation should be absolute'.[40] This was a solution based on classification and the apportionment of value to citizens and their use of the road. Tripp concluded:

> The first step towards solution of the whole problem of the design of streets and traffic requirements must be a critical analysis of the demands of the various classes of road users, and a strict assessment of them in order to determine which are legitimate and which excessive and therefore inadmissible.[41]

To what extent might this culture of segregation and exclusion have been influenced by American ideas of race? A clue lies in the concept of 'cosmopolitanism'. While in Chicago, Tripp asked why there was no enforcement of laws prohibiting pedestrians from crossing except at fixed places. He recalled the reply:

> the Judges (fearful perhaps of incurring unpopularity and loss of votes) refused to convict, unless there were some additional offence. The Chicago Police told me that they thought it would be difficult to enforce such a law 'in a cosmopolitan city like Chicago', but they added that a law had been successfully enforced in Minneapolis where the population is more homogenous.[42]

The term 'cosmopolitan' is telling in this context. At its simplest, the comment may simply have meant that cities with many visitors or 'outsiders' may have had trouble enforcing rules those visitors were not familiar with. There is also, as Peter Norton has eloquently described, a political history of the concept of jaywalking, which may have played a significant role in the judicial reluctance to convict.[43] But there may be a deeper reading of the specific term 'cosmopolitan'. Characterizations of major US cities had for many years been provided by tourist literature and guidebooks written by English and French authors, which, as David Gilbert and Claire Hancock have shown, significantly shaped the European experience of America.[44] In this literature, cosmopolitanism was a distinctly pejorative term, meaning 'monstrous' ethnic diversity, a degenerative state close to the abusive and demeaning American concept of 'mongrelization', although its meanings were complex and variable.

Tripp was not alone in taking part in a sightseeing tour of Chicago's poor neighbourhoods. 'Slumming' – treating mass immigration as a tourist spectacle – was a standard experience in the average European visitor's itinerary in major US cities, with ethnic neighbourhoods in New York in particular acting as 'a kind of museum of humanity, where the customs of the world could be examined', as Gilbert and Hancock put it, and at its heart, they contend, the conception of the 'cosmopolis' as a 'city polluted by diversity' was racist.[45] Tripp's accounts of New York and Chicago fit well into this culture of touristic slumming and the sordid thrill of what one 1924 guide described as 'motley cosmopolitanism'.[46]

The first four decades of the twentieth century were a period in which racial segregation became normal in the USA, not just in the Southern states, which practised extensive and overt spatial segregation in their towns and cities, but also in the Northern states, which encouraged the deliberate ghettoization of urban neighbourhoods and the legislation of 'whites only' restrictions in proliferating New Deal suburbs.[47] In reporting his experiences in Chicago and other American cities, Tripp described local police forces that were consciously aware of – and prejudiced negatively against – minority groups. Black people were 'negroes' and 'negresses' – or, just as commonly, referred to by another well-known but persistently racializing and humiliating epithet. Black men, police infamously held, were likely to be rapists, and working-class white women were fair game; both were sexually promiscuous.[48] Homosexual men were designated 'fairies' or 'pansies', both popular contemporaneous derogations.[49] For the Chicago police, the local population was an assemblage of distinct, identifiable groups – *ethnicities* and *races* – to be categorized and controlled.[50] But segregationist ideas were not restricted to America: they were growing in the UK at this time too, and, by reflecting on the mutability of the concept, we can start to understand the context in which Tripp chose to shape London's traffic scene by segregation. There is no suggestion that *racial* segregation was implied or considered by Tripp in his proposals for London's streets. East India Dock Road was not singled out for railings owing to a particular ethnicity of the local population. It was simply the location of the first installation of what Tripp fervently hoped would be a universal technology across the capital, from Whitechapel to Knightsbridge, and from Commercial Road to Oxford Street. But it is nevertheless notable that he chose to use this term consistently, and he was certainly operating in a segregationist culture of public governance that viewed the classification of people along ethnic, gender and sexual lines as normal.

Paul Rich has examined the situation in Britain in the early twentieth century, decades before the emerging racial conflict following the Second World War and the 'new racism' of the 1970s. Describing segregation in the USA at that time, he observes that it:

> fell substantially into the arena of mainstream political discourse of 'middle opinion', and . . . for a significant period from the 1890s up to the 1940s the concept seemed eminently reasonable even to people of moderate and liberal political persuasion.

Rich notes the 'international nature' of segregation and the 'readily exportable' nature of the term, going on to explore its meanings in Britain.[51] Susan Smith has also written on the politics of race in the UK in this period, describing how ideologies of racism, nationalism and segregationism express a complex network of beliefs.[52] We can observe a nationalism in Tripp's rhetoric of the London police with respect to non-British forces – and this becomes racially inflected, not only when words such as 'cosmopolitan' are deployed, but also in his extensive negative descriptions of racial and cultural diversity. At the local London level, segregationist ideas – the geographical allocation of space according to culturally constructed categories such as 'pedestrian' and 'motorist' – can be seen to be heavily imbued with gender, class, race, age and other embedded prejudices. Thus, although not necessarily overtly racist, we can see in Tripp's localized pedestrian segregation/exclusion an expression of a global movement – what Phillip Gordon Mackintosh has termed 'infrastructural racism', where the 'uncritical organization of public infrastructure abett[ed] and affirm[ed] Victorian and Edwardian conceptions of race'.[53] Importantly, for Smith, such thinking may not have been 'a rigid formula transported between continents, but a system of meanings, norms and expressions, moulding and adjusting to new challenges and changed circumstances'.[54]

In this context, it is hard to read a comment made by Tripp in 1936, expanding on his theme of pedestrian segregation in London, without hearing echoes of the American racial debate. This time he was introducing stronger notions of dominance, coercion and promiscuity, as follows:

> The principle to be aimed at must be to give the wheeled traffic a defined dominance over pedestrian traffic in all roads of an arterial character, but at the same time to provide regulated zones of safety for the pedestrians. In addition, in order to prevent promiscuous crossing by pedestrians, it may be necessary, as I have long advocated, to provide guard rails along main roads . . . If such facilities and safeguards are provided it will pay both vehicular traffic and pedestrians to conform to orderly movement, and they will do so without any form of coercion by the police.[55]

One might substitute 'whites' for 'wheeled traffic' and Tripp's term, 'negroes', for 'pedestrians' and hear a culture of racial categorization and exclusion common to both contexts.

Yet there may be a further link between traffic infrastructure and segregation in the notion of cordons, both physical and social. Anne McClintock, in her work examining the 'degeneration' trope of the late nineteenth century, observes the following:

> In the metropolis, the idea of racial deviance was evoked to police the 'degenerate' classes – the militant working class, the Irish, Jews, feminists, gays and lesbians, prostitutes, criminals, alcoholics and the insane – who were collectively figured as racial deviants, atavistic throwbacks to a primitive moment in human prehistory, surviving ominously in the heart of the modern, imperial metropolis.[56]

She goes on to explain the social purpose of such characterization, stating that 'social classes or groups were described with telling frequency as "races", "foreign groups", or "nonindigenous bodies", and could thus be cordoned off as biological and "contagious", rather than as social groups'.[57]

This cordoning-off activity she considers part of Julia Kristeva's notion of 'abjection', the process by which a social being creates itself by expelling 'impure' elements, which never truly disappear but 'haunt the edges of the subject's identity with the threat of disruption or even dissolution'. Crucially, she explains that, 'Certain threshold zones become abject zones and are policed with vigour: the Arab Casbah, the Jewish ghetto, the Irish slum, the Victorian garret and kitchen, the squatter camp, the mental asylum, the red light district, and the bedroom'.[58]

In 1930s London, we can consider a new 'abject people' and their 'abject zone': pedestrians on the pavements, who, in Tripp's characterizations, were racialized, gendered, sexualized, aged and classed. They formed a group whose disordered presence in the roadway was slowing those people who were driving through the streets. That was their contagion, in the sense of a pollutant or a corrupting influence: pedestrians obstructed the passage of vehicle drivers in a hurry, breaking down Tripp's perceived proper order of the road. Yet pedestrians could only be marginalized, not eradicated, as even the driver was a pedestrian before getting into the car.

Conclusions

The Metropolitan Police's approach to traffic congestion and circulation in the nineteenth century can be characterized as liberal, reactive, reluctant, consensual, localized and fragmented. Yet the first three decades of the twentieth century saw this approach shift, becoming increasingly interventionist, proactive, holistic, coordinated and planned. This mirrored British society's journey from laissez-faire to liberal democratic. In many ways, the Metropolitan Police in the 1930s, under traffic commissioner Alker Tripp, was keener for change on the streets than one might have thought, given its conservative history. On the surface, we can see this partly as a power struggle between the police and other professional groups over who controls the streets and the movement that takes place in them. But we can also see this as the circulation of ideas on policing and social control between major cities around the world.

Alker Tripp's desire to improve the circulation of vehicles on London's streets was wholly intertwined with his wish to reduce fatalities among pedestrians in the roadway – as well as his political goal of taking control of the streets from the myriad local and national bodies who held a stake in their governance. His proposed solution was a London-wide project to cordon off pavements from roadways, with a network of light-controlled pedestrian crossings enabling walkers to cross the roads safely and in a controlled manner, as with the railway network. London's governance structure ultimately frustrated Tripp's plans for a holistic scheme, with only a single thoroughfare – the East India Dock Road – being railed off, and without its traffic lights. Yet, his plan had a more insidious effect and came in part from a more unpalatable sociological belief. Recounting Tripp's study trip to the USA in 1934, months after the end of the Prohibition era and its toxic effect on public attitudes to law and order, this chapter has suggested that distinctly American police cultures of control, together with ideas of racial segregation, were imported by Tripp from places such as Chicago into London. The chapter offers evidence for a policing strategy that could, on the one hand, be seen as a benign attempt to save lives and speed the traffic flow through the mundane technology of the guard rail, or, on the other hand, the expression of a racialized segregationist culture of systematic discrimination and isolation in which pedestrians were configured as impure and contagious, leading to a diseased city.

Notes

1 The National Archives (hereafter TNA): MEPO 2/6748, memorandum, Divisional Road Engineer (London) to Chief Engineer, 17 August 1935.
2 TNA: MEPO 2/6748, file note by Tripp, probably sent to Arthur Dixon (Home Office), 3 May 1935.
3 TNA: MEPO 2/6748, minute, Tripp to Game, 10 January 1938.
4 TNA: MEPO 2/6748, note, Tripp to Select Committee, n.d. but about November 1938.
5 *The Manchester Guardian*, 11 March 1935, 8.
6 *The Manchester Guardian*, 29 May 1936, 10.
7 On the position of pro- and anti-motoring lobbies regarding pedestrians in this period, see Michael John Law, *The Experience of Suburban Modernity: How Private Transport Changed Interwar London* (Manchester: Manchester University Press, 2014), chap. 9; and on age, class and gender issues surrounding pedestrian crossings, see Joe Moran, 'Crossing the Road in Britain 1931–1976', *Historical Journal* 49 (2006): 477–96.
8 *The Times*, 15 January 1935, 14.
9 *Pedestrians' Association Occasional News Letter*, October (1934): 5.
10 *Pedestrians' Association Occasional News Letter*, October (1934): 5–6.
11 *Pedestrians' Association Quarterly News Letter*, October (1935): 9.
12 Living Streets Archive: Minutes of the Pedestrians' Association, 14 January (1936); *Pedestrians' Association Quarterly News Letter*, January (1936): 6–7; *The Times*, 26 November 1935: 11; and, on the interest-group politics of road safety in the 1930s, see William Plowden, *The Motor Car and Politics in Britain 1896–1970* (London: Bodley Head, 1971), chap. 13.
13 Pedestrians' Association, *Ninth Annual Report of the Pedestrians' Association* (London: Pedestrians' Association, 1937), 6–7.
14 Pedestrians' Association, *Tenth Annual Report of the Pedestrians' Association* (London: Pedestrians' Association, 1938), 5.
15 Anonymous, *The Post Office London Directory for 1936* (London: Kelly's Directories, 1936), 388–9; see also Hermione Hobhouse, ed., *Survey of London Volume XLIII: Poplar, Blackwall and the Isle of Dogs* (London: Athlone Press for the Royal Commission on the Historical Monuments of England, 1994), 120–70.
16 *East London Advertiser*, 30 May 1936, 2.
17 TNA: MEPO 2/6748, minute, Tripp to Game, 10 January 1938.
18 Clive Emsley, '"Mother, What Did Policemen Do When There Weren't Any Motors?" The Law, the Police and the Regulation of Motor Traffic in England, 1900–1939', *The Historical Journal* 36 (1993): 357.
19 Plowden, *The Motor Car and Politics in Britain*, chaps. 6, 11–13.
20 Keith Laybourn and David Taylor, *Policing in England and Wales, 1918–39: The Fed, Flying Squads and Forensics* (Basingstoke: Palgrave Macmillan, 2011), 9, 106. The term 'holocaust' applied to road fatalities began gaining popular currency from about 1929, when it was used in the founding speeches of the Pedestrians' Association. See, for instance, *The Times*, 5 November 1929, 9.
21 Alker Tripp, 'Police and Public: A New Test of Police Quality', *Police Journal* 1 (1928): 533.
22 TNA: MEPO2/5937, *Report by H. Alker Tripp, Assistant Commissioner, on Visit to America, October* III (1934): 2.
23 Ibid.: 11–12.
24 Martin Pugh, *State and Society (Second Edition): A Social and Political History of Britain 1870–1997* (London: Arnold, 1999), 62.

25 Ibid., parts II and III and in particular 120–21 and 216–8; see also the essays in Mary Langan and Bill Schwarz, eds, *Crises in the British State 1880–1930* (London: Hutchinson, 1985).
26 David Ascoli, *The Queen's Peace: The Origins and Development of the Metropolitan Police, 1829–1979* (London: Hamish Hamilton, 1979), 219 and 227.
27 *The Times*, 27 March 1939, 14.
28 *The Times*, 23 December 1954, 9.
29 TNA: MEPO 2/5937, memorandum, Hugh Trenchard (Commissioner) to Tripp, 18 December 1934.
30 On the history of the American Prohibition, including its effects, see for instance Norman Clark, *Deliver Us from Evil: An Interpretation of American Prohibition* (New York: Norton, 1976).
31 TNA: MEPO2/5937, *Report by H. Alker Tripp on Visit to America*, II: 6.
32 Ibid., II: 7.
33 Ibid., IV: 1.
34 Ibid., IIA: 5.
35 Ibid., III: 1.
36 Ibid., I: 7.
37 See, for instance, Chief Constable Arthur Bassom in *The Times*, 4 November 1924, 11.
38 Alker Tripp, *Road Traffic and Its Control* (London: E. Arnold, 1938), 136–7.
39 Alker Tripp, 'The Design of Streets for Traffic Requirements', *Journal of the Institute of Transport* 15 (1933): 76.
40 Ibid.: 80.
41 Ibid.: 84.
42 TNA: MEPO2/5937, *Report by H. Alker Tripp on Visit to America*, I: 8.
43 See Peter Norton, 'Street Rivals: Jaywalking and the Invention of the Motor Age Street', *Technology & Culture* 48 (2007): 331–59; and Peter Norton, *Fighting Traffic: The Dawn of the Motor Age in the American City* (Cambridge, MA: MIT Press, 2008).
44 David Gilbert and Claire Hancock, 'New York City and the Transatlantic Imagination: French and English Tourism and the Spectacle of the Modern Metropolis, 1893–1939', *Journal of Urban History* 33 (2006): 77–107.
45 Ibid., 96–8.
46 'Rider's New York City', 1924, quoted in ibid., 87.
47 See Carl Nightingale, *Segregation: A Global History of Divided Cities* (Chicago: University of Chicago Press, 2012), chap. 10; David Freund, *Colored Property: State Policy and White Racial Politics in Suburban America* (Chicago: University of Chicago Press, 2007); Craig Steven Wilder, *A Covenant with Color: Race and Social Power in Brooklyn* (New York: Columbia University Press, 2000).
48 TNA: MEPO2/5937, *Report by H. Alker Tripp on Visit to America*. See for instance IV: 8, where Tripp described a white woman passed out in the street from alcohol: 'I enquired, as we drove off, whether being as tight as that in the Black quarter some negro wouldn't have her before the morning. "Yeah", was the unconcerned reply, "and some of them white women come on purpose for it"'.
49 Ibid., IV: 9. See George Chauncey, *Gay New York: Gender, Urban Culture, and the Making of the Gay Male World, 1890–1940* (New York: Basic Books, 1994).
50 On race and social control see, for example, Matthew Frye Jacobson, *Whiteness of a Different Color: European Immigrants and the Alchemy of Race* (Cambridge, MA: Harvard University Press, 1998); Wilder, *A Covenant with Color*; David Roediger, *Working Toward Whiteness: How America's Immigrants Became White: The Strange Journey from Ellis Island to the Suburbs* (New York: Basic Books, 2005).
51 Paul Rich, 'Doctrines of Racial Segregation in Britain: 1900–1944', *New Community: Journal of the Commission for Racial Equality* 12 (1984): 77.

52 Susan Smith, *The Politics of 'Race' and Residence: Citizenship, Segregation and White Supremacy in Britain* (Cambridge: Polity Press, 1989), chap. 1 (esp. 4–7).
53 Phillip Gordon Mackintosh, 'The "Occult Relation between Man and the Vegetable": Transcendentalism, Immigrants, and Park Planning in Toronto, c.1900', in *Rethinking the Great White North: Race, Nature and the Historical Geographies of Whiteness in Canada*, eds Andrew Baldwin, Laura Cameron and Audrey Kobayashi (Vancouver: UBC Press, 2011), 105.
54 Smith, *The Politics of 'Race' and Residence*, 13.
55 Alker Tripp, 'The Traffic Problem', *Police Journal* 9 (1936): 90.
56 Anne McClintock, *Imperial Leather: Race, Gender and Sexuality in the Colonial Contest* (New York and London: Routledge, 1995), 43.
57 Ibid., 48.
58 Ibid., 71–72.

8 Hurried exchanges

Hybrid office buildings and their uses in the late nineteenth century

Deryck W. Holdsworth

Introduction

In May 1884, during speeches marking the opening of the second New York Produce Exchange, one speaker noted the 'large colony of offices constructed above the Exchange floor, four stories of rooms devoted to mercantile purposes' and observed that 'this office feature was a new departure in Mercantile Exchange building' (Figure 8.1).[1] It was indeed, and this chapter argues that this hybrid exchange-and-office building, an important harbinger of the taller skyscrapers associated with the early twentieth-century downtown, was symptomatic of the

Figure 8.1 The second New York Produce Exchange at the foot of Broadway. The tower of this vast mercantile building was one of the prominent elements of the business district skyline.

Source: Richard Wheatley, 'The New York Produce Exchange', *Harper's New Monthly Magazine* 73 (July 1886): 193.

growing demand for transactional spaces in a newer phase of commodity trading. Volume efficiencies in long-distance commodity chains for the shipping of foodstuffs and industrial resources were brought about by new technologies of telegraph, railroads and steamships. This quickening pace of communication and transportation created a demand for more offices in proximity to key business nodes, to take advantage of the time-sensitive opportunities available during the hours of trading on the exchange.

The 'large colony of offices' conjured up images of hundreds of worker bees or ants[2] and is appropriate for appreciating the 'architecture of hurry': rooms full of clerks and partners busily at work; letters being received and sent; receipts for grain, meat, cheese, fish, hogs or kerosene being filed; orders for shipping or insurance being confirmed. 'Architecture of hurry' is also an appropriate label for a 29,000-square-foot trading floor, a vast space four stories in height, designed to accommodate 5,000 persons at one time (Figure 8.2). Floor plans of the Produce Exchange by architect George Post included 'stands', places along the walls at the edges of the trading floor, for various members whose main address might simply be D3 or C7, rather than rooms in the four floors of office above or in nearby office buildings.[3]

Architecturally, there were distinct areas on the New York Produce Exchange trading floor for bidding associated with different commodities. In contrast to the label of the 'pit' used in Chicago and elsewhere, these sites were called 'rings'

Figure 8.2 The trading floor of the new Produce Exchange, with its many tables for samples and sites for buying and selling. Four floors of offices above made this a hybrid exchange/office building.

Source: Richard Wheatley, 'The New York Produce Exchange', *Harper's New Monthly Magazine* 73 (July 1886): 197.

in New York: three layers of wide wooden platforms up from the floor and then three down, so six layers in total on which a trader needed to be standing in order to participate, and overseen by an exchange official at a high desk who recorded the prices involved in the trade. Nearby was an 'ampitheatre-like "call room"' for auctions.[4] A prominent feature was the Western Union Bulletin Board, which updated out-of-town prices. An article for *Harper's Weekly* about the new exchange noted:

> All the prominent grain and provision firms of Chicago are represented on the floor of the New York Produce Exchange. No black-board in the elegant hall of the Chicago Board of Trade is so eagerly watched as that which, every five minutes, records the prices current in New York.[5]

Constant updates of commodity prices were vital in a fast-moving trading period. The massive blackboards along the walls of the exchange, with men on a gantry updating prices with chalk, also signal this 'architecture of hurry'.[6] Half of the trading floor was given over to sample tables, where a potential purchaser could assess a handful of flour to which a bit of water had been added. As reported by a newspaper summary of activities in the building, the critical hurry here was a matter of minutes, to help decide on a bidding strategy: 'the elasticity of the dough is taken as indicative of the quality of the flour'.[7]

The flowery language spoken at the opening by lawyer Algernon Sullivan pointed to the size and height of the brick-clad Produce Exchange, but captured the Janus-like faces of the building. One looked east: 'Oh! Tower-crowned Trade Hall! Thou standest imperial and complete. Look from thy parapets down the bay where three arms of the ocean meet, and listen to the swash of their flood-tides'; the other looked west, 'across a fruitful continent. The leaves fresh in Spring-life and all the blades bow their heads towards the East in anticipation of their Harvest-Home next Autumn here beneath this gorgeous sky-light'. The skylight was a vast glass ceiling over the trading floor, light enabled by the four floors of offices above being arranged as a hollow rectangle. Access to those floors was provided by nine elevators that carried '21,500 people daily'.[8] And the clock tower on the New Street side of the exchange gave a verticality to this otherwise vast horizontal building.[9] The tower, rising high above the five-story rooftops of warehouses and offices of the wholesale district, was a beacon for both the city and harbor and, for Sullivan, the continent and the oceans. The geographical positionality of the exchange, between a bountiful continent and the world markets across the seas, makes this institution less about produce that would feed New Yorkers, though part of the merchant community was concerned with that regional market of 2 million, and more about orchestrating a throughput to the urban-industrial markets of western Europe.

Only 400 Produce Exchange members had office space in the 190 rooms in the new building, and so the 'architecture of hurry' theme must necessarily extend to the office addresses of another 2,000 members within walking distance, mostly in office buildings south of Wall Street, that sprang up in the turn-of-the-century expansion of lower Manhattan.[10] Several hundred more were based in offices near warehouses across the East River in Brooklyn, west across the Hudson River in Jersey City, and further afield across the Northeast and Midwest, serving firms that wanted access to the trading advantages of the New York exchange.[11]

The 'architecture of hurry' also included the port infrastructure of wharves and warehouses that passed Midwest grain and flour from Erie Canal boats and railroad freight cars to ocean steamers. In addition to the giant grain-storage elevators, a remarkable contraption was a floating marine leg that could move grain from canal barges to ships. In New York's harbor, by 1871, at least 19 floating elevators 'with an average capacity of 3000 bushels per hour transferred grain from canal boats to ocean-going steamers and sailing vessels at their anchorage'. In one instance, reported *Harper's Weekly*, '40,000 bushels of grain were put after six o'clock in the evening, into a steamer that was ready to sail the next day at noon'.[12] Such machine-like mobilities complemented the scurrying of people who tracked these transfers. There were 16 stationary grain elevators, many of them across the East River in the Atlantic Dock in Brooklyn, and others across the Hudson River, where the New York Central Railroad freight cars delivered boxcars full of grain in Weehawken, New Jersey. Each boxcar full of grain was checked by Produce Exchange inspectors, who issued tickets that were filed in the relevant commission brokerage offices. That inspection was critical to the power and grip of the exchange, bestowing its stamp of approval, its sign of trustworthiness of grades and thus worth of the grain, for the fast-moving buying and selling of vast volumes of commodities with precise delivery dates.

This commodity exchange did not emerge from thin air. It must be seen within a trajectory of earlier flour and produce exchange sites and structures and, indeed, among other hybrid exchange/office buildings in New York and elsewhere, in both North America and in Europe, that were emerging in the last quarter of the nineteenth century and whose existence was symptomatic of the new architectures of transactional space that are symbolized by the New York Produce Exchange.

Precursors of hurried exchange

Proximity to key exchange locations has long been a feature of the mercantile city, and the roots of the 'architecture of hurry' are not solely a product of the late nineteenth-century and twentieth-century innovations in transportation and communication.[13] Brief sketches of exchange realms from prior centuries can demonstrate the importance of these precedents.

Bruges, Flanders, the Hanseatic city where Venetian and Genoan galleys came each year for a multi-week trading fair from the thirteenth century, was an important early setting for exchange practices: three hostels of the 'foreign' merchants from Venice, Florence and Genoa were adjacent to an inn run by van der Beurz, from which the term 'bourse' derives, the label used in Europe for the merchant exchange. The Beurzplatz was at the curve of a canal-linked axis of trade from the Waterhalle (New Cloth Hall), the Kraanplatz, where boats were loaded and unloaded with the giant public crane, and the Tolhuis (customs building). Moneylenders did business on one of the bridges over the Kraanrei north of the Cloth Hall, and lawyers set up temporary stalls in the arches of cloisters of St. Donatian's church.[14] In time, distinct streets were also associated with Spanish, English and Baltic merchants.

Similarly, in Venice, the core trading node developed on either side of the Rialto Bridge, which spanned the Grand Canal. On one side was the Fondaco dei Tedeschi, the massive six-story combination hostel and warehouse (1228, rebuilt in 1508), where German and other northern European merchants were required to live, and on the other side the Nobleman's Exchange, built in 1424, an arcaded loggia around the *campo* in front of the Chiesa di San Giacomo di Rialto; nearby passageways were sites for banks and insurance.[15]

However, the bourse as a deliberate architecture has its roots in structures built in Antwerp (1531) and Amsterdam (1611), both of which were models for the Royal Exchange on Cornhill in London; these new major centers for international trading merchants sought the convenience of a deliberately designed structure, rather than the more informal gathering places that had typified Bruges and Venice. The Royal Exchange, rebuilt in 1669 after the first one, of 1571, was burned in the Great Fire of London in 1666, was a massive building with an open-air enclosed quadrangle, 40 yards by 50 yards, which included 'walks' for 27 distinct commodity merchant sectors.[16] In the continental Bourse and the English Exchange, the upper floor was largely devoted to luxury shops, rather than offices, though by 1838 Lloyd's Insurance occupied half of the second floor of the London Royal Exchange, an early sign of the attractions of extreme proximity.

The rhythms of London's trade networks extended the contacts at the Exchange with meetings in nearby coffee houses in the hours before and after the period when the Exchange was in session, vital gathering places for merchants who had their counting-houses in the alleys, lanes, courts and streets of the City. An excellent example can be seen in the activities of the principal of Herries & Co., a firm specializing in Spanish brandy and wheat, but that also traded products to Ireland and to American colonies, with a counting-house in Jeffreys Square, off St. Mary Axe. In a summary of activities from 1766, Herries recorded a journey that, twice a week, on Tuesdays and Fridays, involved going down St. Mary Axe, then turning west a third of a mile along Leadenhall Street and Cornhill. Before going into the Exchange at 2 p.m., he first visited several nearby coffee houses. He went to Tom's Coffee House at 12.30 p.m. for half an hour, then briefly visited three more coffee houses, The Jerusalem, Garroway's and Lloyd's, before going to John's Coffee House at 1.30 p.m., and on to the Exchange; after his time on 'Change', he went back again to John's before returning to his Jeffreys Square counting-house.[17] Many coffee houses had business specialties that echoed those found on the walks inside the Exchange. Some were also transactional spaces, places for auctions by inch of candle, as well as being sites for gossip and intrigue. And adjacent to the coffee houses were many insurance offices and stationers.[18]

Redesigning the perimeters of exchange

Across the nineteenth century, distinctive exchanges emerged for specific commodities, such as for coal, iron, cotton or grain.[19] They were often the largest, and in many cases the tallest, buildings in their downtown areas when built, such as the four-story (plus 109-foot tower) London Coal Exchange of 1849 on Thames Street,[20] or the vast 1862 Corn Exchange in Leeds. They added a further building type to the emergent business district and acted as settings for hurry.

The increasing pull of the exchange floor is notable across the century, and the example of Liverpool is particularly instructive. One early important business node was the Union Newsroom, built in 1800 at the southeast end of Duke Street, which was the location of many of the city's most prominent merchants who had their houses, counting-houses and warehouses along that street.

> [It] stood in the centre of the aristocratic region of trade. The London morning papers arrived early the following morning, and the evening papers the following evening, thus suiting the *habitués* in their way to and from town, and especially the select circle inhabiting the region of Duke Street.[21]

However, by 1875, Liverpool memorialist J.A. Picton observed: 'At the present day . . . the idea of erecting a place of mercantile resort so far distant from the centre of business as the middle of Duke Street would be thought absurd'.[22] That center of business was the Exchange, about a half a mile away, a broad open area around the Town Hall known as the Flags, for its flagstone pavement, referred to as early as 1786 as the 'Rialto where merchants do now congregate'.[23] The Exchange Buildings had been rebuilt in the 1860s, with a suite of meeting spaces, a library and newsroom, even if merchants still preferred to meet outdoors.[24] Picton records:

> The tendency of the present day is to concentrate as much as possible all mercantile operations within a limited area round the Exchange. The hours of business are shorter and more hurried. The newspapers must be close at hand. The local daily newspaper is skimmed over in a few minutes, and even the *Times* is in its main features anticipated by the telegram.[25]

Already by 1875, the pace of business included a commute, the separation of home and work starting to be a feature of middle-class office life, not the combined home/work settings of Duke Street of old:

> The town is a sphere to do business in, to make money or – to lose it; but that done, the omnibus, the steamboat, the railway, whirl off in their thousands to pure air and bright skies, until the dawn of a new day recalls the busy crowds to another struggle in the battle of life.[26]

But when they returned the next business day, 'to be outside the charmed circle narrowly drawn round the Exchange is almost equivalent to being ostracized from business'.[27]

Liverpool's cotton merchants, many of whom rented offices in the Albany Building on Old Hall Street, moved their trading space to a seven-story Cotton Exchange next door to the Albany Building in 1906.[28] Reports of that building's opening noted the telegraph cable links to New York, Bremen and Bombay. They also linked Liverpool with New Orleans, a connection that had long been in place, through New York City, since the era of packet ships such as the Black Ball Line.[29] In New Orleans, too, the pace had quickened. Cotton merchants had small offices west of Canal Street where cotton factors operated by extending loans to planters and then selling their crops. Art historian Marilyn Brown, commenting on the scene that painter Edgar Degas captured in *The Cotton Office* (1873), noted: 'The old-style factoring office was usually a small family outfit that primarily conducted business in spot or transit cotton at a leisurely pace, based on cotton samples on the premise'.[30] However, 'that kind of business quickly became obsolete in the face of large-scale speculation in cotton futures, conducted at a rapid pace through the international system of cotton exchanges that was made possible by the laying of the Atlantic telegraph cable'.[31] New Orleans and Liverpool had now become nodes in a wider, more frantic system, as too were Mobile, Memphis and Savannah, and the New Orleans Cotton Exchange of 1883 became the local manifestation of this new efficiency.

But it was New York City that became the dominant node in the cotton trade, as for decades it had been New York jobbers and financial interests that had extended loans and covered debt for southern cotton planters. The Cotton Exchange in New York, established in 1870, drew in the likes of former Montgomery, Alabama, commission brokers Lehman Brothers after the Civil War and helped make the New York institution a key pivot in this speculative trade network. A nine-story building on a corner site at Beaver and William in New York City, built in 1885 and also designed by Produce Exchange architect George Post, had a galleried exchange room on the second floor and many offices above.[32] It was the other Manhattan commodity exchange that loomed over the five-story wholesale district roofscape, the rounded corner façade with a turreted conical roof being a skyline landmark along with the Produce Exchange tower.[33]

Similar redesigns of the perimeters of trade were evident for the grain and flour trading world, especially with regard to the emergence of hybrid trading-floor-and-office-building exchanges. The Grain and Flour Exchange in Boston, on Milk Street, built in 1892, was very similar to Post's Cotton Exchange in New York, with a curved front and pyramidal roof drawing attention to the three-story trading floor and with offices on the sixth and seventh floor above.[34] In Philadelphia, a 10-story hybrid exchange-and-offices building, the biggest in the city, was constructed by grain merchants in 1895, called the Bourse in homage to the new building of that name in Hamburg, one of Philadelphia's European trading partners; it replaced the earlier Grain Exchange closer to the waterfront.[35] And in Buffalo and Chicago, key nodes in the commodity chain from the American plains to the Atlantic ports, Board of Trade buildings contained prominent trading floors as well as significant office space for brokers and shippers. The Buffalo Board of Trade moved into a seven-story building of 1882, the Corn Exchange occupying a 'Board Room' on the fourth floor, with most of the other 72 offices being grain merchants.[36] The Chicago Board of Trade building of 1885 was a monumental trading hall at the head of LaSalle Street, with a 90-office portion at the back of the block.[37] Similar landmark exchange-and-office structures appeared in Minneapolis and Duluth, responsible for selling and shipping grain from the northern plains, and in Winnipeg, Manitoba, for the marketing of wheat from the Canadian prairies.

In all of these cases, the new trading floor, nested within an emergent skyscraper, was inadequate to house all the commission brokers, lake shippers, railroad freight agents and insurers. Demand for additional office space for these firms was met in nearby skyscrapers as the volume of the commodity chains multiplied in the late nineteenth and early twentieth centuries. Such was the case with the environs of the New York Produce Exchange.

The locational shifts of the New York Produce Exchange

At the southern tip of Manhattan, in colonial and early national periods, East River frontages had been prime business addresses for merchants, and docks with names such as Coffee House Slip, at the east end of Wall Street, and Burling Slip, Old Slip and Coenties Slip round to the foot of Broad Street survived as important moorings, even as landfill repositioned Water Street two blocks in from the river and added Front Street and South Street to the landmass. Merchants with grain and flour interests had their offices in this array of blocks along Water, Front and South Streets, and they met daily at outdoor settings along South Street at the foot of Broad Street, next to where the Erie Canal grain barges moored. Eventually, they purchased an awning to shelter from bad weather and, by 1855, a New York Corn Exchange was developed at No. 17 South Street.[38] Nearby buildings housed offices for grain inspectors and grain-elevator operators, as well as the barge companies.[39]

By 1861, the flour and grain merchants decided to invest in a new, indoor exchange, initially the Corn and Produce Exchange Building, but later renamed the New York Produce Exchange, some 800 feet west, at Whitehall between Pearl and Water; this had a substantial trading floor, but it did not contain any office space (Figure 8.3). It was a crowded place:

> Tables are set as closely together as convenient, and each merchant who wishes has a certain space where, in dainty boxes or bottles, he places samples of his stock – flour, grain, whiskey, lard or petroleum – the name of the brand affixed to each sample.[40]

Assessment of worth came through experienced wisdom:

> You will observe a grave-faced man, with his whiskers all meal, take some flour in the hollow of his hand, pour upon it a few drops of water from a silver tankard, and solemnly work it into dough, which he kneads and pulls and rolls, folds, twists, and worried, judging by the result how good bread it will make.[41]

And these explorations extended to other aspects of produce on offer: 'On the other side of the hall are the grain exhibits; and here men go round with their pockets full of wheat and oats and nibble, nibble, nibble until you think they have become granivorous'.[42] Once decisions to purchase were made, and prices agreed on, this might trigger the chartering of a floating elevator, communication with the steamer that was going to take the grain overseas, and contact with insurance brokers. Proximity to transportation and insurance services was vital, and some large grain brokers had them close at hand.

The dominance of flour and grain merchants in the South Street/East River area is evident in Produce Exchange membership data from 1875.[43] The five-story Herrick Building on South Street near Coenties Slip housed 12 members, all in the grain sector, and 34 other members had offices in four adjacent buildings. Most noteworthy was David Dow & Co., one of the earliest merchants to develop a significant hinterland linkage, with schooners operating on the Great Lakes and

Figure 8.3 The earlier Produce Exchange on Whitehall between Pearl and Water Streets. The crowded trading floor (below) was a frantic site of buying and selling.

Source: Ernest Ingersol, 'The Lading of a Ship', *Harper's New Monthly Magazine* 55 (September 1877): 485.

branch offices in Chicago and Duluth, as well as other places. Seventeen different merchant names are listed as representing David Dows in the New York Produce Exchange, indicating a significant financial entry fee by the firm. Their office address was 20 South Street, next door to the first Corn Exchange.

Other buildings in lower Manhattan acted as anchors for the transactional networks. The earlier Merchant Exchange on Wall Street was now used as the Custom House, and thus was an important close-by space for the daily activities of many of the import/export merchants. At the eastern end of Wall Street, most of the tea-and-coffee import merchants in New York could be found, plus some specializing in sugar. The strength of the export market meant that shipping interests were an important element of the Produce Exchange, many with offices closer to Battery Park at the foot of Broadway, along what was known as Steamship Row, and likely a feature of the decision to move the exchange westward. So too the Barge Office, at the foot of Whitehall, which was a vital cog of the Hudson River and Erie Canal transportation element of this commodity chain.

Memberships more than doubled in the decade after the Civil War, as the western plains and other regions produced increased volumes of foodstuffs. The space of the 1861 exchange was soon inadequate. With its construction begun by 1881 and its opening in 1884, the new (i.e. second) Produce Exchange at the foot of Broadway contributed to a significant reshaping of lower Manhattan in the next few decades. A new 'steamship row' developed for shipping concerns renting offices in 15- to 20-story skyscrapers on the west side of Bowling Green and, by 1907, a new federal Custom House, adjacent to the Produce Exchange. Clusters of new office buildings up Broadway and Broad Street were tenanted by hundreds of Produce Exchange members. A far more complex array of foodstuffs was now being traded, but the major portion of the overall membership of the exchange, and for tenants in the four floors of offices above the trading hall, still reflected that core grain and flour business.

The prominent early firm of David Dows & Co. was now listed as being in Room 102, Produce Exchange, in membership lists. But an advert in the Chicago Board of Trade annual report for 1886 helps clarify the contours of that firm's broader geographical footprint:

> David Dows & Co. Commission Merchants, 87 Board of Trade, Chicago; Produce Exchange, New York; Williamson Block, Duluth. Headquartered in New York, 20 South St., with four resident partners, plus branches in Chicago, Duluth, St. Paul, and Baltimore. They are the largest Eastern receivers of flour and grain, via lake and canal from Chicago to New York. Their Chicago warehouse and elevator facilities can handle and store the largest consignments, and in New York their 'stores' have the capacity for three million and a half bushels of grain.[44]

A similar geographical reach characterized Milmine, Bodman & Co., which had bought memberships for 12 different members of the firm and was based in Room 491, Produce Exchange. They too had offices in Chicago, where part of the firm originated. The 'hurry' theme is evident in a 1907 note that survives on their Produce Exchange office letterhead, from George Milmine to his brother Herb:

> I have not left the office before 6:30 yet this week and when one starts writing and figuring at 9:30 and keeps it up for nine hours, just as tight as they can go, by evening they are pretty well used up, which has been my case. We have been very busy especially in grain and stocks, and the brunt of it has fallen on me. The Market has been crashing steadily since last Friday, until the majority of people consider it bottomless, – a pit engulfing the unsuspecting widows and orphans.[45]

The colony of offices was a stressful place, for sure, and not just during the hours of trade on the exchange floor below.

The complexity of all the interrelated activities that were associated with the Produce Exchange is evident in the summary of membership identities offered by its president, E.J. Carhart, in 1911:

> An indication of the great variety of interests represented on its floor is found in the fact that an attempt to make up a classified business list of its present membership resulted in a list of eighty-four headings, and one of these was miscellaneous with 202 members included therein; 297 members are identified with the flour industry; 256 with grain; 374 are receivers and shippers of general commodities; 133 are connected with transportation interests, 91 of these being identified with the steamship trade, showing the great importance of the export trade on the floor; 131 are provision men, 88 are concerned in the various lines of oil, 64 deal in stocks and bonds, 54 are bankers, 50 are insurance men, 38 are brewers, 37 are grocers, 63 are executives of corporations, etc. Hay, seed, feed, elevating, inspectors, weighers, measurers, tallow, greases, fertilizer, naval stores, millers, custom brokers, lawyers, bakers, coffee, lumber, vinegar, dried fruits, beans and peas are all included in the list, and so it goes on.[46]

The annual reports of the Produce Exchange listed 68 pages of members' addresses and their sectors, 80 percent in a New York office, though some firms with multiple memberships for their partners confirm the links to flour, grain and meat concerns in Chicago and Milwaukee.[47] Each of them had their business stories, and their geographies. David Dows & Co., or Milmine, Bodman & Co., get noticed because of the multiple memberships listed. But, even if a firm just had a single membership, that one name fronted an often impressive array of staff, part of the vast tentacles of trade that were interwoven across the continent and Atlantic. Three examples are profiled here.

John Gledhill, of the Co-operative Wholesale Society, Manchester, England, had an office in Rooms 420–22, Produce Exchange. 'The Co-op' was the place where working people in industrial Britain shopped. Started initially in Manchester, with offices in Liverpool, the better to link with its eight Irish sources of produce, a New York office was established in 1875. Its company history by Redfern noted:

> The importance of having opened the New York branch is daily experienced. We are thus enabled to ascertain the state of the market for cheese, bacon, lard, and grain, both there and in Liverpool, almost at the same time, and thus are enabled to determine when we can purchase with advantage.[48]

With 69 butter and cheese merchants listed in the Produce Exchange, 57 lard merchants and 742 in grain and flour, this was certainly an important place to be.

William Lunham (member no. 2829) and Walter Moore (member no. 591), Room 461 Produce Exchange, ran a significant business for ocean freight brokering, forwarding, marine and fire insurance and banking. An advert for their services in a trade magazine in 1915 shows 30 people sitting at desks in this one office. The scope of their business network is evident in their declaration: 'We book grain from New York, Philadelphia, Baltimore, Newport News, Norfolk, New Orleans and Galveston. We are in a position to handle grain at New York, Philadelphia and Baltimore'.[49] Their advert also stressed that they had an office in Buffalo, solely for grain. Their head office was in Great Saint Helens Street in London, which, like the Jeffreys Square address of a counting-house discussed earlier, was just off St. Mary Axe in the heart of London's business district. By this time, the street was the location for the Baltic Exchange, built in 1903, the main trading place for the shipping trades in the city.[50]

R. Braga Rionda, of the firm Czarnikow-McDougall Co. Ltd., Export Merchants, was listed as being at 112 Wall Street, which was at the eastern end of the street, near Coffee House Slip. Czarnikow-McDougall had been founded in 1891 as the New York-based North American branch of Czarnikow & Co, of 29 Mincing Lane, London. Spanish-born Rionda helped the Scot McDougall import Cuban sugar into New York, and this Wall Street locale of tea and coffee importers was a logical place for a sugar importer. The Czarnikow company history points to a far wider world, their London office in Mincing Lane containing 16 departments, each in a different small room.[51] John Summerson notes that the reshaping of the City in Victorian London started first in the streets around the Royal Exchange, as banks and insurance companies built prominent new office headquarters, but then ripples of rebuilding added speculative capacity purely for offices in an area where once merchants had both residences and counting-houses.[52] In both London and New York, then, as in other leading trade nodes, dozens of employees were helping to orchestrate flows of a product across wide regions.[53] In a longer study, a similar web of strands can be woven for many other members, involved in marketing a wide range of commodities or confirming transportation options, all within a quarter of a mile of an exchange floor.

The New York Produce Exchange Building, once rising above the five-story roofscape near the tip of Manhattan, was soon surrounded by a forest of new skyscrapers. The Kemble Building and the aptly named Merchant Building, along Whitehall between Stone and Bridge just south of the Produce Exchange, both rented office space to dozens of exchange members, as did the Welles Building and Produce Exchange Bank Building across Beaver to the north. On the west side of Broadway across Bowling Green, the 12-story Washington Building of 1887[54] and others anchor-tenanted by shipping lines further framed the Exchange. One-year rental contracts for office space led to a merry-go-round of office relocation of many firms who responded to new locations and perceived new status of a prominent new office tower. There was little stability in addresses in a comparison of membership addresses across 1875, 1889 and 1903.

Though the Produce Exchange stayed in that location until it was demolished in 1957, the Cotton Exchange was more footloose. Since first organized in 1870, the exchange moved four times in 50 years, the last move to a new 25-story building in the early 1920s, with the trading floor on the 19th floor that would 'run through three stories to the roof. Uninterrupted elevator service will have to be maintained because of the necessity of the members having prompt access to the trading floor at all times during the hours of transacting business'.[55] Much of the rest of the tower was rented for offices of member firms.

Wheatley's essay on the Produce Exchange in *Harper's Weekly* summarized the focus of the exchange in an interesting sentence: '"What shall we eat, what shall we drink, and wherewithal shall we be lighted?" are the three questions with whose pleasant solutions the New York Produce Exchange charges itself'.[56] The third, 'Wherewithal might we be lighted?' seems somewhat puzzling for a Produce Exchange. Yet kerosene, increasingly, was the answer, rather than candles made of whale oil. Pennsylvania oil was the new source of lighting, and the Cleveland oil merchant John D. Rockefeller was a prominent merchant in the distribution of this product. In 1875, Rockefeller was recorded in the Produce Exchange membership lists at 140 Pearl Street, one of five members at that address and one of 25 oil merchants on Pearl more broadly. Rockefeller moved from Pearl to 26 Broadway in 1883, a few buildings north of the Produce Exchange, initially to a 10-story building that grew to 16 in 1895, as his Standard Oil empire developed. Most of the petroleum interests abandoned the Produce Exchange for the Petroleum Exchange, a bit north on Broadway and closer to the Stock Exchange that many see as synonymous with Wall Street and the arena for industrial stocks. Rockefeller's office needs expanded, and, between 1921 and 1928, the earlier 26 Broadway building was replaced by a huge 31-story office complex, the Standard Oil Building, which faced the Produce Exchange across Beaver Street; its illuminated pyramidal roof was a beacon for shipping, much as Algernon Sullivan had hailed the 'Tower-crowned Trade Hall' almost four decades earlier.[57] A significant portion of Standard Oil wealth was by then connected to gasoline, not kerosene, and thus to a new medium of hurry, the automobile. The financial and industrial world north of and along Wall Street, and the shipping and foodstuffs world south of Wall Street were now part of a broader 'downtown' New York, but one where commodity nodes such as for produce, cotton, and coffee were still important.

Commodity exchanges throughout the Midwest occupied hybrid office/exchanges in the late nineteenth and early twentieth century, and, by 1903, many members of the New York Produce Exchange that had addresses in these Midwest cities were listed specifically in rooms in these hybrid spaces.[58] The foodstuff commodity chains were made ever more efficient by the merchants working out of these buildings.

The writing on mobilities and hurry often focusses on the worlds of schedules, for trains, busses, streetcars and subways, and the cultural practices of commuters using them. Commuters going to work by streetcar from the Garden District in New Orleans to offices near the Cotton Exchange, or on the suburban London Metropolitan Railway corridor to Mincing Lane or St. Mary Axe, or by elevated railway to work at or near the Produce Exchange in New York all arrived at stationary structures. There, at the end of the hurry, stands the office building or exchange. Mobility is something that takes place before and after the business day. This chapter suggests that, when we look inside those colonies of office spaces, where countless thousands of messages move millions of items around the world daily, time is of the essence, for ships to catch a tide, or for a market to be profited from, the hurry continued. The idea of geographical mobility of business needs to be expanded to include these stationary buildings – housing undulating hurriers inside. They appear static, but everything they contain is dynamic.

Notes

1 Algernon Sullivan, speech. *Ceremonies on Leaving the Old and Opening the New Produce Exchange, May 5th and 6th, 1884* (New York: The Art Interchange Press, 1884), 39–40.

2 A visitor to Liverpool's Exchange in the 1870s watched 'a crowd of merchants and brokers swarming and humming like a hive of bees on the floor of the vast area below'. Quoted by Tristram Hunt, *Ten Cities that Made an Empire* (London: Allen Lane, 2014), 395.

3 George B. Post Architectural Record Collection, New-York Historical Society, New York.

4 Sarah B. Landau and Carl W. Condit, *Rise of the New York Skyscraper, 1865–1913* (New Haven: Yale University Press, 1996), 118.

5 Richard Wheatley, 'The New York Produce Exchange', *Harper's New Monthly Magazine* 73 (July 1886): 207.

6 'Instantaneous quotations are shown on its blackboards in wheat, corn and oats from Chicago, Toledo, St. Louis, Kansas City, Minneapolis, Duluth and Winnipeg; in provisions and oils from Chicago; and cables are shown from Liverpool, Paris, Antwerp, Berlin, Budapest and Buenos Ayres.' E.R. Carhart, 'The New York Produce Exchange', *The Annals of the American Academy of Political and Social Science* 38 (1911): 219. Frank Norris, in his 1903 novel *The Pit*, notes a slightly different set of places connected with Chicago's Board of Trade: 'And all the while, Liverpool, Paris, Odessa, and Buda-Pesth clamoured ever louder and louder for the grain that meant food to the crowded streets and barren farms of Europe' (New York: Doubleday, Page), 270. For a detailed account of the ways that the exchange system worked in Chicago, see William Cronon, *Nature's Metropolis: Chicago and the Great West* (New York: Norton, 1991), chapter 3.

7 Anonymous, 'The New York Produce Exchange: Wide Scope of Its Operations and History of Its Growth', *New York Times*, September 22, 1901, 41.

8 Wheatley, 'The New York Produce Exchange', 204.

9 The spire of Trinity Church at the end of Wall Street had long identified that part of downtown Manhattan, and the tower of the Exchange, visible down Broad Street, was a commercial above the five-story wholesale and office district. For a thoughtful essay

on the late nineteenth-century tensions between a horizontal and vertical aesthetic for the city, see William A. Taylor and Thomas Bender, 'Culture and Architecture: Aesthetic Tensions in the Shaping of New York', in *In Pursuit of Gotham: Culture and Commerce in New York*, ed. William A. Taylor (New York: Oxford University Press, 1992), 51–69.

10 Deryck W. Holdsworth, 'Morphological Change in Lower Manhattan, New York, 1893–1920', in *Urban Landscapes: International Perspectives*, eds J.W.R. Whitehand and P.J. Larkham (London: Routledge, 1992), 114–29; Gail Fenske and Deryck W. Holdsworth, 'Corporate Identity and the New York office building, 1895–1915', in *The Landscape of Modernity: Essays on New York City, 1900–1940*, eds David Ward and Olivier Zunz (New York: Russell Sage, 1992), 129–59.

11 Membership lists were published in the annual reports of the Exchange and included street addresses, down to room numbers in office buildings.

12 *Harper's Weekly*, May 20, 1871. A floating elevator is pictured on p. 486 in a detailed chronicling of 'The Lading of a Ship' (Anonymous, 'The Lading of a Ship', *Harper's New Monthly Magazine* 55 [September 1877]: 481–93).

13 For a broad overview of the wider European trading world, see Fernand Braudel's magisterial summary of 'the instruments of exchange', in *The Wheels of Commerce: Civilization and Capitalism, 15th–18th Century, Volume 2* (New York: Harper & Row, 1982), 81–113.

14 James M. Murray, *Bruges, Cradle of Capitalism, 1280–1390* (Cambridge: Cambridge University Press, 2005), 63–81.

15 Mark Girouard, *Cities and People: A Social and Architectural History* (New Haven: Yale University Press, 1985), 85–112; Patricia H. Labalme and Laura S. White, eds, *Venice, Città Excelentissima: Selections from the Renaissance Diaries of Marin Sanudo*, trans. Linda L. Carroll (Baltimore: Johns Hopkins University Press, 2008), 11; Richard R Goy, *The Building of Renaissance Venice: Patrons, Architects, and Builders* (New Haven: Yale University Press, 2006), 57–78.

16 See Natasha Glaisyer, 'Merchants at the Royal Exchange, 1660–1720', in *The Royal Exchange*, ed. Ann Saunders (London: London Topographical Society, 1997), 199–205.

17 Jacob M. Price, 'Directions for the Conduct of Merchant's Counting House, 1766', *Business History*, 28 (1986): 134–50. See also Daniel Defoe, *The Anatomy of Exchange-Alley: or, a System of Stock-Jobbing* (London: E. Smith near the Exchange-Alley, 1719), 35, for a similar, even earlier, account.

18 Bryant Whitehall, *London Coffeehouses: A Reference Book of Coffee Houses of the Seventeenth, Eighteenth and Nineteenth Centuries* (London: George Allen & Unwin, 1963); Aytoun Ellis, *The Penny Universities: A History of the Coffee-house* (London: Decker & Warburg, 1956). For a detailed property-by-property map of the variety of land uses near Change Alley in 1748, see 'A New & Correct Plan Of All The Houses Destroyed And Damaged By The Fire Which Began In Exchange-Alley, Cornhill, On Friday, March 25th, 1748', *The London Magazine, or, Gentleman's Monthly Intelligencer, 1748* (accessed September 1, 2017), http://mapco.net/cornhill/fire.htm.

19 Nikolaus Pevsner, *A History of Building Types* (Princeton: Princeton University Press, 1979), chapter 12, 'Exchanges and Banks'.

20 'In its heyday, as many as five or six hundred merchants, colliery agents and factors attended, on three afternoons a week' (Elspet Fraser-Stephen, *Two Centuries in the London Coal Trade: The Story of Charringtons* [London: Gurwen Press, 1952], 44).

21 James A. Picton, *Memorials of Liverpool Historical and Topographical, Vol II* (London: Longmans, Green, 1875), 270.

22 Picton, *Memorials of Liverpool*, 268.

23 Quentin Hughes, *Seaport: Architecture and Townscape in Liverpool* (Liverpool: Bluecoat Press, 1993), 41.

24 Joseph Sharples and John Stonard, *Built on Commerce: Liverpool's Central Business District* (Swindon: English Heritage, 2008), 43.

25 Picton, *Memorials of Liverpool*, 269.
26 Picton, *Memorials of Liverpool*, 269–70.
27 Picton, *Memorials of Liverpool*, 142.
28 Hughes, *Seaport*, 50–53. For a fine illustration of the array of offices and cotton sample rooms for commercial buildings on Dale Street and Exchange Street East, see the Goad's fire-insurance map of 1884 reprinted in Sharples and Stonard, *Built on Commerce*, 12.
29 Robert G. Albion, *The Rise of New York Port, 1815–1860* (Boston: Northeastern University Press, 1984 reprint of 1939). Albion describes the way that, in 1825, the New York cotton trader Jeremiah Thompson worked with Liverpool traders Cropper, Benson & Co. in the 'biggest orgy in cotton speculation' (114–15).
30 Marilyn R. Brown, *Degas and the Business of Art: A Cotton Office in New Orleans* (University Park, PA: The Pennsylvania State University Press, 1994), 35.
31 Ibid., 35.
32 Landau and Condit, *Rise of the New York Skyscraper*, 127–31.
33 Kenneth J. Lipartito, 'The New York Cotton Exchange and the Development of the Cotton Futures Market', *The Business History Review* 57 (1983): 61.
34 On Boston, see George M. Cushing, Jr., *Great Buildings of Boston: A Photographic Guide* (New York: Dover, 1982), 6–7. A delightful 1907 booster advert for the Boston Chamber of Commerce 'Bigger Better Busier Boston' campaign features the prominent Flour and Grain Exchange Building at the center, with boxcars of wheat and beef on the left, and Liverpool-bound ships loading at grain elevators on the right; https://en.wikipedia.org/wiki/John_F._Fitzgerald (accessed September 1, 2017).
35 Anonymous, *The Philadelphia Bourse, with Sketches of the Hamburg Börse and Manchester Exchange, 1891* (Philadelphia: Allen, Lane & Scott, 1891), 31. Philadelphia Bourse Records, Renters' Cards 1896–1912, Boxes 20–26, Temple University, Special Collections Research Center, Philadelphia PA.
36 On Buffalo's Board of Trade Building, see *Western New York History: As We Were* (accessed September 1, 2017), http://wnyhistory.com/portfolios/businessindustry/corn_exchange/part1/corn_exchange_pt1.html.
37 Daniel Bluestone, *Constructing Chicago* (New Haven: Yale University Press, 1991); see also Gretta T. Roman, 'The Reach of the Pit: Negotiating the Multiple Spheres of the Chicago Board of Trade Building in the Nineteenth Century', unpublished doctoral dissertation (Penn State University, 2015); Cronon, *Nature's Metropolis*.
38 The date and address are recorded in an 1864 court case, *Roberts vs. White*, related to a party wall with an adjacent structure. See Anthony B. Robertson, *Report of Cases Argued and Determined in the Superior Court of New York*, Volume II (Albany, NY: W.C. Little, 1867), 426–9.
39 See 'South and Broad Streets, 1867', Plate 11, in Mary Black, *Old New York in Early Photographs, 1853–1901, 196 Prints from the Collection of New-York Historical Society* (New York: Dover, 1973), 13. An image from *Harper's Weekly*, November 15, 1873, 'Scene on the Great Flour Dock, Coenties Slip, East River, New York', shows inspectors gauging the contents of a barrel.
40 'The Lading of a Ship', 484.
41 Ibid., 484.
42 Ibid., 485.
43 *Annual Report of the Produce Exchange of New York*, 1875.
44 *Chicago Board of Trade*, 1886, 109.
45 *Across the Generations: Bodman Family*, 2002 (accessed September 3, 2017), www.smith.edu/libraries/libs/ssc/atg/bodman/btextgmb-hbletterw.html.
46 Carhart, 'The New York Produce Exchange', 218.
47 For example, the Hecker-Jones-Jewell Milling Company was an amalgam of New York area flour mills, but it was very much connected to Minneapolis flour concerns. The company – including Hecker, Jones and Jewell themselves - had 14 memberships and operated from a spread of four different offices on the second floor of offices above the trading floor.

48 Percy Redfern, *The New History of the C.W.S.* (London: Dent, 1938), 577.
49 'View of Main Section of New York Office of Lunham & Moore', part of an advert for Ocean Freights, *Exporters' Review* 18 (May 1915): 22.
50 Stuart Burch, 'An Unfolding Signifier: London's Baltic Exchange in Tallinn', *Journal of Baltic Studies* 39 (2008): 451–73. The group had its origins in the Virginia and Baltick coffee house on Threadneedle Street in 1744. Its focus was tallows, oils, flax, hemp and seeds.
51 The Czarnikow company history includes images of the London office at 21 Mincing Lane. Photos of directors in their offices include 32 people in 16 small offices, labelled as departments for Canada, the Levant, refined sugar, cash, produce, correspondence, accounts and insurance, and coffee samples, plus a larger general office with at least 26 staff; www.czarnikow.com/about-czarnikow/history (accessed April 9, 2017). Mincing Lane was an address for many tea and coffee importers, and Mark Lane, one street to the east, was an earlier site of the Corn Exchange. The Commodities Trading House had been built in Mincing Lane in 1811, its neoclassical façade standing out from the otherwise similar four-story counting houses along the street. Tea auctions were held in what was later known as the London Commercial Salerooms, and tea merchants established offices in and around the street during the nineteenth and early twentieth centuries. For a superb summary of that Mincing Lane world in the early twentieth century, see David Kynaston, *The City of London Volume II, Golden Years 1890–1914* (London: Chatto & Windus, 1995), chapter 15, 'Merchants and Others', especially pages 258–60.
52 John Summerson, 'The Victorian Rebuilding of the City of London', *The London Journal* 3 (1977): 163–85; and *The Architecture of Victorian London* (Charlottesville, VA: University of Virginia Press, 1976), 52–4.
53 Morton Rothstein, 'Centralizing Firms and Spreading Markets: The World of International Grain Traders, 1846–1914', *Business and Economic History* 2nd series, 17 (1988): 103–13.
54 'At its ultimate height, 258 feet from the sidewalk to the top of its lighthouselike cupola, the Washington Building surpassed even the Produce Exchange tower.' Landau and Condit, *Rise of the New York Skyscraper*, 127.
55 The quote about the new skyscraper is from Anonymous, 'The New York Cotton Exchange', *The Edison Monthly* 14 (1922): 171.
56 Wheatley, *Harper's New Monthly Magazine* 73 (July 1886): 204–5.
57 Landau and Condit, *Rise of the New York Skyscraper*, 133–4.
58 For example: 19 Board of Trade Building, Buffalo; 24 Produce Exchange, Toledo; 7 Board of Trade Building, Chicago; 29 Chamber of Commerce, Minneapolis.

9 Shelter from the hurry?

Hospitality in Montreal, 1836–1913

Sherry Olson and Mary Anne Poutanen

Introduction

The situation of Montreal on the St Lawrence River furnishes a perspective on the continental sweep of population and commodity flows in the years 1840–1914. Ranking among the top 10 metropolitan centres of North America, its population surged every 20 years: from 7,000 households in 1840 to 15,000 in 1860, 30,000 in 1880, and 70,000 in 1900. Dredging in the 1840s made Montreal the head of steam navigation on the river, and the Victoria Bridge, completed in 1859, defined a pivotal position in the rail network and spurred its transformation from a commercial into an industrial metropolis whose merchant bourgeoisie would mobilize further investment in transcontinental links (1885). As in other metropolitan centres, concentration of wealth and power increased, reaching in the 1890s the extremes we acknowledge today.[1] Commercial and industrial expansion meant dramatic increases in movement and messaging, as reactions to surfeit and shortage, hustle and slump. Although hotels can be understood as privileged sites of urban hospitality, we consider a wide array of related services that varied in urgency from hour to hour, day to day, and season to season.

From its founding as a French colony and mission into 'Indian' territory, an incurable cultural diversity was a factor in the power of attraction of Montreal, as well as in its political volatility. The nineteenth-century surges of city-building were synchronized across the continent with surges of immigration. As the city funnelled streams of immigrants into Canada, it experienced recurrent crises of speculation, immiseration, and contagion, making the city always 'first respondent' to the needs of the homeless, the unemployed, the unemployable, and the abandoned. By early May 1831, *The Vindicator* reported 296 vessels arrived with 14,500 emigrants, and anticipated 40,000 by the end of the navigation season – a number comparable to the resident population. Relief meetings were held in inns, and money was raised for 'forwarding destitute emigrants' elsewhere by wagon or canal boat.[2] Cholera followed. In the summer of 1847 and again in 1849, arrivals from Ireland approached 100,000, followed by typhus and cholera. As settlement of the Canadian prairies proceeded, Montreal profited from both a westward flow of migrants and an eastward flow of wheat (1896–1914). Massive. Turbulent. Impatient. Expectant. By 1901, the diversifying stream included vacationers, travelling salesmen, and government agents, trainloads of Chinese labourers southbound for railway construction in Mexico, Italian sojourners northbound from New York City for railway construction on the Canadian Shield, Quebecers westbound for the Klondyke, and intermittent streams of Doukhobor refugees, girls imported from Scotland for domestic service, and Rumanian and Lithuanian Jews who would recreate a garment industry in Montreal.

In the broadest sense, the hospitality industry drew upon the entire city. The whole of Montreal was a great inn in which some people spent a few hours or a few days, and others a lifetime. We argue that the pressures of urban hospitality presented a recurrent ethical challenge: 'I was a stranger, and you took me in'.[3] From ancient times, trade and news were compelling reasons for encounters between stranger and host, and a sacred law of hospitality mitigated the inherent risks: on the doorstep, each party agreed not to attack the other and sealed the pledge by an exchange of gifts. Succour for the stranger was valued as a good deed (the Buddhist *satkara* or the Jewish *mitzvah*), and song and story reinforced the idea of a guest who might be an angel in disguise. At the moment Montreal was transferred from French to British control (1760), three generations before the capitalist expansion we are observing, the Encyclopedists were presenting hospitality as the greatest of virtues, universally revered, extending a reciprocal liberality beyond kin and friend to the whole of humanity:

> la juste mesure de cette espèce de bénéfice dépend de ce qui contribue le plus à la grande fin que les hommes doivent avoir pour but, savoir aux secours réciproques, à la fidélité, au commerce dans les divers états, à la concorde & aux devoirs des membres d'une même société civile.[4]

The contradiction of threat and obligation persisted; it was intensified in the metropolis, where the host as well as the guest confronted an ever-greater proportion of strangers relative to familiar figures, and an ever-wider gulf between wealth and beggary.

An argument in lieu of theory

Concentrating on the urban context, we focus on three aspects of hospitality: its grounding in domesticity, the fault-line of gender that traversed it, and its role in the percolation of information. The inn, hotel, or boarding house was a home away from home: the traveller needed to find substitutes for most of the services normally received 'at home', self-provisioned from home-based tools and furnishings, or provided by a family member, a servant, or a domestic animal. This meant recreating rapidly, in foreign territory, a 'pocket of local order' centred on the temporary home.[5] The trip itself added urgency: the guest arrived dirty, exhausted, stressed, and loaded with baggage, sometimes ill or injured. Hazards of storm or vehicle might surprise both guest and host at any hour. Will there be room in the inn? For travel to Montreal, winter was advantageous, but gave priority to shelter and warmth, captured in the verses from the medieval Norse:

> Fire is needed by the newcomer
> Whose knees are frozen numb;
> Meat and clean linen a man needs,
> Who has fared across the fells,
> Water, too, that he may wash before eating,
> Hand cloth's and a hearty welcome,
> Courteous words, then courteous silence,
> That he may tell his tale.[6]

In domestic households of nineteenth- and early twentieth-century Canada, as well as the countries from which immigrants and travellers were coming, most of these services – the cooking, laundry, bed-making, cleaning, mending, and minding of stoves and fires – were routinely assigned to a wife, daughter, hired or slave woman. Classical and Near Eastern hospitality provided first the bath or washing of the feet. Rebecca watered the traveller's thirsty horses. As so few of the travellers were in fact angels, they expected also liquor, tobacco, entertainment, and sex. The gender roles implicit in the domestic model created an enormous rift: on the demand side, the desires of men predominated; on the supply side, the services of women; the imbalance created considerable risk of abuse of power, exploitation, misunderstanding, violence, and ambiguity.[7]

Beyond the warmth of fire, the stranger's most urgent need was information. Messenger services and brokerage would compensate for services of the usual near-home partners, but every communication was an act of trust. Wherever information was vital, there was also misinformation, disinformation, and disguise. In the home away from home, domestic proximity made third parties or near-strangers privy to personal information, so that the guest was peculiarly vulnerable to surveillance or indiscretion that might reveal the tryst, the speculation, the valuables carried, the military or diplomatic objective. The roles of women are rarely addressed in studies of the information economy,[8] but emerge boldly in literature. The importance of trust is apparent in the displays of hospitality in Homer's *Odyssey*: the observation of a guest by a servant, recognition of a guise, and the risks incurred by a lie.

Location preferences for centre city and riverfront, crossroads and street corners, point to the function of the inn as a node of information, but nominal sources such as the census, reporting 'residents', understate the presence of short-term travellers and their contacts with locals. Because cities are places of intense connectivity, the challenge of the urban object is as a 'connectome'. In the neurologist's terms, how are we connected? How does the structure of connections change? How does it make us what we are? To put it in the architect's terms, 'the fundamental correlate of the spatial configuration is movement . . . Encountering, congregating, avoiding, interacting, dwelling, conferring are not attributes of an individual, but patterns or configurations, formed by groups or collections of people'. We can reframe it from a geographical viewpoint: 'Geography is all about connections. No connections, no geography'.[9] From the historian's perspective, changes in the pattern of connections alert us to the dynamics of encounter and the urgency of response.

Some of the contradictions are effectively treated by Jacques Derrida's essays on the stranger and Linda McDowell's analysis, based on interviews with the personnel of a present-day chain hotel in London (UK), addresses the 'embodiment' of labour and its performative aspects: 'doing gender', 'doing ethnicity', and the intersection between them.[10] As historical treatments of these themes are sparse,[11] it may be more profitable for us to look first into our own experiences as jet-lagged hotel guests and our own upbringing in the culture of hospitality. Or we might re-read the *Odyssey*, the *Edda*, or the *Torah*. Wherever we look, the rules of hospitality required offering the welcome, the bath, the blessing, and the meal, before exacting the story from the stranger: 'Who are you? Who are your parents? Where do you come from? What is your need?' Penelope is a figure of the risks incurred in hospitality, the idealized solidarity of the couple, the critical façade of respectability, and the ever-present ambiguity.[12]

Four cases

Pertinent information is scattered. Nineteenth-century hotels and inns came in all sizes and catered to a variety of 'niche' markets. Although travellers' accounts are numerous and entertaining, they display vigorous biases. The municipal tax-roll, attentive to taxation of liquor outlets, offers soundings for 1830–1914: outlets increased in proportion to population. But we find no comprehensive, continuous, and reliable statistical sources, as a consequence of the total disinterest of census-makers in activities that were domestic in location, 'informal' (not fully monetized), seasonal, part-time, or perceived as temporary. The standard sources are therefore inadequate for taking the measure of the hospitality industry and its female personnel. To unroll a tapestry, we have instead seized upon four cases grounded in exceptional local sources: the records of a small inn operated by the O'Brien couple in 1824–49; the career of Madame Angélique Saint-Julien, from 1842 to 1860; registers of Henry Hogan's 'St Lawrence Hall Hotel' for 1886–7; and business records of 'The Windsor' at the height of its eminence, in 1902–10. The four enterprises overlapped in time, and for each case we have surrounded the core document with 'pinpoints' of information from newspapers and print advertising: glimpses that acknowledge the potential of such sources.[13]

Mr and Mrs O'Brien

Eliza McDugald managed the inn three blocks from the riverfront, while her husband, Bartholomew O'Brien, carried on activities as a broker of silver and coin. His success as a broker made possible his legacy of £1,000 for the creation of a refuge for 'out of place servants' and subsequent conservation of a carton of documents.[14] Although Mrs O'Brien could not sign her name, she had her own bank account and status as a *marchande publique*. After her husband's death in 1849, she continued to operate a boarding house for 18 years longer, with the same set of furniture and kitchen gear, in the same rental premises, but without a liquor licence.[15] Her decision to remain in the hospitality business was not unusual: many Montreal widows continued the family business, renewing liquor licences in their own names.

The inn was within a two-block walk of the steamboat landing, canal-boat landing, and stagecoach depots in what is today known as 'Old Montreal'. It was one of a hundred inns in the city, and an inventory of 1843 provides unusual material detail. Entering the main floor through the small parlour, a guest would choose among the 15 pairs of slippers. The largest stove reigned here, with its 19 lengths of pipe and five elbows, and a gas lamp illumined the bookshelves, the drop-leaf table of cherry, and the big roller-map of Canada. In the larger parlour beyond, the guest would settle into the sofa, the big upholstered rocker, or one of 17 other chairs, to enjoy the fire, carpets, and silver chandelier. In a homelike décor, reproductions of St Anthony and the Holy Virgin got along with portraits of George Washington and Daniel O'Connell, and with popular prints – *Negro Caricatures* and *Sailor's Wedding*. In the two upper stories, each of the seven guest rooms had its double bed and table with pitcher and bowl. One room was distinguished by a bed with white curtains and a child's crib with turned rails; elsewhere, there were extra folding beds or bench-beds. Mrs O'Brien's purchase of 10 featherbeds is recorded, at 40 shillings each.[16]

The image of a 'family hotel' implied a degree of cleanliness, personal service, and warmth. The innkeeper herself played out a visible performance of respectability and asserted a code of behaviour. Out front, she welcomed travellers and neighbours and served food at the shared table. Behind the scenes, with the help of a boy and a girl servant, she oversaw the cooking and cleaning and washing. Much of the work was carried out in the three large service rooms of a lower story or 'lighted basement'. Turnover of help was frequent: the day book mentions a departure or arrival on average once a month of another Biddy, Maggie, or Annie. Their beds (a servant boy specified in one room, a servant girl in another) were surrounded by buckets, washtubs, boilers, smoothing irons, churn, and rat trap. The menu was responsive to the seasons: in summer, Mrs O'Brien kept a cow, and in winter she bought turnips and carrots by the bushel. The meals were plain and substantial – a dinner of 'turkey tongue' or fish with hominy; for dessert, apple pie. As with Odysseus or Odin, the O'Briens' guests anticipated the after-dinner drink – strong beer or temperance beer, gin sling, brandy, or milk sling – exchange of stories, and, on occasion, a little music from the Irish carpenter-fiddler.

In a city of only 7,000 families, the O'Briens received several hundred guests per year, most of them for a single night, a few for a week or a month, 90 per cent of them men. Some had to be cared for in sickness: when a man died of smallpox, the innkeeper wrote to the family to recover the 3 weeks' board and outlays for candles and visits of the priest. O'Brien's niece, who nursed him, caught the disease but recovered. What distinguished this inn from the others was the seasonal clientele of raftsmen. Having sold their firewood in Montreal, the high-quality timbers in Quebec City (for export or ship-building), and even the substance of the raft, they were returning upstream with money in their pockets and a yearning for a night of comfort, after a season in the forest and on the river. Destinations recorded in guest registers from 1836–48 display a hinterland important to Montreal merchants and

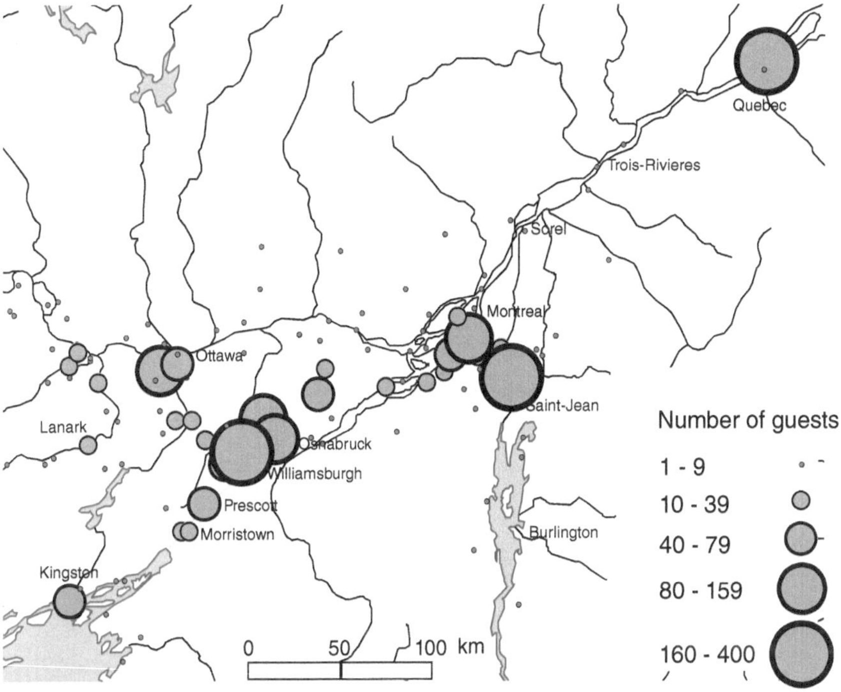

Figure 9.1 Destinations of guests of O'Brien Inn, 1836–48.

Source: McCord Museum, O'Brien papers, Guest register.

investors (Figure 9.1), and, from small packets of letters dated in the 1820s, we can actually observe the westward progress of timber cutting. The registers, together with activities mentioned by O'Brien in a more personal 'line a day' journal, provide eloquent testimony of the seasonality of the entire urban economy. In winter, when guests were few, as shown in Figure 9.2, there was time to keep a journal, catch up with correspondence, entertain the nephews and sisters-in-law, and go to the races or the theatre.

Despite the political turbulence of the 1830s, the city's role as capital of the Union of the Canadas (1843–9), and massive arrivals of famine refugees in 1847 and 1849, the O'Brien documents are near silent with respect to politics. Other sources confirm his role as a conduit of information, charity, and enterprise. Piecing together the memos provides coherent evidence of the intensity of communications: networking among family members, among neighbours, and among tavern-keepers citywide. Pocket account books uncover further networks. Numerous small loans, many of them tagged 'No interest – Irish', were repaid a day or two later. O'Brien made some more substantial 30-day or 60-day loans to commission merchants at the legal interest rate (5 per cent per annum), and in the 1840s he was extending an ongoing 'line of credit' to a score

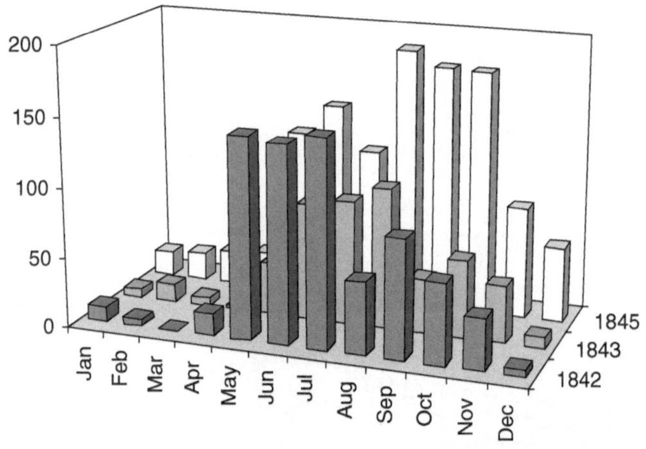

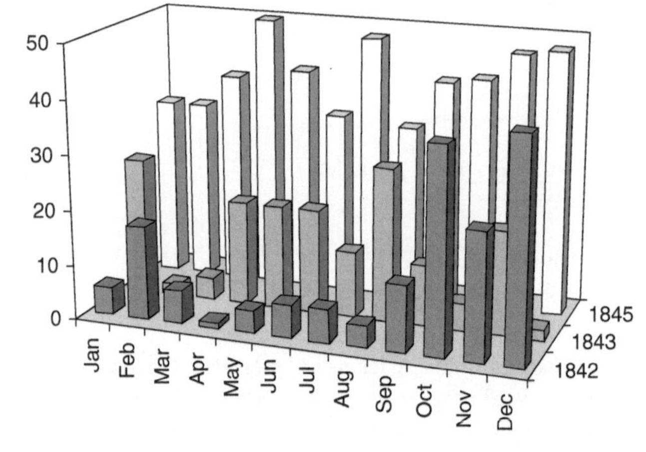

Figure 9.2 Activities at the O'Brien Inn, 1842–7, summarized by month of the year. Numbers of entries are compiled from the guest register of the inn (top) and from O'Brien's personal day-book (bottom) for the years 1842, 1843, and 1845.

Source: McCord Museum, O'Brien papers.

of local artisans and manufacturers, all of them Irish Catholic immigrants of his own generation – those who arrived before the famines of the 1840s. They included a rope-maker, butchers, ironworker, mason, plasterer, and carpenters.[17] Everyone depended on credit, and credit was an act of trust. The bankruptcy of John Kelly, an experienced carpenter-builder, reduced the O'Brien legacy by half, but thickened the story, weaving their network into the politics of the city. The Irish tavern-keepers were key fundraisers for the construction of St Patrick's Church, and in 1843 they charged Kelly with keeping a watch on its progress. In nearby Près de Ville, he was at that moment building houses for a dozen families, most of whom were among O'Brien's borrowers. As the building boom swelled in the 1840s, Kelly obtained promising contracts from wealthier Protestant merchants and architects in adjacent Beaver Hall Hill and set his sights on the contract for roofing the new Bonsecours market and city hall. The requisite political connections stretched across the 'great rift' [of religious affiliation] and entangled Kelly in the electoral manipulations of Mayor James Ferrier. The hundreds of loans, contracts, and guarantees of both Kelly and O'Brien appear to have been negotiated entirely in the inn, on the street, or in the front room of a home. O'Brien's role, remarkably discreet, illustrates the way the innkeepers were knitting together various social networks, extending into the hinterland of farmers, traders, and suppliers of timber. His personal accumulation of 2,500 newspaper editions (inventoried by his nephews) is suggestive of the urban thirst for news, the eagerness with which the city was pushing forward its telegraph lines and protesting the sluggishness of its postmaster. New technologies of information and more intense exchange enhanced the value of face-to-face conversations.

Madame Saint-Julien

Madame Angélique Saint-Julien operated a hotel reputed internationally for its elegant accommodation and attentive service. As she achieved unusual success and visibility, we have traced her 23-year career through research in Quebec's rich genealogical sources. She had learned the hospitality trade first hand from her parents, who were innkeepers at Pointe-aux-Trembles, a village on the King's Highway 15 km east of the city. Following her marriage to Antoine-Timothé Saint-Julien in 1835, she began her career as a publican independent of her husband. In 1842, she was operating a boarding house at the very centre of town, two blocks from the steamboat landing, the parish church, and the courthouse. Three years later, she rented another building on the same street and advertised her 80-room Canada Hotel as 'a large, commodious house, spacious yard, good stabling'.[18] The 1851 renewal application confirms that she had held a liquor licence for at least 8 years, and the next year she leased the newly rebuilt 100-room Donegana Hotel, located on the north-west corner of Notre-Dame, 'the most fashionable street in the city'. By 1857, Saint-Julien had added a new wing, refurbished the drawing and sitting rooms, and refitted all the bedrooms 'with every luxury and convenience'.[19] The hotel was advantageously situated near the Champs-de-Mars parade ground, where military bands entertained residents and tourists alike with daily musical and marching spectacles. Within a 5-minute stroll from the hotel (now 150 rooms), guests could shop at fashionable stores or attend the theatre. An 1858 bill of fare publicized the hotel's extensive menu and wine list and the current play at the Theatre Royal.[20]

Madame Saint-Julien underscored in advertisements that the Donegana was a home away from home:

> Has all the quietness and domestic comforts of a private house . . . visitors will feel themselves quite at home while resident in it and their comfort and convenience will be as much attended to and consulted, as if they were living in their own families.[21]

The hotel was considered the equal of the Astor Hotel in New York, even before she took over,[22] and an experienced manager and a large staff catered to the many needs of a well-off clientele. The hotel also hosted balls, meetings, and charity events. Saint-Julien publicized her several businesses under her own name and annually renewed the tavern licence independent of her husband, who was occupied as a boarding-school master. The couple had no children. Antoine-Timothé Saint-Julien died in 1863 at the age of 49, and it appears that Madame Saint-Julien left the city's hospitality service not long after his death.

Advertisements of the 1840s and 1850s emphasized the three great assets of location, hours, and comestibles. 'The Bar is constantly supplied with the best of Wines, Ales, Liquors and Cigars.' 'Snacks at all hours.' Activities ebbed and flowed to the rhythms of the transportation system. Madame Saint-Julien arranged conveyance of guests to the steamboat landing, railway station, and stages servicing Albany (via Burlington), Upper Canada, and Quebec City. The workforce served meals to coincide with the arrival and departure of steamships and trains.

Critical to attracting well-off guests were the small comforts, prompt service, 'social skills', and advertisements aimed at particular clienteles. From the first, women were decision-makers about art and decoration, and the female clients, whether accompanied, alone, or with children, expected certain accouterments of respectability; they included separate sitting rooms and a staff versed in discretion, social distance, and knowing one's place. Madame Saint-Julien understood this perfectly, meticulously attending to the hotel's culture and the constraints embedded in the liquor-licencing process – never jeopardizing this amenity.

Whatever the size and rank of the hotel, directories of the 1840s suggest frequent turnover of management, moves between installations, and recycling of high-access corner sites. All of the inns and boarding houses, but especially those operated by women, were vulnerable to accusations of being 'disorderly houses' and to any gossip that they condoned illicit sexual behaviour in the establishment. The city offered alternatives: some women in the sex trade achieved notoriety, but few attained a high standard of living. Concentrated in popular-class neighbourhoods with access to military barracks and riverfront, they aimed at a low profile to ensure tolerance by their neighbours.[23]

Hogan's St Lawrence Hall Hotel

Henry Hogan's, located at 13 Great Saint James Street since 1851, served as headquarters for the British contractors and engineers engaged on the Grand Trunk Railway and bridge – the vital line that one of his guests described as '1,200 continuous miles of discomfort from Chicago to Quebec and Portland, the western and eastern termini'.[24] This flamboyant guest, Captain Rhys, performed at Nordheimer's Music Hall, and his travel narrative sums up for us the social geography of Montreal:

> The principal hotels, of which there are but two, are enormous edifices, the others mere pot-houses. The principal streets, of which there are but two, are wide and handsome, the rest mere lanes. The people of whom there are but two, the rich, holding high estate, the poor, holding nothing. There is nothing *middling* in Montreal.[25]

The layout of the 150-room hotel confirms the domestic model. In an account from the agitated period at the end of the US Civil War (1865), another of Hogan's guests leads us from the bustle of public spaces fronting the street – the height of hurry – into progressively more private and protected spaces. He noted the exercise of control over entrances and internal traffic, analogous to the penitentiary superintendent's eye over his panopticon layout, or to the alert ear of the housewife in her kitchen centrally located in a little Montreal duplex.[26] The 'enormous' hotel produced, however, some novel semi-public spaces and a more rigorous segregation by gender, with the ladies' sitting room and music room, the men's smoker and billiard room. These spaces differentiated the architecture of hurry, allowing people to group and re-group, to observe or overhear. The sitting room and smoker offered distraction from delay and retreat from the intensity of encounters. Hubbub was highest in the communications spaces. Hogan's reception desk, with its 'incessant roll of clamour for beds and bills', was flanked on one side by a post office and newspaper stall, on the other by the barbershop: 'from dawn until dinner, a succession of unshorn Yankees – a Yankee never shaves himself – submitting their cheeks to the barber's razor, and their ears to his latest news'.[27]

In the 1860s, Hogan's was indeed the antenna for the latest news. The city's cotton textile industry fostered sympathy with the Confederacy, and local élites, including the mill owners and officers of the British garrison, shared fashions, hunts, balls, and seasonal visits to resorts in the Great Smokies and the Adirondacks, seaside Newport, and mineral springs at St Johnsbury, White Sulphur Springs, and Saratoga Springs.[28] Harbouring Confederate diplomats, expatriates, and spies, Hogan's was reputed to be the only hotel in the Canadas to serve mint juleps. On 18 October 1864, actor John Wilkes Booth checked into Hogan's, bringing with him a wardrobe of velvet suits, plumes, and doublets and a box of stage swords and pistols. It was the eve of the St Albans border raid, which threatened British neutrality: 'A diplomatic uproar ensued, filling the Hall with judges, lawyers, detectives, and civil and military authorities of every description'.[29] Booth, as thirsty for news as O'Brien had been, responded to reports of Confederate successes by 'scattering small silver pieces round among the newsboys and bell boys'.[30] Wartime considerations had tripled the size of the British garrison (ordinarily about 800 men) and attracted to the city other boisterous guests, not all of them with deep pockets: deserters from the Union Army, draft evaders, Canadians eager for the enlistment bonus, crimps organized to bring in recruits drunk or sober, and, persisting after the end of the war, Irish Fenian agents.[31]

Hogan's guest registers allow us to compile large samples of places of origin of his guests for a summer month 20 years later, 5 July through 3 August 1886, and a 6-week winter season, 15 December 1886 through 3 February 1887 – the basis for Tables 9.1 and 9.2. Since O'Brien's time, the 'iron network' of rail and telegraph had extended the city's catchment of travellers to the central place system of the continent. Canada's transcontinental had just opened (1885), and the railway president had already selected a site in the Rockies for the first of a string of grand hotels across Canada, but no impact was yet apparent. The January sample includes the week-long winter carnival to which the city welcomed 50,000 spectators to enjoy a nocturnal siege of the ice palace by 'armies' of hundreds, exhibiting what Thorstein Veblen referred to as modern vestiges of barbarian male prowess.[32] 'Women were encouraged to freely participate in tobogganing provided a man was in command, and velvet seats could be added to make the activity more comfortable and more "feminine".'[33] Hogan's registers reflect the persistent massive gender imbalance. The samples comprise a population of 1,495 parties arriving in the summer season, 1,764 parties in the winter season. Two-thirds of the groups travelling in summer came from neighbouring regions of the St Lawrence Valley (as shown in Figure 9.1); in winter, three-quarters. Family groups including women and children amounted to one-quarter of the parties

Table 9.1 Samples of guests at Hogan's St Lawrence Hall Hotel, 1886–7

	Number		*Percentage*	
	Summer	*Winter*	*Summer*	*Winter*
Number of parties				
Man alone	1,055	1,541	70.6	87.4
Men 2 or more in party	73	85	4.9	4.8
Couple	148	63	9.9	3.6
Couple with family	49	26	3.3	1.5
Man and Miss or daughter	21	4	1.4	0.2
Woman with son or children	14	10	0.9	0.6
Women 2 or more in party	47	12	3.1	0.7
Woman alone (Miss)	44	8	2.9	0.5
Woman alone (Mrs)	44	15	2.9	0.9
All parties	1,495	1,764	100.0	100.0
Number of people (estimated)				
Men	1,473	1,778	75.0	89.8
Women	414	157	21.1	7.9
Children	71	44	3.6	2.2
Servant or nurse	7	0	0.4	0.0
Total persons	1,965	1,979	100.0	100.0

Source: McCord Museum, St Lawrence Hall Hotel Registers (P093), 5 July through 3 August 1886 and 15 December 1886 through 3 February 1887.

Note: Numbers of persons are slightly underestimated as a result of ambiguous entries 'and children', or unspecified occupants of a second room.

Table 9.2 Places of origin of guests arriving at Hogan's St Lawrence Hall Hotel, 1886–7

Region of provenance	*Individuals*		*Percentage*	
	Summer	*Winter*	*Summer*	*Winter*
Montreal	76	128	3.8	7.3
St Lawrence watershed*	1258	1348	63.3	76.6
Rest of Eastern US	419	148	21.1	8.4
Maritime Canada	32	52	1.6	3.0
Western Canada & US	133	49	6.7	2.8
Transoceanic	69	35	3.5	2.0
Total†	1987	1760	100.0	100.0

Source: McCord Museum, St Lawrence Hall Hotel Registers (P093), 5 July through 3 August 1886, and 15 December 1886 through 3 February 1887.

* Ontario, Quebec, and upper New York State
† For 275 additional guests the place of origin could not be determined.

in the summer holiday season, in winter to fewer than one in ten.[34] The male-dominant clientele continued to challenge hotel mangers to maintain the reputation of an environment safe and attractive to women and 'family'.

A single enterprise displays only a portion of the flows into the city, of course, and we cannot assume that the map of residences of people arriving in Montreal matched the array of places to which Montreal residents dispersed. In the 1890s, Hogan expanded his establishment to 250 rooms, installed an elegant passenger elevator, and, with electric incandescent light in the halls and public rooms, claimed to be 'the most attractively lighted hotel in the Dominion'. His coaches and porters were 'in attendance' at all the arrivals of trains and steamers. Soundings in the registers of that decade point to patronage of theatre companies, music clubs, bowling teams, football and lacrosse teams – clienteles that compensated for the loss of an élite who now favoured the newer and larger Windsor Hotel, situated 'ringside' on Dominion Square where the ice palaces were built and the first international hockey tournaments held.

The Windsor

Designed to promote and exploit tourism, the Windsor had opened 28 January 1878 on an uptown site convenient to the new railway station, and executives of both railway companies retained offices in the hotel and residential apartments for top brass. Integration of the railway system of North America created a social time warp. Long-distance service was decisive in justifying night work, Sunday operations, and 'standard' time.[35] Telegraph service, faster steamships, snow-ploughs, and ice-breakers also contributed to more reliable schedules, effecting a regime change in the seasonality of everyday life. The compulsive summer hurry and winter relaxation evident in O'Brien's journal was reversed, and the registers of the Windsor suggest bursts of agitation year round: the flurry of hurry to enjoy a weekend spectacle in the city, and hectic stopovers of those who travelled in luxury – with their children, servants, wardrobes, dogs, and horses – to a summertime social season at seaside, lakeshore, or riverside. The Windsor's request for special treatment to operate its barbershop on Sunday mornings reflected a long-running debate in Montreal between Protestant sabbatarians and Catholic sociability.

Minutes of the syndicate of owners in 1877–81 and 1902–7 provide pinpoints of information about clients, suppliers, employees, and lawyers who faced the ethical problems of commercial hospitality.[36] The syndicate advertised in newspapers, pamphlets, and books targeting readers in New York, Ontario, and Quebec. Members' sensitivity to profit reached into the minutiae of the everyday. They denied Mr Walter's request for a reimbursement of $8 for his missing umbrella and appealed a judgment to pay Mr Greene $250 for his loss of two trunks. (When the higher court upheld the decision, the syndicate had to issue a cheque to Mr Greene for $701.[37]) They sought experienced waiters from New York and contracted Chicago decorators, Mitchel and Halbach, to paint the dining room at a cost of $4,500.[38] To settle Mrs Austin's outstanding bill of $1,500, the secretary of the syndicate engaged in a protracted correspondence with her son and the Boston lawyer who managed her trust.[39]

These documents, then, capture the risks inherent in the travel sector: delays, uncertainties, weather, predation and violence, inefficiencies in the connections between modes of transport (baggage lost or stolen), and failures of communication. In the literature of modernity, emphasis on steam and speed has led us to underestimate the waiting times, the frequency of disconnects, and levels of resistance (see Pooley and Pooley in this volume).[40] 'Time is money.' The investors in the hotel syndicate, as well as the railway and steamship agents, were tuned to that; for their employees and the travelling public, the tighter schedules and interconnections intensified awareness of the alternative values: paid time or free time, his time or her time, your time or my time.

The Windsor, by catering to the rage for conspicuous leisure and conspicuous consumption, sharpened the distinction between front-stage and back-stage actors. The grand charity balls and civic dinners it hosted dramatized the gap between those entitled to be served in a hurry and those expected to scurry at their behest. The employer's capacity to confine the time–space of its entire staff can be seen in the wake of a smallpox epidemic that deterred tourists: in June 1885, the hotel vaccinated its employees and their families, quartered the unmarried employees in the hotel, and set up clubrooms, male and female. 'Then they were shut up.' Six months later, the 25 waiters refused to be quarantined any longer; they 'walked out', returned to give guests their dinners, and left again after lunch. The manager dismissed the headwaiter for 'deserting service'.[41]

The 1901 Bell telephone book, among its 524 entries, listed 20 full-fledged hotels and identified distinct departments of the Windsor. From census samples, we obtain some clues to the labour force. Extraction of 100 male 'hotel keepers' in the city (of an estimated 300) shows their year's earnings distributed around the mode of $1,000, three times the labourer's. At St Lawrence Hall, Hogan reported, in addition to his wife and daughter and 35 guests in long-term residence, 55 employees housed on the premises. The Windsor reported 90, with a more complex hierarchy: among the women a head housekeeper, a laundry head, and a superintendent of scrubbers. The women scrubbers and laundresses were kept out of sight, whereas male waiters in the dining room continued to display their cheek and agility in the midday rush. The team of 12 male cooks and steward all received double the average working-class wage, and the two head cooks were paid more than the cashier and his chief clerk.

Although a fire in 1906 destroyed almost 100 guest rooms, renovations and a new wing doubled the number of rooms, from 368 to 750. Management's concern with baths, telephones, and elevators identifies the systems challenge of technologies that embedded the hotel in the larger urban system of cabling, smoke and soot, water supply, sanitary sewers, waste disposal, streetcar traffic, and snow clearance.

Among the largest events staged in Montreal was the Eucharistic Congress of 6–11 September 1910, a week-long religious spectacle of global significance, believed to have attracted 100,000 visitors to the city.[42] The reception committee of the Archdiocese published a pamphlet with the Windsor at the top of the list ($4 a night), boarding houses at $1–$2 a night, and eating places in the vicinity of its several centres of activity, of interest to clerics, nuns, or laymen, French- or English-speaking. The arrangements point to the elasticity of the hospitality sector, enrolling the city's numerous convents, boarding schools, and presbyteries to house 2,000 priests. All the steamships, sailing vessels, and small boats turned out to escort the Papal Nunzio; a dress code was prepared for his reception by the Mayor, and the Windsor was the site of the banquet and ball given in his honour by the provincial and federal governments. The Nunzio and the Archbishop departed to a midnight mass, while the élite of Montreal continued dancing into the wee hours. In anticipation of the grand 7-hour procession, the Archdiocese articulated its concern with respectability, urging all the clergy to participate and not be seen looking on from windows and balconies.

Taste and service in a large hotel like the Windsor were driven by the circulation of the one-tenth-of-one per cent who frequented also the Waldorf Astoria, the resort at Saratoga Springs, and the Canadian Pacific Railway's 'chateaux' at Banff and Quebec City, but pressure increased in due proportion for more basic kinds of hospitality. At Windsor station, a block from the hotel, the cellars were renovated to fit up two 80-foot dormitories to receive immigrants.[43] The immigration service employed translators, but the trainloads of Chinese were still locked in for transshipment. Syrian entrepreneurs were boarding and deploying 950 immigrant peddlers.[44] The city jail received farm labourers in winter, and a night refuge was created adjoining the central police station. Close to 4 per cent of the city's residents lived in institutions operated by religious women; several of them sheltered more than 1,000 persons: the insane, orphans, the aged and chronically ill, unmarried expectant mothers, illegitimate infants, 'protected' or 'delinquent' girls, and women prisoners. At St Bridget's, the institution created by O'Brien's bequest, the Grey Nuns provided 60,000–70,000 nights of shelter to out-of-place servants each year.

Conclusion

Examination of the hospitality sector must, as we have shown, extend well beyond the inn or hotel itself, and the case studies leave us unsatisfied. In terms of innovations, to what extent were successive generations of hotelkeepers innovators of high-rise 'solutions'? Did they lead or follow? Did they aggravate the hurry or provide effective shelter for the guests? From a wealth of trivia – notarized acts, court records, work accidents, and fires – we extracted 'pinpoints' of information that specify the way 'respectability' cast a veil over the roles of women as managers, proprietors, and hostesses, and as backstage labour, low paid and vulnerable. Often incidental to the *fait divers* reported, they are erratic, but are they biased? If we treat them as random occurrences, such tidbits have potential to illuminate both the contradictions of hurry and the contradictions of hospitality.

Among the three aspects we selected for observation – information, gender, and domesticity – the case studies suggest interaction. Innkeepers, large and small, played a vital role in accelerating the flow of information through the connectome of the urban economy, wired for extracting the resources of the continent. The largest enterprises stole the limelight, but the attraction of each particular hotel demanded a homelike environment. The domestic services of women conveyed a market image, assured the guest's sense of security, and enabled a comfortable sociability. The flow of information, the entrepreneurial opportunities, and the personal sense of 'being at home' all depended on the construction of trust between strangers: the ancient ethic of hospitality.

If the argument is correct, it invites consideration of present-day challenges. We see today swifter journeys, taller buildings, and dispersal of commercial short-term 'homes' outside urban centres – at airport or expressway, along coastal margins and ecological frontiers. Women's participation in tourism has increased, as has participation of family groups, unchaperoned youth, and loners in older age sets. By carrying their own instruments of communication, individual travellers are today less dependent on an innkeeper and are prepared to deploy and extend rapidly their pocket of local order. The supply of services is less visible than in the nineteenth century, with greater differentiation of front-stage and back-stage roles of women. The wage differential persists, however, with disturbing implications for the ongoing process of change. Any reform of the hospitality sector in its 'industrial' manifestations – multinational managements, seasonal flows to lower latitudes, transnational sex trades, and mass movements of refugees – depends on reduction of income inequality and remodelling of domestic relations in the home. The ethical challenge persists essentially as the Encyclopedist expressed it, '*L'hospitalité* est la vertu d'une grande âme, qui tient à tout l'univers par les liens de l'humanité'. In older and simpler words, 'Whosoever is in need let him come and eat'.[45]

Notes

1 Thomas Piketty, *Le Capital au XXIe Siècle* (Paris: Seuil, 2013).
2 *The Vindicator*, 21 June; 5 and 19 July 1831.
3 The Bible, King James Version, Matthew 25:35.

4 '[T]he just measure of this type of beneficence depends on what contributes the most to the great end that men must have as a goal, namely reciprocal help, fidelity, exchange between various states, concord, and the duties of the members of a shared civil society' (Louis, chevalier de Jaucourt, 'Hospitalité', in Denis Diderot and Jean d'Alembert [eds], *Encyclopédie ou Dictionnaire raisonné des sciences, des arts et des métiers, par une Société de Gens de lettres*, vol. 8 [Paris: 1765]: 316, ARTFL Encyclopédie Project, University of Chicago, http://artflsrv02.uchicago.edu/cgi-bin/philologic/getobject.pl?c.7:1154.encyclopedie0416.3220096 For translation by Sophie Bourgault in *The Encyclopedia of Diderot & d'Alembert Collaborative Translation Project*, University of Michigan Library, http://hdl.handle.net/2027/spo.did2222.0002.761 [accessed 8 May 2017]).

5 Kajsa Ellegård and Bertil Vilhelmson, 'Home as a Pocket of Local Order: Everyday Activities and the Friction of Distance', *Geografiska Annaler* 86B (2004): 281–96.

6 *The Hávamál: With Selections from Other Poems of the Edda, Illustrating the Wisdom of the North in Heathen Times*, ed. and trans. D.E. Martin Clarke (Cambridge: Cambridge University Press, 1923). See also Sërpa Alto, 'Commercial Travel and Hospitality in the Kings' Sagas', *Mirator* 10 (2009): 31–43.

7 For examples of the way the risk inherent in reception of the stranger was transferred to women, see Genesis 19:1–9 and Judges 19:23–30.

8 James Gleick, in his book *The Information* (New York: Vintage, 2011), is exceptional in his attention to gender, but only 6 per cent of the persons mentioned are women.

9 Sebastian Seung, *Connectome: How the Brain's Wiring Makes Us Who We Are* (Boston: Houghton Mifflin Harcourt, 2012); Bill Hillier, *Space Is a Machine* (Cambridge: Cambridge University Press, 1996): 152, 31; Peter Gould, 'Dynamic Structures of Geographic Space', in *Collapsing Space and Time: Geographic Aspects of Communication and Information*, eds S.D. Brunn and T.R. Leinbach (London: Harper Collins, 1991): 4.

10 Jacques Derrida, *'De l'hospitalité': Anne Dufourmantelle invite Jacques Derrida à répondre* (Paris: Calmann-Lévy, 1997). The article by Linda McDowell, Adina Batnitzky, and Sarah Dyer, 'Division, Segmentation, and Interpellation: The Embodied Labors of Migrant Workers in a Greater London Hotel', *Economic Geography* 83 (2007): 1–25, provides entry points to a wider discussion of the persistence of categorical inequalities.

11 Noteworthy are studies of the intersection of race and class in Pullman cars and nineteenth-century 'watering places', and the intervention of 'the Irish innkeeper' worldwide: S.J. Mathieu, 'North of the Colour Line: Sleeping Car Porters and the Battle against Jim Crow on Canadian Rails, 1880–1920', *Labour/Le Travail* 47 (2001): 9–42; Eliza Potter, *A Hairdresser's Experience in High Life*, ed. Xiomara Santamarina (Chapel Hill: University of North Carolina Press, 2009). For the burgeoning discussion of present-day 'hospitality management', see Barry O'Mahony, 'The Role of the Hospitality Industry in Cultural Assimilation: A Case Study from Colonial Australia', in *Hospitality: A Social Lens*, eds Conrad Lashley, Paul Lynch, and Alison J. Morrison (Amsterdam: Elsevier, 2007): 89–100.

12 See Barbara Clayton, *A Penelopean Poetics: Reweaving the Feminine in Homer's Odyssey* (Lanham: Lexington Books, 2003); Agathe Thornton, 'The Homecomings of the Achaeans', in *People and Themes in Homer's Odyssey* (Dunedin: University of Otago in association with Methuen, London, 1970): 1–15; and Steve Reece, *The Stranger's Welcome: Oral Theory and the Aesthetics of the Homeric Hospitality Scene* (Ann Arbor: University of Michigan Press, 1993). They identify 18 'hospitality scenes' in the *Iliad* and the *Odyssey*, with 38 ordered elements.

13 Mary Anne Poutanen, 'Due Attention Has Been Paid to All Rules: Women, Tavern Licences, and Social Regulation in Montreal, 1840–1860', *Histoire Sociale/Social History* 50 (May 2017): 43–68.

14 St Patrick's Church donated the documents to the McCord Museum.

15 Eliza McDugald was first married in 1816, in Newcastle on Tyne, and was abandoned by her husband, John Silbit, soon after arrival in Lower Canada (Bibliotheque et Archives nationales du Québec (hereafter BAnQ), Tutelles, 3 June 1851). The annexed

act of notoriety (Repertory of Isaacson, 19 December 1850), signed by merchants Henry Mulholland and William Workman, makes no mention of events in the intervening 27 years: a son, the second marriage (to O'Brien), the £50 annuity provided in his will, or the £500 entitlement from their marriage contract.

16 McCord Museum, O'Brien Papers (PO83), inventory taken by O'Brien and his nephews Dan and Ed Kingsbury in April 1845; Day Book of Bartholomew O'Brien.

17 See Sherry Olson, 'Silver and Hotcakes and Beer, Irish Montreal in the 1840s', *Canadian Ethnic Studies* 45:1/2 (2013): 179–201. On Kelly's bankruptcy, BAnQ-M, TP11, S2, SS10, SSS1, Dossiers des faillites.

18 BAnQ-M, Cote: P 1000, D880, Licences de tavernes, 17 April 1851.

19 *The Stranger's Guide through the City of Montreal*, 1857 (Salter and Ross, 1857).

20 *The Donegana Hotel, Bill of Fare*: *Montréal, Sunday, August 29, 1858* (Montréal: J. Potts, Herald Office, 1858).

21 *The Stranger's Guide through the City of Montreal*, 1857.

22 'Important from Canada', *Colonial Times*, 11 January 1850; Edgar Andrew Collard, 'Donegana's and the Charity Ball', *Montreal Gazette*, 15 January 1972.

23 Mary Anne Poutanen, *Beyond Brutal Passions: Prostitution in Early Nineteenth-Century Montreal* (Montreal: McGill-Queen's University Press, 2015); Sherry Olson, 'Feathering Her Nest in Nineteenth-Century Montreal', *Social History/Histoire sociale* 33 (May 2000): 1–35.

24 Captain Horton Rhys, *A Theatrical Trip for a Wager! Through Canada and the United States* (Vancouver: Alcuin Society, 1966), 123.

25 Ibid., 111. The two principal streets were, of course, Notre-Dame and St James, chosen by both Saint-Julien and Hogan.

26 George Tuthill Borrett, *Letters from Canada and the United States* (London: J.E. Adlard, 1865), 30–37. He mentions the charge at $2 per day, including bed. See also Janet Semple, *Bentham's Prison: A Study of the Panopticon Penitentiary* (Oxford: Clarendon Press, 1993); on the centrality of the kitchen in the Montreal dwelling, see François Dufaux and Sherry Olson, 'Reconstruire Montréal, rebâtir sa fortune', *Revue de la Bibliothèque et archives nationales du Québec* 1 (2009): 44–57.

27 Borrett, *Letters*, 32. Jon Sternass, *First Resorts: Pursuing Pleasure at Saratoga Springs, Newport, and Coney Island* (Baltimore: Johns Hopkins University Press, 2003), credits the innovation in scale of hotel construction to the Boston Exchange Coffee House, the City Hotel in Baltimore (1826), and the Tremont House in Boston (1829).

28 Sternass, *First Resorts*; Louise B. Roomet, 'Vermont as a resort area in the nineteenth century', *Vermont History* 44:1 (1976): 1–13; Michel Prévost, *La Belle époque de Caledonia Springs* (Hull: Lettres Plus, 1997). Our innkeeper, Bartholomew O'Brien, stocked the mineral waters from springs at Varennes, Québec, and Caledonia Springs, Ontario, and visited the resorts in the 1840s.

29 Terry Alford, *Fortune's Fool: The Life of John Wilkes Booth* (New York: Oxford University Press, 2016): 186. He would assassinate Lincoln on 14 April 1865 in Washington, DC.

30 'Interviews with Henry Hogan, When Wilkes Booth Was in Montreal', *Montreal Star*, 6 December 1902; see also 12 February and 8 March 1902; cited by Alford, *Fortune's Fool*.

31 Robin Winks, *Canada and the United States: The Civil War Years* (Montreal: McGill-Queen's University Press, 1998), 192–200; Barry Sheehy, *Montreal City of Secrets: Confederate Operations in Montreal during the American Civil War* (Montreal: Baraka Books, 2017); David A. Wilson, 'The Fenians in Montreal, 1862–68: Invasion, Intrigue, and Assassination', *Eire-Ireland* 38 (Fall/Winter 2003): 109–33.

32 Thorstein Veblen, *The Theory of the Leisure Class* (New York: Macmillan, 1899).

33 Sylvie Dufresne, 'Le Carnaval d'hiver de Montréal, 1803–1889', *Revue d'Histoire Urbaine* 11 (February 1983): 26. Carnivals were held in 1883, 1884, 1885, 1887, and 1889. The railway system is well described in *Montreal Star*, 9 September 1886.

34 Volumes at the McCord Museum, Saint Lawrence Hall Hotel Fond (P093), are of varied dates, with outside limits 16 August 1858 through 19 January 1905. Library and Archives Canada R6035-0-1-E reports possession of 38 volumes, outside limits 1851 through 1903; Guests' origins (usually the town of residence) are not reported in the 1860s volumes. The 1880s volumes lend themselves to more detailed analysis, such as carried out by C. Weaver, D. Fyfe, A. Robinson, D. Holdsworth, D. Peuquet, and A.M. MacEachren, 'Visual Exploration and Analysis of Historic Hotel Visits', *Information Visualization* 6:1 (2007): 89–103; *idem*, 'Visual Analysis of Historic Hotel Visitation Patterns', *IEEE Transactions on Visualization and Computer Graphics* 12 (2006): 35–42.

35 Jarrett Rudy, 'Do You Have the Time? Modernity, Democracy, and the Beginnings of Daylight Saving Time in Montreal, 1907–1928', *Canadian Historical Review* 93 (December 2012): 531–54. Integration was achieved in bursts: east of the Mississippi about 1852, along the St Lawrence and Great Lakes corridor 1872, and the transcontinental links 1885.

36 McCord Museum, Fond PO92/A, Windsor Hotel, Minutes Syndicate, 3.24-27/C-01-03, Box 10. The holdings of the McCord include registers for the years 1877–81 and 1902–7 and diverse plans.

37 Minutes, 7 April 1903, 26 January 1904, and 31 January 1905.

38 Minutes, 7 April 1903, 8 and 11 December 1903.

39 Minutes, 7 April 1903.

40 Andreas Bernard, *Lifted: A Cultural History of the Elevator* (New York: New York University Press, 2014).

41 *Montreal Star*, 6 January, 25 February, 10 March, 5 April 1886. The headwaiter appealed and lost.

42 Guy Laperrière, 'Le congrès eucharistique de Montréal en 1910: une affirmation du catholicisme montréalais', *SCHEC, Études d'histoire religieuse* 77 (2011): 21–39; Robert Rumilly, *Histoire de la Province de Québec*, Vol. XV (Montréal: Bernard Valiquette, 1945); *XXIe Congrès eucharistique international*, Montréal (Montréal: Beauchemin, 1911).

43 *La Presse*, 27 September 1901.

44 *La Presse*, 6 March 1901.

45 '*Hospitality* is the virtue of a great soul that cares for the whole universe through the ties of humanity' (Louis, chevalier de Jaucourt, 'Hospitalité').

Part III

Practices of mobility

10 Pedestrianism, money and time

Mobilities of hurry in George Gissing's *The Private Papers of Henry Ryecroft*

Jason Finch

Mobilities, hurry and literary texts

Novels and works of life-writing typically considered 'literary' provide in-depth case studies in the geographical mobilities of modernity seen from individuals' points of view. They neither unproblematically represent reality nor act as 'representations' operative only within text and ideology; they indicate actuality, but in a way both complex and limited. The objective of this chapter is to highlight bodily and financial mobilities in George Gissing's 1903 book *The Private Papers of Henry Ryecroft*, concentrating on physical mobilities of pedestrianism driven by monetary *need*, and to facilitate dialogue between historical geographers and literary scholars, in the process. The study builds on literary scholars' recent interest in the often 'frenetic' mobilities of the Victorian city.[1] Such work has directly asked what a turn towards mobilities as an object of attention in the social sciences, and particularly in human geography,[2] would mean for literary studies, where a growing groundswell of interest in literary-spatial matters is noticeable from the 1990s on. Among literary-spatial studies of Victorian literature, James Buzard has examined both the literal geographical paths of tourism and the role of literary authors such as George Eliot in producing 'metropolitan autoethnography' explaining a territory to its own inhabitants.[3] Charlotte Mathieson, more recently, has spatialized Dickens's *Bleak House* (1853) in the world beyond Britain by tracing its characters' moves and relating these to a social-sciences discourse of mobilities emerging in the 2000s and 2010s that focuses on the multiplicity of human 'scales and modes of movement', putting the small-scale alongside the global.[4]

Gissing's writing in general exhibits an extreme sensitivity to the material pressures of the modern age. His renowned novel of the modern literature business *New Grub Street* (1891) exemplifies this. In *New Grub Street*, the writer Jasper Milvain prospers thanks to his discernment of the market's shifting tastes, while his contemporaries Edwin Reardon and Harold Biffen sink towards poverty and death, largely because of their naïve view that what one must do is write a great novel. As is apparent from *New Grub Street*, Gissing's writing can seem to be reducing everything to a monetary account and therefore to be, in one sense of the word, materialistic. At the same time, his writing repeatedly presents characters who radically reject and protest against a money-driven modernity, often via the motif of disinterested classical scholarship that is opposed to writing for the market as Milvain does.

Ryecroft is a fictionalized memoir that itself might be regarded as an architecture of hurry, structured around the temporal gap between its time of writing, around 1900, and that of the youth both of its author, George Gissing, and his autobiographical protagonist, Ryecroft, in London, around 1880. Writing in rural seclusion, Ryecroft proclaims his past position as an urban pedestrian: 'For more than six years I trod the pavement, never stepping once upon mother earth – for the parks are but pavement disguised with a growth of grass'.[5] These were times, he says, when public transport was beyond his means: 'In those days I hardly knew what it was to travel by omnibus. I have walked London streets for twelve and fifteen hours together without ever a thought of saving my legs, or my time, by paying for waftage' (35). In youth, then, Ryecroft pushed through a zone – the pavement, the carriageway for foot passengers – which has been described as one of 'anarchy', frequently '[c]ongested or obstructed', in places divided from the area of wheeled traffic by a 'verge' less respectable than either, the territory of street traders.[6]

Gissing could seem an archetype of the struggling writer, longing to write principled, lofty and uncommercial texts, but prevented from doing so by poverty. The pedestrianism fits into this view of him: the poor writer must trudge from editor to pupil and home again in order to put bread – or even lentils – on the table. Virginia Woolf sums it up sarcastically in her 1932 essay on Gissing:

> The writer has dined upon lentils; he gets up at five; he walks across London; he finds Mr. M. [a wealthy pupil] still in bed, whereupon he stands forth as the champion of life as it is, and proclaims that ugliness is truth, truth ugliness, and that is all we know and all we need to know.[7]

Woolf herself never knew the financial hardship experienced by Gissing, who, as Frank Swinnerton writes, 'lived among the poor [. . .] by reason of the most lamentable necessity' for many years.[8] The view of Gissing publicized by Swinnerton and Woolf both romanticizes and caricatures him, as the text of *Ryecroft* indicates. The youthful Ryecroft's London walks are not all acts of necessity driven by poverty. He recalls 'an August bank holiday, when, having for some reason to walk across London [. . .] unexpectedly' he found himself 'enjoying the strange desertion of great streets'. This, he writes, was an occasion when the defamiliarized city, empty and lit by the sun's 'noontide radiance', 'seemed to fill my veins with life' (88).

Ryecroft's memories are precisely localized in London. In the country lanes near where he now lives, he experiences 'by a freak of mind' the wish to be back in the London of his youth, amid 'the shining of shop-fronts, the yellow glistening of a wet pavement, the hurrying people, the cabs, the omnibuses' (213). It is not some beautiful corner of the lost city that enters Ryecroft's mind when he reminisces, but somewhere extremely mundane: 'the High Street of Islington, which I have not seen for a quarter of a century, at least; no thoroughfare in all London less attractive to the imagination'. What he recalls fondly is youth, as he recognizes, but he also recalls the true town-dweller that he once was, finding 'pleasure in the triumph of artificial circumstance' – light, warmth, human noise – 'over natural conditions' (214). He recognizes too that, were he to be 'thrown back into squalid London' needing to support himself, he would simply go out to work (215). Sometimes, he may summon up an urban image likely to be broadly unchanged 25 years on, as with Islington High Street. At another moment, he knowingly conjures the 'dear old horrors' of a London scene now gone: 'Some of the places, I know, have disappeared. I see the winding way by which I went from Oxford Street, at the foot of Tottenham Court Road, to Leicester Square' (23–24). The 'labyrinth' Ryecroft remembers passing through, lingering outside a pie shop when he had not a penny for food, disappeared with the construction of Charing Cross Road, completed in 1887.[9] As in the perspective on London developed in *Howards End* by Forster's character Margaret Schlegel, the march of building and demolition here encapsulates the onward rush of the city. Yet, unlike Margaret, profoundly disturbed by the impermanence this change points out to her, Ryecroft seems quite at ease with the knowledge that, in details, the city has altered.

Work on nineteenth-century literary mobilities by Buzard, Mathieson and others focuses on canonically Victorian figures such as Dickens and George Eliot, and on the genre of the novel.[10] *Ryecroft*, in contrast, has a complex position on the frontiers of both the Victorian and the novel. Gissing's satirical and brutally self-examining mid-period naturalist novels, such as *The Nether World* (1889) and *New Grub Street* (1891), and his 1890s fictional analyses of suburbs and gender relations are, in the 2010s, more widely read. *Ryecroft* is hard to place generically. It sits oddly at the end of both Gissing's career and Victoria's reign in a sub-period too early to be easily classifiable as 'modernist'. When contemporary Gissing scholarship took off in the 1970s, *Ryecroft* was denigrated as 'a piece of bogus autobiography and wish fulfillment', guilty of 'self-indulgent, self-caressing sentimentality'.[11] Since the 1970s, with a few exceptions, *Ryecroft* has largely been ignored.[12] But, in the early decades of the twentieth century, *Ryecroft* was one of Gissing's most admired and successful works, reprinted no less than 13 times before the First World War. A review for the *Week's Survey* in 1903 described it as 'one of the most distinguished books written in the last ten dull years, years of an outstanding intellectual stagnation, brought about [. . .] by the commercialization [. . .] of literature'.[13] In the 1920s, it was exported to newly independent European countries, including Finland and Estonia, by inclusion in sets of books representing modern English literature bought for university libraries by a forerunner of the British Council.[14]

Ryecroft could be understood in the terms proposed by Max Saunders as a work of 'autobiografiction', a genre widely practised in England between the 1880s and the First World War. Autobiografictions develop a complex, layered relationship between fictionality and the telling of recalled truth.[15] In Gissing's case, this happens via the figure of Ryecroft, whom the reader is clearly intended to compare, even conflate, with Gissing himself. The invitation is offered in a preface during which a certain 'G.G.' claims to be presenting an edition of 'three manuscript books', which he discovered among Ryecroft's papers after the latter's death (x). The reader is invited to see through 'G.G.', recognizing that, although these initials transparently figure George Gissing, it is the character Ryecroft who actually conveys Gissing's experiences and opinions. Such complexity goes beyond even the intricate autobiografictional moves outlined by Saunders in his account of the book, which connect Gissing with other writers' presentations of him and of themselves. In fact, Ryecroft is a thoroughly ironized figure, and the joke may indeed be on readers who admire Ryecroft, thinking that in him they meet the true Gissing.

In the present day of *Ryecroft*, the supposed time of writing of the reflections discovered by 'G.G.', the protagonist lives in retired comfort far from the city; the book is built around the contrast between this present and his city-bound youth. Ryecroft has decelerated, but only thanks to money generated in the economic foci of Victorian Britain. He remains umbilically connected to London through the trains that bring him books in Devon. These, and the regular postal deliveries they make possible, multiple times every day, would also enable a writing career to be conducted from far outside London. Technical improvements or augmentations introduced in modernity had the potential also to introduce new hurry and stress: running to catch the post, for example.[16] When he makes Ryecroft reminisce about his past involvement in what this volume identifies as architectures of hurry, Gissing demonstrates phenomenological aspects of the accelerated movements brought about in the new urbanity of the nineteenth century. Ryecroft's retrospective of 1870s and 1880s London is founded on his pedestrianism as an impoverished, newly arrived immigrant in what was then the world's biggest city. Inverting the shuttling movement from home to workplace and back again typically associated with employees, the young Ryecroft's freelance work as a badly paid writer is remembered as keeping him seated in the garret room where he sleeps. Walks forth from here are only possible in intervals of leisure provided by the completion of work and the receipt of money. Yet his walks are not, in the decades past to which he looks back, leisured in the sense that they later become when he is able to retire to rural England. For one thing, they are confined to the city; for another, they are undertaken in a great hurry; and furthermore they are also often closely related to his working life in another portion of the book trade:

> There came a day when I was in funds. I see myself hastening to Holywell Street (in those days my habitual pace was five miles an hour), I see the little grey man with whom I transacted my business – what was his name? – the bookseller who had been, I believe, a Catholic priest, and still had a certain priestly dignity about him. (34)

As in Islington High Street and in the 'labyrinth' later occupied by Charing Cross Road, the localization rewards attention. Although mourned today as a 'picturesque' site of 'immense character', in the late nineteenth century Holywell Street, lined with decayed, pre-Fire, seventeenth-century buildings, had a reputation as one of the 'worst streets in London'.[17] By the time of Gissing's death in 1903, also the year of publication of *Ryecroft*, Holywell Street was already doomed, it and the area around it slated to be removed as part of the construction of Aldwych and Kinsgway, a broad new road development lined by public buildings commemorating the imperial present.[18] As such, Holywell Street in itself represents a signifier for the fast-moving, past-obliterating present in this passage.

E.M. Forster's notion of an 'architecture of hurry', adapted for the title of the present volume from Chapter XIII of *Howards End*,[19] would not have surprised Gissing. He died 7 years before *Howards End* was published in 1910, but many times in his writing of the 1880s and 1890s Gissing thematized hurry, the subjective sense of pressure brought about by human time–space compressed by mechanization and the application to them of precisely calculated monetary values.[20] Nowhere is the centrality of hurry to Gissing's writing, and Gissing's importance to an understanding of the experience of hurry, clearer than in *Ryecroft*. Hurry seems stranger when viewed from a distance of decades and hundreds of miles, the distance, together with his later financial security, that provides the older Ryecroft with his perspective on a hurried, financially pressured youth.

The book is punctuated by references to the passage of time. These include reflections on the varying speeds with which time is experienced as passing during different phases of life, and verbs and time adverbials highlighting Ryecroft's habitual rapidity of movement as a struggling young man in London. Apart from 'hastening to' Holywell Street because he is afraid that another purchaser may have spotted the volume he wants, Ryecroft remembers how he 'hastened back' to a squalid Islington garret overlooking the canal through thick fog, following meals at a 'City Road coffee-shop' (34; 27). He later speaks of 'speeding home' from Battersea Bridge, having been inspired by a sunset visible from 'the old picturesque wooden bridge', and turning the experience into copy that he 'straightway' sent to a newspaper, then finding the very next day that the newspaper has printed his sketch. It will pay him 'a couple of guineas' for a piece of writing he 'enjoyed' doing (193). Time and money are thus linked in *Ryecroft* in ways that are multiple, complex and fraught.

Techniques for reading pedestrianism in modernity

For architectural and even urban historians, the streets of London exist primarily as landscapes of buildings, as planned and platted conceptions, or as resulting from a succession of property transactions.[21] But the streets along which the ageing Ryecroft imagines his younger self trudging might be better understood by means of the varied and multiple human uses of and passages through them. In Ryecroft's London past, he is presented by Gissing as forever walking, but also forever static. This is because, while in London as a hack writer, money prevented him from ever leaving the vast city. Contrasting with that trapped immobility in London is Ryecroft's later existence in the present tense of the book. This is leisured and situated outside London, in the English countryside. Financial sufficiency has enabled Ryecroft to travel in his later years as he never could earlier, both in rural England and abroad in continental Europe (188–90; 258–62). Reduced hurry can thus be combined with greater range, and hurry can emerge in modernity as a quality of financial insufficiency, as part of a life-stage devoted to the gathering of money.

Techniques needed to take account of pedestrianism in *Ryecroft* include those developed in studies of urban modernity – specifically as it was enacted in London – and in work on Gissing. Theorists of urban modernity, examining the gendered walker of the nineteenth-century city, have focused on leisured pedestrians capable of taking their time precisely because their days were not required to be spent at a specific place of employment, and nor were they chained to home by an occupation paid as piecework. Under the influence of Walter Benjamin, during the 1970s and 1980s, Marshall Berman and Richard Sennett drew attention to the figure of the (typically male) strolling, loitering observer represented in nineteenth-century Parisian writing and art as the *flâneur*; Janet Wolff, meanwhile, began critiquing the gendered exclusions and assumptions represented in such a figure.[22] By the 1990s, Judith Walkowitz, in the words of Wolff, was indicating 'the new spaces available for women in a redefined public domain in London in the 1880s'.[23] In the light of these critical discussions, it is worth stating that Ryecroft is not merely male; he seems unable to conceive of anyone female other than in the role of a housekeeper, such as '[t]his poor woman who labours for me in my house'. He repeatedly uses the locution 'the man' (in 'any toiling man', or 'a young man fresh from class-rooms', to take only two examples) to refer to those active in his world as other writers or putative readers. The urban poor he conceives of in an entirely masculine shape via 'muscular toil, the swinking of the ruder man' (43, 19, 37, 176). Attitudes similar to Ryecroft's are latent in much of Gissing's fiction.

Among the sparse readings of *Ryecroft* offered this century, Kevin Swafford relates the book to the late nineteenth-century Aesthetic Movement and specifically the idea attributed to Gissing here that decorous mourning, 'artistic integrity and sincerity of expression', as presented in this particular later work, are what he saw as the summit of his own literary achievement.[24] Alternatively, a more directly material approach to Gissing could build on information now available about his income during his first 12 years as a writer.[25] The vagaries – the mobilities – of wealth are important to a reading of *Ryecroft*. According to 'G.G.', in the playful preface to Ryecroft's papers:

> At the age of fifty, just when his health had begun to fail and his energies to show abatement, Ryecroft had the rare good fortune to find himself suddenly released from toil, and to enter upon a period of such tranquility of mind and condition as he had never dared to hope. On the death of an acquaintance, more his friend than he imagined, the way-worn man of letters learnt with astonishment that there was bequeathed to him a life annuity of three hundred pounds.
>
> (ix)

Fortune, for Ryecroft, can be understood in two parallel senses: in the sense of good fortune or chance, and in the financial sense of 'a fortune', specifically meaning the amount of money on which one could live leisurely for life. Gissing lived and wrote in an era when wealth in Britain was more broadly distributed than previously, but still held by a very small minority and transmitted, above all, by the instrument of the legacy and the writing of the will, thus insulating legatees from the dangers of a pre-welfare society.[26] Identified by 'G.G.' as 'the way-worn man of letters' (ix), Ryecroft has carried out hard work over an extended period of time and bears the marks of a traveller on a journey that has been extensive in spatial terms. Within this environment, Ryecroft has been freed by good fortune from the necessity of either working or living in London. As such, he has in effect become 'respectable' for life at a time when many members of the commercial and professional middle class in Britain struggled to do just that.[27] Without spouse or children, Ryecroft is not obliged, say, to live on considerably less than 300 pounds a year so as to save a legacy for others. Instead, he is free to enjoy and dispose of this income, approximately double the three pounds a week that, Gissing himself told his brother Algernon in 1883, 'would suffice to all my needs'.[28]

The fast (masculine) business walk: discursive histories of pedestrianism

Ryecroft's pedestrianism does include moments of stillness and peace, such as the August Bank Holiday walk in the sun in which he is able to appreciate the city at rest, temporarily empty, but these are exceptions. Typically, his pedestrianism is not that of a leisured flâneur: it involves physical hurry, bustle and the actual getting of money. It exists in the streets of London at a time when the poor still overwhelmingly travelled on foot. Things changed after the establishment of an extensive electric tram network, with tickets priced for the working classes, at the end of the nineteenth century.[29] But Ryecroft chiefly walks the main roads and in the more sedate, newer residential areas surrounding the West End of London. This zone should not be conflated with the noisy, smelly, plebeian street environment that Gissing described on the northern fringes of the City of London in *Workers in the Dawn* (1880) or south of the River Thames in *Thyrza* (1887).[30] These specific topographies originate in Gissing's personal experience and in the actual social geographies of 1880s London, but Ryecroft's pedestrianism also has textual antecedents. Gissing wrote an important book about Dickens and has sometimes been seen as a late Dickensian writer, one of the last of the three-volume realists in English fiction, as a new dichotomy between powerful demotic voices and art fiction emerged in the 1890s.[31] Alternatively, as for Peter Keating, he can seem a figure on the boundary between earlier and later modes of social-problem fiction.[32] In the 1890s, Gissing was still writing novels energetically. But the Gissing scholar M.D. Allen looks to the twentieth century when calling him something different, 'the first of the very greatest Dickensians'. A deliberately

Dickensian stance, after all, was a nostalgic position only emerging at the end of Gissing's life and just afterwards, for example in the writings of G.K. Chesterton, a patriotic, insular and politically conservative move.[33]

Walking in *The Private Papers of Henry Ryecroft* is seen as, ideally, leisured, yet also as a proper and decent aspect of English country life. Ryecroft nowadays need not go anywhere. He announces this early in the first section of the book ('Spring' – the four sections are the four seasons):

> The exquisite quiet of this room! I have been sitting in utter idleness, watching the sky, viewing the shape of golden sunlight upon the carpet, which changes as the minutes pass, letting my eye wander from one framed print to another, and along the ranks of my beloved books.[34]
>
> (3–4)

Ryecroft's rate of movement may have declined, but in the course of the book he details numerous country walks taken during the current phase of his life, supported by his annuity. These include both daily walks out from his Devonshire cottage, in all seasons, and trips to other parts of England on which he has walked between stays in inns. In Ryecroft's present, what Adrian Poole calls 'a perfectly defined distance between Self and the World' (that of the financially comfortable man in his study, going for walks when he chooses) has been established.[35] But Ryecroft oscillates between accounts of these country walks (interspersed with reflections on the present-day world) and memories of his past. Earlier in life, his existence in motion was inescapably meshed into the squalor and struggle of 'the streets', of existence in 'a great town' (10, 175). Ryecroft's identity as a literary character includes his middle-aged crankiness and his reactionary views on democracy and gender, facets of personality explaining why Gissing critics since the 1970s have been so shy of talking about him. But, like Georg Simmel's famous essay published in Berlin in the same year *Ryecroft* was published in London, Gissing's book is a study connecting 'The Metropolis and Mental Life'.[36]

In *Ryecroft*, leisure and work are both repeatedly conceptualized in relation to pedestrianism. As revealed by Roger Milbrandt's study of Gissing's finances in the 1880s, tutoring in the West End of London involved him in a complex network of pedestrianism as he moved from one house to another across main roads and the boundaries of districts.[37] Later in the book, in Chapter VI of 'Winter', Ryecroft's meditative, sitting position at teatime seems something earned, just as the (male) agricultural or industrial worker is entitled to sit beside the fire in his armchair when he has completed a day's labour (215–17). Here, Ryecroft highlights 'One of the shining moments of my day [. . .] when, having returned a little weary from an afternoon walk, I exchange boots for slippers', reflecting that, 'it is while drinking tea that I most of all enjoy the sense of leisure' (215). He has his housekeeper, who serves him his tea with care, bring him the tea table at the fireside, 'so that I can help myself without changing my easy position'. Then she retreats, most of her day's work done, 'to enjoy her own tea, her own toast, in the warm, comfortable, sweet-smelling kitchen' (217). But, in an earlier stage of his life, the walk was not the day's work but the leisure that succeeded the day's work. In Chapter V of 'Winter', immediately before the Devon teatime scene, when Ryecroft feels his pang of nostalgia for Islington High Street 25 years earlier, he sees himself setting out from lodgings after a day of writing in a garret room, 'walking with the quick, light step of youth' to the theatre (213). And the contrast is not just between the walk now and the walk then, but between different experiences of tea-drinking now and then. Tea, Ryecroft remembers, had earlier to be 'gulp[ed] down', 'hurried, often harassed, by the thought of the work I had before me', the tea-drinker 'quite insensible of the aroma, the flavour' of the tea (215). The teacup, too, could be an architecture of hurry.[38]

Walking to Holywell Street at 5 miles an hour for a book, Ryecroft is not a flâneur, nor a rural stroller, nor again an urban commuter-on-foot. His fast, masculine and businesslike walks, undertaken as part of his life in the book trade, have precedents in Dickens. Ralph Nickleby, uncle to the titular character of *Nicholas Nickleby* (1839), is a moneylender based on the unfashionable side of the West End in Golden Square, east of Regent Street. Early in the novel, Ralph announces that he is walking from there into the City of London (where he himself has dealings with richer financiers) that morning. His clerk, Newman Noggs, must, if a certain letter arrives, come and meet him on foot on the way back: 'I shall leave the city about that time and walk to Charing Cross on the left-hand side of the way; if there are any letters, come and meet me, and bring them with you'.[39] Ralph epitomizes the fast-moving early Victorian man of business, and not in a good way. He is contrasted with the Cheerybles, merchants whose mobility is not that of fast pedestrianism but of personal stasis lost in a 'quiet, little-frequented, retired' square in the City of London, small and unmodernized by railings or gravel, while their house makes fairy-tale fortunes doing business across the sea with Germany.[40] Ralph's business life, by contrast, is defined by moves between the West End, where his customers who aspire to fashion are and on the fringes of which his own house lies, and the City of London, the source of money for him, where he must go to raise it. Another Dickens creation is *Bleak House*'s Mr Bucket, a detective with masterly knowledge of London's characters, secrets and byways. He exerts pressure via the lawyer Tulkinghorn on Mr Snagsby, a law stationer who relies on Tulkinghorn's business, to join him on a walk of 'a couple of hours'' duration, to and from the notorious slum of Tom All Alone's in search of information from the boy crossing sweeper, Jo.[41] Ryecroft's own business walking in his remembered London years involved walks to and from editors, booksellers and pupils. It is precisely from such movement that Ryecroft has retreated. His later country walks fall into a different category. All of these walks exemplify mobilities articulated in modernity, specifically that of the gigantic London of the nineteenth century, pulsating with commerce, in which most must work for money for as many hours as necessary, and some are freed from this need by the possession of enough capital.

Feminist cultural studies of walking in the Victorian city have pointed out that leisured walking and observation were by no means confined to men, as the theory of the flâneur seemed to imply.[42] But the working pedestrianism of nineteenth-century women remains under-examined. Ralph Nickleby, arriving at the lodgings of his brother's widow on the Strand just east of Charing Cross, addresses his niece Kate: '"Now," he said, taking her arm, "walk as fast as you can, and you'll get into the step that you'll have to walk to business with, every morning." So saying, he led Kate off, at a good round pace, towards Cavendish Square'.[43] Given the chance, Ralph would terrorize others by means of the walk.[44] His aim is to terrorize Kate via a gendered pedestrianism. For Ralph, Kate's genteel femininity suggests that he can walk harder, faster, and further than she, and that she will fear being left unprotected in the street as a result.[45] Kate, in the novel, proves herself able to face this challenge, however, just as she is able to fight off the sexual predator Sir Mulberry Hawk. The massively multiple toil of the city resounds through *Nicholas Nickleby*, even more than through *Bleak House*, and at every stage it is tied up with rapid pedestrianism. The literary business, for Gissing's Ryecroft and Jasper Milvain, in the 1880s and 1890s, is entirely analogous with the more openly rapacious business which Dickens presented Ralph Nickleby operating half a century earlier. And yet there are differences, for, whereas Newman Noggs, Kate and Snagsby are all oppressed by a business superior (Ralph or Tulkinghorn) through the instrument of the business walk, Ryecroft does his alone.

These examples raise the question of whether the fast business walk, carried out in more or less of a hurry, ought to be read via notions of gender, or class, or both. Given that *flânerie* as a gendered activity has been deconstructed by feminists such as Wolff, why consider the sort of walk that Ralph and Ryecroft, and indeed Kate, take for monetary reasons as masculine at all? One answer would be that Ralph brutalizes Kate precisely by forcing her to enter a male-dominated street environment regularly, as a male commuter does. Numerous Victorian commentators, as Lynda Nead shows, were concerned that women walkers in the streets of London, if they did not fall victim to assaults on their physical bodies, would repeatedly have their being as women attacked by the visual images that would confront their eyes, for example in the notorious bookshop windows of Holywell Street, near where Kate and her mother lodge.[46] Still, in 1903, Gissing is able to have Ryecroft recall his business walks of youth in a way that separates them entirely from questions of decency and respectability, focusing instead on matters financial and existential, ways that perhaps would not yet at that stage have been available to a woman writer.

Town and country pedestrianism: walks for work, walks for leisure

Gissing presents a number of parodies and distortions of the business walks of Ralph, Newman, Kate, Bucket, and Snagsby in his own Ryecroft, ironized protagonist and hater of the modern business world. On one such walk Ryecroft goes to and fro along the City Road, Pentonville Road, Euston Road and Marylebone Road of today (the mid eighteenth century's first London bypass, called the New Road), to pick up a fine set of Gibbon's *Decline and Fall of the Roman Empire* on sale at a bookstall near Portland Road (today's Great Portland Street) station and cart them home to an Islington room east of the Angel (34–35).[47] So heavy and bulky are these volumes that Ryecroft has to carry them in two goes, making the journey there and back twice. The copy of Gibbon is his reward to himself paid for by missed meals and by his own static hackwork in his Islington garret. The physical toil involved in the walk is stressed: it leaves Ryecroft 'perspiring, flaccid, aching – exultant' (35). Another walk too is a sort of failure that is also a success. It extends from another rented room, this one on Hampstead Road north of Tottenham Court Road, to the Knightsbridge home of a wealthy pupil being tutored and back again, because the pupil is unwilling to rise for his class, and yet he pays for the class nevertheless.[48] The hack teacher's dream: paid and not having to teach, because the pupil cancelled late. Walking then back to Hampstead Road and working, fired with energy, the whole day long on the writer-scholar's real work, for which the teaching is mere necessary drudgery paying for bed and board.

Sometimes work is seated and leisure pedestrian; sometimes it is the other way around. Both of these London walks, as recalled by Ryecroft, enact and mimic, not the relationship between businessman and clerk as in Dickens's pairings of Ralph and Newman, Tulkinghorn and Snagsby, but, in a distorted fashion, the walks of the labouring man. Carrying the books along the New Road, then trudging back and retrieving the second load because he cannot carry the whole lot at once, Ryecroft mimics the lowest sort of delivery man or even a dray horse ('[s]ometimes I added the labour of a porter to my fasting endured for the sale of books'). Yet he does so in the pursuit of what he calls 'literature': writing, the experience and understanding of which liberate its reader from the pursuit of gain, embodied in the works of the father of English free thought, Gibbon. To achieve this noble goal, Ryecroft is prepared to make sacrifices: 'To possess those clean-paged quartos I would have sold my coat' (34). As in many ways, Ryecroft thus simultaneously rejects and participates in modern commercialism and acquisitiveness. His ardour for the object of his desire, the self-denial he is prepared to experience for it, resembles nothing more than the relentless struggle for money of a young Andrew Carnegie or John D. Rockefeller.[49] Yet Ryecroft's patriotism and conservatism compel him to desire, not money, but social standing within a British class system fetishizing the figure of the gentleman. On the trip from Hampstead Road to Knightsbridge, Ryecroft establishes himself as a gentleman rather than a boor for the simple reason that, walking to work and finding no work there, he is still paid. This differentiates him from London labouring figures such as the docker who arrives too late at the dock gates and walks home empty-handed.

Other London walks are remembered that involve beauty and art as a momentary release from the seated drudgery of the garret. Unlike Newman and Kate in *Nicholas Nickleby*, Ryecroft is a freelance, but, like them, he is still dominated by the ticking of the clock, if only because he must arrive on time to tutor pupils and deliver set numbers of words to editors for tight deadlines. Ryecroft recalls 'walking London streets by night, penniless and miserable', and hearing music from open windows in Eaton Square, 'one night when I was going back to Chelsea, tired, hungry, racked by frustrate passions. I had tramped miles and miles, in the hope of wearying myself so that I could sleep and forget' (135).[50] Here, as with the wealthy Knightsbridge pupil who slept in and cancelled the class for which he was still paid, Ryecroft is thankful: 'for an hour or so I revelled as none of the bidden guests could possibly be doing' because of the free concert in the street, which he, unlike the invited guests (he assumes), can truly appreciate (135). Another instance is his August Bank Holiday encounter with an empty and seemingly pristine London. For once unhurried, Ryecroft feels differently about the buildings of the city, finding 'a charm in the vulgar vista, in the dull architecture which I had never known' (88). London without people, either the raucously boorish poor or the smugly boorish wealthy, becomes aesthetically beautiful. 'Deep and clear-marked shadows' have more visual power 'when they fall upon highways devoid of folk'. This happens in the course of yet another walk that is not leisured, nor again straightforwardly connected to work.

In Ryecroft's present retirement, cushioned by money, to walk is a choice. Walking can, however, become the day's activity, balanced by the rest that follows physical exertion. As such, walking, for the retired, country-dwelling Ryecroft, can mirror the working day of the person who (unlike the older Ryecroft but like his younger self) must work. So Ryecroft, when 'a little weary', rests from this physical exertion by sitting in his chair waiting for 'the soft but penetrating odour which floats into my study with the appearance of the teapot' (215). And so, in the present tense of the book, apart from reminiscing about the past and responding splenetically to what he reads, he presents his central daily activity as walking in the different seasons of rural England. Here is the close of 'Autumn':

> Yesterday I passed by an elm avenue, leading to a beautiful old house. The road between the trees was covered in all its length and breadth with fallen leaves – a carpet of pale gold. Further on, I came to a plantation, mostly of larches; it shone in the richest aureate hue, with here and there a splash of blood-red, which was a young beech in its moment of autumnal glory.
>
> I looked at an alder, laden with brown catkins, its blunt foliage stained with innumerable shades of lovely colour. Near it was a horse-chestnut, with but a few leaves hanging on its branches, and those a deep orange. The limes, I see, are already bare.
>
> To-night the wind is loud, and rain dashes against my casement; to-morrow I shall awake to a sky of winter.
>
> (202)

Ryecroft can thus mark the passage of time in a fashion that seems free of hurry. And yet the hurry of his past always shadows his unhurried present.

A different sort of walk in the present is that of the past as it is projected in memory, a memory Ryecroft indulges precisely because of the money that enables him now to walk, sit or reflect when he wants. In the remembered past, Ryecroft had two roles, as aspirant writer and intellectual in 1880s London: in one with the walk as relief from labour, in the other with the walk as part of the drudgery.

First, he carries out hack writing that is badly paid but necessary for survival. Because of his poverty and lack of a position, this is carried out, not in a study or office, but in the same single room, cellar or 'the squalidest garret', sometimes 'pestilential' (26), where the writer sleeps. Then, walking out in the evening to the theatre (227) or 'in funds', marching to the book stall to buy his coveted Gibbon (34–35), the physically sedentary toil at the desk has its opposite when the work is done in escapes into the streets. These are themselves limited to London because there is insufficient money to leave London (and the lodgings are too bad to leave precious books in for any length of time). Second, like Gissing in the phase of his career highlighted by Woolf, the young Ryecroft often walked to work tutoring pupils in their homes. The walk would be from the cheapest lodgings in proximity to the homes of the rich, between those according to a timetable of lessons, and back again. The tutoring and the writing are also alike in that they are both freelance work paid for by multiple better-off customers, rather than employment requiring attendance at a place of work for set hours, but both keep Ryecroft on a leash; they restrain his mobility, because he cannot be very far at any time from pupils or editors.

Modernity begets conservatism

Although Ryecroft himself is a splenetic opponent of modernity in guises as varied as London, the contemporary inn, literary commerce, the education of children from working-class families and universal suffrage, *Ryecroft* the book is entirely a product of modernity and, far from denying or ignoring the changes of Gissing's lifetime, it calls attention to them. This side of *Ryecroft* has been compared to the concern in Thomas Hardy's novels with what Angel Clare views in Tess Durbeyfield as 'the ache of modernism'; it could equally well be compared, with its gnomic numbered subsections and reflections that are personal to the risk-taking limit of egotism, to Friedrich Nietzsche's *The Gay Science*.[51] In these books, Gissing and Nietzsche strive towards the formal departures today associated with the high modernism of the 1920s. In Gissing's *Ryecroft*, as in both Hardy and Nietzsche, the experience of modernity can be combined with both a massively expanded time range and a sense of bitterness or a callous indifference in relation to the present moment.[52]

For Gissing, via Ryecroft, literature as a business or profession seems to be squalid. He claims to write at a time when an interest in 'the processes of "literary" manufacture and the ups and downs of the "literary" market' have taken over from actual discussions of literature, clearly alluding to industry and the stock market (194–5). On numerous occasions, Ryecroft refers slightingly to writing that pays by the word, or to the profits made by publishers and hangers-on of the publishing industry, or to writers suited to coping with the brutality of that industry, such as 'A big blusterous genial brute of a Trollope', or Dickens, a 'shrewd and vigorous man of business', contrasting them with himself, 'in all practical matters [. . .] idle and inept' (195, 30). Perhaps the most striking example of Ryecroft's caustic view of his own former occupation comes when he recalls that, on a dark winter morning, 'sitting up in the dark, I heard men going along the street, roaring news of a hanging that had just taken place'.

> 'Execution of Mrs' – I forget the name of the murderess. 'Scene on the scaffold!' It was a little after nine o'clock; the enterprising paper had promptly got out its gibbet edition. A morning of mid-winter, roofs and ways covered with soot-grimed snow under the ghastly fog-pall; and whilst I lay there in my bed, that woman had been led out and hanged – hanged. I thought with horror of the possibility that I might sicken and die in that wilderness of houses, nothing above me but 'a foul and pestilent congregation of vapours'. (212)

Through Ryecroft, Gissing bemoans the fact that writing is a commercial business, associating this fact with the frightening accelerations of urban modernity and contrasting it negatively, in a manner clearly reminiscent not just of the conservative English Romanticism of Wordsworth and Ruskin but of much older modes,[53] with rural retirement. Despite his assertions to the contrary, Ryecroft forever sees literature in terms of its market materiality, in terms of its business, and the long-distance, high-speed walk becomes the physical figure for this, its symbol, its talisman. Ryecroft claims to be no businessman, no Ralph Nickleby. Yet he emerges as a deformed and decayed figure, lurking in his Devonshire bower. He complains about the state of the English inn, while benefitting from the mysterious inheritance (in effect, his benevolent patron, his Maecenas, to continue the Horatian analogy) that has freed him from the slavery of the pen and the canvasser offering him typewriting services, an inheritance perhaps derived from industrial or commercial sources. He praises the poor benighted servant he has in Devonshire, as well as the railway post, which swiftly brings him books from London. Ryecroft may be an older man, but, like numerous youthful lead characters in the young Gissing's fiction – notably Arthur Golding in *Workers in the Dawn*, Julian Casti in *The Unclassed*, Gilbert Grail in *Thyrza*, Adela Waltham in *Demos*, Sidney Kirkwood in *The Nether World* and Reardon in *New Grub Street* – he is both wise and a fool. The puzzle in each case relates to modernity and working for money: no one can avoid this unless money makes the avoidance possible; yet, perhaps as Ryecroft himself realizes, it is just as noble to charge around a dirty and rapacious city grabbing money, as it is to contemplate a beautiful fire in rural seclusion.

Conclusion

Ryecroft's move through the seasons, within the frame narrative provided by 'G.G.', is a move towards the close of his own life. Just before the end, he mulls over the clichéd expression '[t]ime is money' and turns it around, transforming it into 'money is time' (262). Here he finds 'a precious truth', regarding the 'glorious fire crackling and leaping' suddenly as an example of the 'material comfort' with which in his latter days he is blessed: 'What are we doing all our lives but purchasing, or trying to purchase, time?' What is the hard-working City man doing, whether working honestly or, like Ralph Nickleby, exploiting others, but trying to retire early?[54] This chapter has offered a contribution to the 'historical geographies of mobility' announced by Tim Cresswell, by illustrating how sedentary positions of stillness and pedestrian motion could both, in this era of print for mass consumption, railways, and competitive examinations as the gateway to membership of a new middle class, be associated with either leisure or work. It has opened up experiential dimensions of these moves and obstructions by examining feelings about them committed to paper and, crucially, dramatized through the device of a semi-fictional character reflecting on the relationship between past and present.[55] The present collection's contribution to the literature of mobility lies in the ability of 'hurry' as a concept to capture aspects of modern subjectivity peculiarly well. *The Private Papers of Henry Ryecroft*, this chapter has argued, is imbued with feeling-laden instances of hurry, each brought about by one-off, unique conjunctions of money and time.

Ryecroft is a 'socioeconomic outsider', at least in the early part of his career, yet sociological readings of Gissing have by and large overlooked this book.[56] Offensive as many of the opinions expressed in it are, *Ryecroft* needs to regain its place among the earliest non-fictional studies of urban modernity, alongside works by Simmel and Ferdinand Tönnies. It is a profoundly literary work in its complex, knowing games with authorship and editing, games played with an audience whose particular sophistications as readers have largely disappeared from sight, and these complexities are among the mobilities of the urban environment that produced it. Precisely through its rendering of felt experience, which is to say its literariness, *Ryecroft* adds considerably to an understanding of 'the rhythms of everyday life' in nineteenth-century London,[57] and as such makes the case that historical geographers and literary scholars need to interact more profoundly.

Hurry appears in the architectures of London: those that are strangely made beautiful on August Bank Holiday for Ryecroft, wandering alone through an empty city; those of Islington High Street, which at times he cannot separate from his own hurried youth. Gissing's special interest for students of urban modernity is found in the peculiar relish his writing expresses for scenes that might seem merely squalid or boring: in the life he gives to a world of hurry, without overly aestheticizing it. As such, although it might be tempting to condemn Gissing as an egotistical depressive who was becoming anachronistic in an era running into modernism, to do so would be a mistake. He represents a major cultural strand: the anti-town spiritual urge was remarkably strong throughout Britain and London's era of material pre-eminence, as evidence from Dickens's Wemmick (in *Great Expectations*) to the Hampstead Garden Suburb of Henrietta Barnett and Raymond Unwin indicates. Gissing is a powerful writer of town who is also always completely alienated from town, and as such represents quite precisely one British relationship to the urban. His account of the complex refigurations of the time–money relationship, additionally, is vital among accounts of the moment in modernity being produced in many contexts during the last decades of the nineteenth century, notably in Germany and the USA, as much as in England.

Notes

1 Katharina Boehm and Josephine McDonagh, 'Urban Mobility: New Maps of Victorian London', *Journal of Victorian Culture* 15 (2010): 184–200, 187. See Charlotte Mathieson, '"A Moving and a Moving On": Mobility, Space, and the Nation in Charles Dickens's *Bleak House*', *English* 61 (2012): 395–405; Josephine McDonagh, 'Space, Mobility, and the Novel: "The Spirit of Place is a Great Reality"', in *A Concise Companion to Realism*, ed. Matthew Beaumont (Oxford: Blackwell, 2010), 50–67.

2 Tim Cresswell, *On the Move: Mobility in the Modern Western World* (London: Routledge, 2006); Tim Cresswell, 'Mobilities II: Still', *Progress in Human Geography* 36 (2012): 645–53; Tom Hall and Robin James Smith, 'Stop and Go: A Field Study of Pedestrian Practice, Immobility and Urban Outreach Work', *Mobilities* 8 (2013): 4–5; Mimi Sheller and John Urry, 'The New Mobilities Paradigm', *Environment and Planning A* 38, (2006): 207–26; Mimi Sheller, 'Moving with John Urry', *Theory, Culture & Society* (2016), www.theoryculturesociety.org/moving-with-john-urry-by-mimi-sheller/.

3 James Buzard, *The Beaten Track: European Tourism, Literature, and the Ways to 'Culture', 1800–1918* (Oxford: Oxford University Press, 1983); James Buzard, *Disorienting Fiction: The Autoethnographic Work of Nineteenth-Century British Novels* (Princeton: Princeton University Press, 2005), 8–14.

4 Mathieson, '"A Moving and a Moving On"', 396. For another effort to spatialize *Bleak House*, based on a north–south geographical line running through England, see Jason Finch, *Deep Locational Criticism: Imaginative Place in Literary Research and Teaching* (Amsterdam: Benjamins, 2016), 57–62.

5 George Gissing, *The Private Papers of Henry Ryecroft* (London: Constable, 1921). Page references hereafter in parentheses.

6 James Winter, *London's Teeming Streets 1830–1914* (London: Routledge, 1993), 100–1.

7 Virginia Woolf, 'George Gissing', in *The Common Reader Volume II*, ed. Andrew McNeillie (London: Vintage, 2003), 221–22.

8 Frank Swinnerton, *George Gissing: A Critical Study* (London: Martin Secker, 1912), 61; Peter Keating, *The Working Classes in Victorian Fiction* (London: Routledge, 1979), 57.

9 Jerry White, *London in the Nineteenth Century: A Human Awful Wonder of God* (London: Vintage, 2008), 56. This area contained numerous ramshackle pubs and small halls where 'Soho radicalism' was focused (see Sarah Wise, 'The Eclectic Hall, Headquarters of Soho Radicalism', *History Workshop Journal* 83 [2017]: 289–300, doi: 10.1093/hwj/dbx009). Political meetings of this sort are important in Gissing's 1880 debut novel, as are lowlife drinking scenes set in the same neighbourhood (George Gissing, *Workers in the Dawn*, ed. Debbie Harrison [Brighton: Victorian Secrets, 2010], 478–92). The 'labyrinth' south of Crown Street, which became the northern section of Charing Cross Road, can be seen clearly on Edward Weller's 1868 map (http://london1868.com/weller41.htm).

10 Even the recent interest in formerly more obscure figures such as Margaret Harkness and Amy Levy, as keys to aspects of everyday gendered experience formerly overlooked, largely remains within analysis of the novel.

11 Gillian Tindall, *The Born Exile: George Gissing* (New York: Harcourt Brace Jovanovich, 1974), 41; Adrian Poole, *Gissing in Context* (Totowa, NJ: Rowman & Littlefield, 1975), 204.

12 Exceptions: M.D. Allen, '"Feeble Idyllicism": Gissing's Critique of *Oliver Twist* and Ryecroft', *Gissing Journal* 43 (July 2007): 26–32; Max Saunders, *Self-Impression: Life-Writing, Autobiografiction, and the Forms of Modern Literature* (Oxford: Oxford University Press, 2010); Kevin Swafford, 'Mourning, Pleasure and the Aesthetic Ideal in *The Private Papers of Henry Ryecroft*', *Gissing Journal* 38 (July 2002): 1–13.

13 Cited by Swafford, 'Mourning, Pleasure and the Aesthetic Ideal', 12–13.

14 Copies of the 1921 printing of *Ryecroft* in the library of Åbo Akademi University, Finland, and of the 1930 printing of the book in the library of the University of Tartu, Estonia, share this origin.
15 Saunders, *Self-Impression*, 129–31.
16 See the reflections in the introductory essay of this volume on multiple posts and daily editions of newspapers around 1900.
17 Philip Davies, *Lost London: 1870–1945* (Croxley Green: English Heritage, 2009), 22, 117; Patricia E.C. Croot ed., *A History of the County of Middlesex Volume XIII: The City of Westminster Part I* (Woodbridge, UK: Boydell & Brewer for the Institute of Historical Research, 2009 [Victoria County Histories]), 133.
18 Lynda Nead (*Victorian Babylon: People, Streets and Images in Nineteenth-Century London* [New Haven: Yale University Press, 2000], 168) notes that, for nineteenth-century observers of a progressive bent, Holywell Street 'represented a London that was rapidly disappearing, but not fast enough'. Thinking of architectures of hurry, the street and those around made up a literal blockage of circulation in the city. Visitors to London were jokingly said to have disappeared between Lincoln's Inn Fields and the Strand because of this knot of old, decaying thoroughfares. Conversely, as Nead (*Victorian Babylon*, 184) observes, Holywell Street before its demolition was 'a place of stasis and reverie' for ordinary Londoners, notably women, because of the multitude of shops in it that sold paintings and illustrated books and displayed these in their windows: a temporary respite from the city of hurry.
19 E.M. Forster, *Howards End* (London: Penguin, 2000), 93.
20 As discussed in the introductory essay to this volume.
21 On Aldwych, Simon Bradley and Nikolaus Pevsner, *London 6: Westminster* (New Haven: Yale University Press, 2005), 330–3; F.M.L. Thompson, *Hampstead: Building a Borough, 1650–1974* (London: Routledge, 1974).
22 Walter Benjamin, *The Arcades Project*, trans. Howard Eiland and Kevin McLaughlin (Cambridge, MA: Harvard University Press, 1999); Marshall Berman, *All That Is Solid Melts into Air: The Experience of Modernity* (New York: Simon & Schuster, 1982); Richard Sennett, *The Fall of Public Man* (Cambridge: Cambridge University Press, 1974); Janet Wolff, 'The Invisible "*Flâneuse*": Women and the Literature of Modernity', *Theory, Culture & Society* 2–3 (1985): 37–46; Janet Wolff, 'Gender and the Haunting of Cities (or, the Retirement of the Flâneur)', in *The Invisible Flâneuse: Gender, Public Space and Visual Culture in Nineteenth-Century Paris*, eds Aruna D'Souza and Tom McDonough (Manchester: Manchester University Press, 2006), 18–32.
23 Judith R. Walkowitz, *City of Dreadful Delight: Narratives of Sexual Danger in Victorian London* (Chicago: University of Chicago Press, 1992); Wolff, 'Gender and the Haunting of Cities', 20.
24 Swafford, 'Mourning, Pleasure and the Aesthetic Ideal', 12.
25 Roger Milbrandt, 'How Poor Was George Gissing? A Study of Gissing's Income between 1877 and 1888', *Gissing Journal* 43 (October 2007): 1–17.
26 Lesley Hoskins, Samantha Shave, Alastair Owens, Martin Daunton and David R. Green, 'The Death Duties in Britain, 1850–1930: Evidence from the Annual Reports of the Commissioners of the Inland Revenue' (History of Wealth Project Working Paper 1, 2014), https://historyofwealth.files.wordpress.com/2013/06/working-paper-1.pdf; Martin Daunton, *Wealth and Welfare: An Economic and Social History of Britain, 1851–1951* (Oxford: Oxford University Press, 2007).
27 David R. Green and Alastair Owens, 'Geographies of Wealth: Real Estate and Personal Property Ownership in England and Wales, 1870–1902', *Economic History Review* 66 (2013): 848–72. Gissing himself never got enough money with his pen – let alone from any mysterious legacy – to retire from the business of writing daily for money, although he gradually became more prosperous from the late 1880s onwards. His annual income always fluctuated; in the 1890s, when he was a well-established writer, it was similar

to that of a better-paid clerical worker, rather than that of a middle-class professional such as a clergyman or solicitor.

28 Milbrandt, 'How Poor Was George Gissing?', 10.

29 A network still in its infancy in the year that Gissing died and *Ryecroft* was published (Jerry White, *London in the Twentieth Century: A City and Its People* [London: Vintage, 2001], 12–13; Winter, *London's Teeming Streets*, 100–17).

30 Gissing, *Workers in the Dawn*, 5–8, 48–56; George Gissing, *Thyrza*, ed. Pierre Coustillas (Brighton: Victorian Secrets, 2013), 60–62; see Keating, *Working Classes in Victorian Fiction*, 58–63. See also Richard Dennis, '*Thyrza's* Geography', in Gissing, *Thyrza*, 560–7.

31 H.J. Dyos, 'The Slums of Victorian London', in *Exploring the Urban Past: Essays in Urban History by H.J. Dyos*, eds David Cannadine and David Reeder (Cambridge: Cambridge University Press, 1982), 129–53, 139.

32 Keating, *Working Classes in Victorian Fiction*, 53–92.

33 M.D. Allen, '*Bleak House* and *The Emancipated*', *Gissing Journal* 43 (October 2007): 17–27, 18.

34 M.D. Allen ('"Feeble Idyllicism"') claims to have found specific Dickensian roots for the rural portion of *Ryecroft* in *Oliver Twist*, but does not demonstrate this via specific verbal or thematic parallels. Taken together, the specifics of 1890s nature-writing (e.g. C.J. Cornish, *Wild England of To-Day and the Wild Life in it* [London: Thomas Nelson, 1895]; see Jason Finch, *E.M. Forster and English Place: A Literary Topography* [Turku, Finland: Åbo Akademi Press, 2011], 122) plus the general idea of the country as opposed to the town represented by the English West Country, as in at least two separate Dickensian cases (see below, n. 54) form more convincing analogues for Ryecroft's accounts of his walks and life in the English countryside.

35 Poole, *Gissing in Context*, 206.

36 George Simmel, 'The Metropolis in Modernity', in *The Blackwell City Reader*, eds Gary Bridge and Sophie Watson (Malden, MA: Blackwell, 2010), 103–10.

37 Milbrandt, 'How Poor Was George Gissing?', 11.

38 I am grateful to Richard Dennis for this observation about the transformation of Ryecroft's tea-drinking (and more generally for his expertise on London and on Gissing).

39 Charles Dickens, *Nicholas Nickleby*, ed. Paul Schlicke (Oxford: Oxford University Press, 1990), 9.

40 Dickens, *Nicholas Nickleby*, 468, 558–66.

41 '"And if you have no real objection to accompany Mr. Bucket to the place in question", pursues the lawyer, "I shall feel obliged to you if you will do so"' (Charles Dickens, *Bleak House*, ed. Norman Page [Harmondsworth: Penguin, 1971], 328).

42 Walkowitz, *City of Dreadful Delight*, 22–24, 41–80; Wolff, 'Invisible "Flâneuse"'. On (largely middle-class) women walkers in the streets, note Nead (*Victorian Babylon*); Deborah Epstein Nord, *Walking the Victorian Streets: Women, Representation and the City* (Ithaca, NY: Cornell University Press, 1995). Deborah Parsons, *Streetwalking the Metropolis: Women, the City and Modernity* (Oxford: Oxford University Press, 2000) concentrates on observational walks by female writers. In late Victorian and Edwardian fictions, for example Amy Levy's *The Romance of a Shop* (1888) and Arnold Bennett's *The Old Wives' Tale* (1908), respectable working women are sometimes placed within the boundaries of the shop, over which they can have some control. See Elizabeth F. Evans, 'We are Photographers, Not Mountebanks! Spectacle, Commercial Space, and the New Public Woman', in *Amy Levy: Critical Essays*, eds Naomi Hetherington and Nadia Valman (Athens, OH: Ohio University Press, 2010), 25–46. Women's pedestrian commutes as a multitudinous aspect of everyday life, rather than a cause for outrage or moralizing, or even a statement of independence, would reward more attention.

43 Dickens, *Nicholas Nickleby*, 121.

44 The ferocious speed at which Ralph himself walks is more than once emphasized by Dickens (e.g. Chapters 10, 44).
45 Dickens, *Nicholas Nickleby*, 127.
46 Nead, *Victorian Babylon*, 182–9.
47 Gibbon was crucial to the development in England of a modern view of the classical past, free-thinking and sceptical about all religions, Christianity as much as Paganism. In narrating its decadence and decline, Gibbon chose to examine the era when the Roman Empire was hurrying towards its demise, a choice that was resonant in the era of the *fin de siècle* and 'decadence' during which Gissing wrote. And yet, for Ryecroft, and presumably for Gissing himself, Gibbon also represents an escape from the hurry of modernity: in the massive volumes of the folio *Decline and Fall*, the opposite of small, cheap, portable reprints designed for reading on trains, in the single-minded concentration that reading them requires, and the sense of transport to another world that reading them gives.
48 The distance from the middle of Hampstead Road (running north–south) to the middle of Knightsbridge (running east–west) is approximately 3 miles. Assuming this, and Ryecroft's vaunted pace of 'five miles an hour', the walk there and back would have taken him 72 minutes.
49 The young Gissing, after an early disgrace in Manchester and poverty in London, went as far as Boston and Chicago seeking his fortune, but it did not work out.
50 The young Ryecroft (like the young Gissing in the account given by Milbrandt in 'How Poor Was George Gissing?') was always able to work for a living and support himself to the extent of having a bed for the night. In this sense, his tramp differs from that of the homeless who 'carry the banner' in Jack London's *The People of the Abyss* (New York: The Macmillan Company, 1903), 113, who must 'walk the streets all night', having failed to secure a bed in the cheapest lodging houses of the city. *The People of the Abyss*, like *Ryecroft*, was first published in 1903. I am grateful to Phillip Gordon Mackintosh for pointing out the comparison.
51 Swafford, 'Mourning, Pleasure and the Aesthetic Ideal', 1. Friedrich Nietzsche, *The Gay Science*, ed. Bernard Williams (Cambridge: Cambridge University Press, 2012).
52 On massive temporal expansions in modernity and its art, see Charles M. Tung, 'Baddest Modernism: The Scales and Lines of Inhuman Time', *Modernism/Modernity* 23 (2016): 515–38.
53 *Ryecroft* is to a considerable extent a formal imitation of the *Epistles* of Horace put into a modern setting.
54 Or perhaps, like another peculiarly unpleasant Dickensian businessman, Mr Vholes, 'supporting an aged father in the Vale of Taunton – his native place' (Dickens, *Bleak House*, Chapter 37), buying time for another who has himself invested time in the London money-maker. In all three cases, the Vholeses', the Nicklebys' and Ryecroft's, the other of hard-nosed London is the West Country, often used discursively as an epitome of rural England.
55 Cresswell, 'Mobilities II: Still'.
56 Bart Keunen and Luc De Droogh, 'The Socioeconomic Outsider: Labour and the Poor', in *The Cambridge Companion to the City in Literature*, ed. Kevin R. McNamara (Cambridge: Cambridge University Press, 2014), 99–113, 108.
57 Alastair Owens, Nigel Jeffries, Karen Wehner and Rupert Featherby, 'Fragments of the Modern City: Material Culture and the Rhythms of Everyday Life in Victorian London', *Journal of Victorian Culture* 15 (2010): 212–25.

11 'We're going to move . . . I can't rush backwards and forwards, I'll go mad – I am sure of it.'

Representations of speed and haste in English life-writing, 1846–1958

Colin G. Pooley and Marilyn E. Pooley

Introduction

The feeling that there is insufficient time in the day to complete all the things we consider necessary, and seeking to hurry between places and to rush tasks that might be more pleasurable (and executed more efficiently) if done more slowly are normal human experiences. Although the need to hurry and compress many activities into a limited period of time has undoubtedly been felt by some throughout human history, it has also been argued that processes of urbanization and 'modernization' of society, economy, transport and communications – in other words, the development of architectures of hurry – have increased the pace of life and led to new stresses and strains in urban living through a process of 'time–space compression'.[1] It is also argued that one of the key characteristics of such time–space compression is the unequal power relations that exist within society: those with the least power experience the most deleterious effects of pressure and hurry, and fail to share in the benefits of a more connected society.[2] Such processes may be observed at a range of scales, from the global to the local.[3] The processes of time–space compression have been particularly associated with the development of a late twentieth-century globalized economy and society, where global interconnectedness reduces (though does not eliminate) the constraints of time and space, but similar processes may be observed in the past, as transport speeds increased and many communities became more easily connected to a wider world.[4]

In this chapter, we examine the speeding up of time and space through the eyes of selected travellers who lived in Britain during the century after 1840. Whereas accounts of changes in transport technology, urban structure and workplace regimes are relatively common and of long standing,[5] there are far fewer accounts of how such shifts in economy and society were experienced by travellers. Although the scope of our investigation is necessarily limited by the sources available, we assess the ways in which past travellers engaged with new, faster transport technologies, the situations that created stress when travelling, and the role of time–space compression in producing such feelings. We also examine selectively the experiences of men and women of different classes and in varied parts of the country, together with any evidence for change over the period studied. Most of our diarists lived for at least part of the recorded time in urban areas, and all travelled to and interacted with towns and cities. We argue that, by the mid nineteenth century, rural and urban areas were already closely connected, and that new architectures of travel and hurry which had developed first in urban areas could have an impact on the lives of most travellers, wherever they lived. Evidence is partial and cannot be deemed conclusive, but it does provide pointers to the ways in which past processes of time–space compression were experienced by individual travellers in Britain.

Anyone living in Britain from the 1850s to the 1930s would have seen massive changes in the transport available to them, although the extent to which this could be accessed varied by location, social class and gender. In the mid nineteenth century, the principal means of travelling from place to place were by road (on foot, on horseback, by cart or by carriage); by water (on inland canals and rivers or on coastal steamers); or increasingly by rail, which, by 1850, connected most major towns and cities, although smaller communities in rural areas were for the most part not linked into the national rail network until much later in the century. During their lifetime, they would have witnessed urban transport transformed by the development of first steam and then electric trams and, towards the end of their life, witnessed the expansion of travel by bicycle, motor bus and private car. In the early twentieth century, vehicles powered by horses, humans and motors all vied with each other for space on the streets of Britain's towns and cities.[6] Transport history is often presented as a linear story, with new developments (such as motor cars) displacing older forms of transport. However, in practice, old technologies and means of travelling usually persisted alongside the new, and many people remained excluded from access to the fastest and most modern technologies and infrastructures.[7] In this chapter, we seek to examine how selected informants engaged with this changing transport landscape in the century after 1840.

The analysis we present is informed by contemporary mobility theories as developed initially by Mimi Sheller and John Urry, and subsequently refined by many scholars.[8] We draw especially on the following five characteristics that are central to much research on mobility: first, the assertion that mobility in all its forms (of people, goods and ideas) is so embedded in everyday life that it is a major factor that shapes society, economy and culture, rather than being a product of it; second, the belief that, for an individual traveller, mobility has meanings that extend well beyond the simple act of moving from one place to another – the journey itself has significance and meaning; third, the suggestion that experiences and, especially, expectations of mobility have changed over time, so that assumptions of speed and relatively unrestrained movement became strongly embedded within most societies by the late twentieth century; fourth, the way that mobility research has focused on the impacts and influences of new sites of mobility, such as international airports or motorway service stations; finally, and most crucially, the ways in which human movement is experienced by an individual and the impacts that it has on their everyday life. In this chapter, we seek to utilize some of these concepts in a historical context.

Using life-writing to study mobility

Contemporary mobility research is often conducted through the use of in-depth interviews, accompanied journeys and travel ethnographies, thus constructing a detailed picture of the ways people engage with everyday travel.[9] Clearly this is not possible in a historical context but, by using life-writing, especially diaries or letters written contemporaneously, it is possible to gain some insights into the experience of everyday travel. Life histories and autobiographies, though usually more readily available, are more problematic in that they were mostly constructed long after the events recorded and thus may be particularly influenced by the distortion of memories over time.[10] Much the same is true for oral histories that may be used for the more recent past.[11] For these reasons, we focus only on diaries written, so far as we are aware, on an almost daily basis. A selection of nine diaries is used, drawn from some 50 items of life-writing that we have studied so far. Brief details of the diarists are given in Table 11.1, and further contextual information is provided in the text.

Diaries are, of course, not without their difficulties of interpretation, and these have mostly been well rehearsed elsewhere.[12] By definition, diaries were written only by those with sufficient leisure time and literacy, many were destroyed later in life, and there are no ways of knowing what criteria (if any) were used in decisions to write or retain a diary. Similarly, we do not know what diarists omitted from their record of daily events, but it is reasonable to assume that unusual events were recorded more often than the mundane. This is a problem for the investigation of everyday travel, as much of this was indeed repetitive and mundane in nature. Diaries appear to be kept more often by women than by men (men were more likely to write autobiographies), and are most common for adolescent girls and young women. It is thus likely that the demographic profile of the authors of surviving diaries is skewed. In our research, we have avoided using diaries of the rich and famous that may have been written with a view to future publication, but have concentrated on those written, so far as we can tell, with no thought to any future readership. Analysis of any life-writing is time consuming. Most diaries consulted were hand-written manuscripts, often hard to read and very slow to transcribe. However, there is no substitute for the careful reading of each diary, recording those items relevant to the research in hand while also building a contextual picture of the life and times of each diarist.

Reactions to new and faster forms of transport

When viewed from the perspective of an individual traveller, newness is a relative concept structured by age, gender, social group and location. The fact that a form of transport existed does not mean that an individual traveller had access to it. Thus someone growing up in a remote rural area in Britain might have been aware of the development of the railway network, but may not actually have

Table 11.1 Brief details of the diaries used in the study

Name of diarist	*Occupation*	*Date of birth*	*Location during diary*	*Dates of diary*	*Location of source*
John Leeson	House proprietor	1803	London	1846–65	Bishopsgate Institute archive (GDP/8)
John Lee	Apprentice draper	1842	Lancashire	1859–64	Private collection
James Bennetts Williams	Tin/copper mine worker	1856	Cornwall and Liverpool, then South America	1883–7	Bishopsgate Institute archive (GDP/57)
Mary Anne Prout	None. Limited domestic duties	1861	Cornwall	1882	Bishopsgate Institute archive (GDP/58)
Ida Berry	None. Limited domestic duties	1884	Manchester	1902–7	Bishopsgate Institute archive (GDP/28)
Freda Smith	None	1887	London, Oxfordshire, Northumberland and elsewhere	1904–14	Bishopsgate Institute archive (GDP/99)
Gerald Gray Fitzmaurice	Lawyer	1901	London and elsewhere	1926–7	Bishopsgate Institute archive (GDP/52)
Annie Rudolph (Rudoff)	Art school and assists in father's shop + domestic duties	1905	London	1923 (+ later summary)	Bishopsgate Institute archive (GDP/31)
Gillian Caldwell	School, college, assists in parents' hotel, foreign travel agent and secretarial work	1937	Eskdale (Cumbria), Zurich and Edinburgh	1952–8	Bishopsgate Institute archive (GDP/1)

travelled on a train until later in life, when they moved to a larger community with rail connections. Even if a form of transport was in theory available, it does not mean that it was used by all: in Britain, motor cars remained the preserve of the wealthy during the first half of the twentieth century, and most people, despite seeing cars on the road, would have travelled rarely if ever in a private motor vehicle (see Mackintosh in this volume). Access to different forms of transport could also be strongly gendered, as both bicycles and motor cars were much more likely to be available to, and used by, men than women. Social norms and societal constraints meant that many women did not use some of the fastest and most convenient forms of transport as quickly or easily as their male counterparts. Reactions to new forms of transport could also vary with age and previous experience: for instance, the way in which someone in their 80s reacted to and engaged with the coming of the railway possibly was quite different from that of a younger contemporary who had grown up during the development of rail travel. Though there is clear evidence of variations in travel mode by factors such as age and gender in the twentieth century,[13] there is little direct information on how and when individual travellers encountered and engaged with different forms of transport. Although clearly highly selective and limited in scope, life-writing can provide one avenue through which these themes may be explored at an individual level for periods prior to the present or recent past when oral and survey data may be utilized.

All the diarists whose travel has been analysed for this chapter made some use of railway travel. Although it did not supplant all older means of travel (for instance by boat, on foot or by horse power), it did provide a fast, convenient means of travelling over longer distances that appeared to be enthusiastically embraced by those who encountered it. This was the case for diarists living in both urban and rural areas (although those living in the countryside usually had to undertake a longer journey by other means to reach the nearest railway station), for both men and women, and for diarists of all ages and at all times, from the 1850s to the 1950s. From its early development, travel by train appeared to be welcomed, with appreciation of its speed and convenience and few concerns about either its impact on the environment or any risks from travelling on the railway.

John Lee was born in 1842 and kept a surviving diary from 1859 to 1864. He lived initially in north-east Lancashire and then on Merseyside, working as a draper's assistant. In common with most life-writing, his diary probably does not record all his everyday journeys, but it is clear that the railway was his normal means of travel whenever he moved outside his local area. As he grew up, he would have seen the railway network develop in East Lancashire, and travelling by train around Lancashire and across the Pennines to Yorkshire was a regular feature of his diary, with on average about seven such journeys recorded explicitly each year. Most such travel was related to leisure, longer work-related travel or visits to relatives, but all journeys appeared unremarkable and were recorded in a matter-of-fact way that suggested that they were a routine part of his everyday life. The following two extracts from 1860 are typical: 'Went by the first train to Burnley to get some of my school books &c out of my large box';[14] 'walking, visiting friends, then train to Skipton, to

Burnley for agricultural show'.[15] John Leeson (born 1803) was much older than Lee and, as a house proprietor in London, lived a much more affluent lifestyle in the capital. However, despite these differences of age, social class and location, his diary (kept 1846–65) shows an equally comfortable relationship with the railway. Leeson was a young man, living at home with his parents in London, when the rail network first developed. He would have already established a pattern of travel, and would have had to adjust himself to this new and faster means of transport. However, by the 1840s, travel by train was established as a normal part of life for himself and his family for trips outside London, mostly to visit relatives in Norwich and for leisure travel. In 1847, he even noted that his 76-year-old mother travelled by train (probably for the first time) to visit family in Norwich, commenting specifically that she enjoyed the experience. Two extracts again illustrate the matter-of-fact way in which travel by train for leisure and family visits was recorded: 'Left London and I went by Railway from Euston Square to Derby and Ambergate, Matlock, to Buxton, got there at 5;'[16] 'Left London with Mrs Leeson and went by Eastern Union Railway to Norwich, very foggy'.[17] These examples from the mid nineteenth century suggest that travel by train was quickly adopted by people of different ages and social backgrounds, living in very different parts of the country. Unsurprisingly, later diarists were equally comfortable with rail travel when it provided the most convenient option.

From the late nineteenth century, the bicycle could provide quick and convenient travel over relatively short distances, and it could be used as a substitute for walking (where it had the advantage of speed) or public transport (where its main advantages were privacy and flexibility). Before the 1920s, bicycles were too expensive for most families on low incomes, and cycling was mainly undertaken by the middle classes (and above) for leisure purposes.[18] Although cycle use was greater among men than women, our diary evidence supports other studies showing that, from the late nineteenth century, many young women also cycled and enjoyed the speed and independence that it gave.[19] Ida Berry (born 1884) lived in south Manchester and kept a surviving diary from 1902 to 1907. She and her sister were keen cyclists, going out most weeks in summer and regularly undertaking rides of some 15 km, with occasional longer rides of up to 80 km. Almost all such trips were for leisure. Cycling had multiple attractions for Ida and her sister: they enjoyed the exercise and fresh air, revelled in the speed they could attain on downhill stretches, often (though not always) had the company of male friends, sometimes interrupted the ride with a walk, and frequently incorporated tea and cakes at a convenient café into their excursions. Two extracts from Ida's diary of 1906 illustrate these points: 'Glorious day, just like summer. We brought the bicycles down and went for a lovely ride through Cheadle, Gatley and Northenden';[20] 'Maud and I cycled to Alderley Edge after tea, it was a glorious ride we scorched home'.[21] Cycling seemed to give Ida and her sister a degree of freedom that they may not have gained so easily elsewhere at that period.

Freda Smith (born 1887), whose surviving diaries date from 1904 to 1914, lived with her parents in London, but frequently stayed with her paternal grandmother and aunts at Britwell Salome in rural Oxfordshire. For Freda, cycling (always with a relative or other chaperone) provided a convenient and speedy means of transport in the countryside, as well as a source of exercise, allowing her to socialize with family friends or to access services in larger settlements within a radius of about 8 km of Britwell House, although she also walked or was driven (by carriage or car) to such locations at other times. Freda cycled in all seasons; for example, in February 1904, she wrote: 'M and I bicycled to Watlington in a great hurry to buy goloshes for this afternoon's hockey as the lawn was very soft'.[22] Occasionally, Freda undertook longer cycle rides for leisure purposes (though nowhere near as long as some undertaken by Ida Berry). In August 1904, having bought a new bicycle in London, she 'was seized by a sudden desire to spend Sunday at Windsor'; she and Aunt G travelled from Britwell to Windsor: 'then bicycled by Datchet, Old Windsor and in the Park', before staying the night at the White Hart; next day they 'bicycled to Virginia Water. Lovely ride through the Park . . . well worth the total of 10½ miles'.[23] In both instances, there is the suggestion that these journeys were undertaken on an impulse, and that any hurry was as much about the process of decision-making as it was about the journey itself.

Engagement with private motor cars was much rarer for our diarists during the early years of motoring: even those from relatively affluent families rarely travelled by car; none drove regularly themselves, and so they relied on other family members or friends if they were to travel by car. In common with bicycle use cars were used mainly for leisure activities in the early decades of the twentieth century in Britain. Drivers and passengers revelled in the kinaesthetic feelings of speed that could be achieved on the open road once early restrictions were lifted, and the private motor car opened up new horizons and destinations for leisure and pleasure travel. They were also often associated with romance. Travel by car did not have to have a purpose: going for a ride was itself sufficient reason to drive.[24] For instance, while staying at some family friends' house party at Neasham Hall, near Darlington, in 1907, Freda Smith and a group of other young adults 'suddenly at 4.26 . . . motored off to Harrogate in his car (William Cox's) a perfect Siddeley. I never had such a ripping run . . . 30 in abt 1¼ hours. Tea at the Majestic. Left H. 6.30 home at 8'.[25] By then, Freda had been sufficiently interested in cars and driving to have had a 'Motor lesson' while staying at another house party in Burton-on-Trent earlier that week, and by 1907 travel in the cars of Freda's older relatives, or of her contemporaries, was a frequent and normal means of transport for her to and from social events, both in rural areas and in London.

There has been a formal age restriction on driving in Britain since the 1930s,[26] and driving a motor vehicle is the only form of private transport that is restricted in this way. This means that there is a constant flow of fresh generations of young people for whom the experience of driving is new. In this case, engagement with the new is unrelated to when the technology was introduced, but is controlled by the age at which someone is legally allowed to engage with it (although once cars became common, most young people would have experienced motoring as a passenger, and some would have driven informally before they were legally able to take a test). For much of the twentieth century, access to a private motor vehicle has been especially important in rural areas, because public transport options were more limited than in towns, and this is well illustrated in the diary of Gillian Caldwell (born 1937), who grew up in the remote valley of Eskdale in the English Lake District. Gillian was 15 years of age when the diary began in January 1952, but, aged 17, she left home to work (briefly) in Zurich and then more permanently in Edinburgh. Although Gillian had a number of transport options in Eskdale at this time, including a narrow gauge, mainly tourist, railway and buses that could connect (often poorly) with the national rail network, her father (who ran a hotel and restaurant) had a car, and, whenever possible, Gillian persuaded either her father or various male friends with cars to provide her with lifts. By this stage, access to a car for everyday travel was increasingly viewed as essential in a remote rural environment such as Eskdale, though it was still largely a male preserve. Nonetheless, the car also retained a strong leisure and romantic association for Gillian and her friends, with feelings of freedom and speed clearly associated with pleasure and excitement. When at home in Eskdale, Gillian and her (mainly) male friends regularly drove out to local beauty spots such as Wastwater, taking a plentiful supply of alcohol, to sit by the lake chatting and drinking. For instance, in July 1955, Gillian was back in Eskdale for the weekend and described a day touring in the Lakes, with plentiful stops at inns along the way, finished off as follows: 'we dropped into Eskdale just in time for dinner. Pa . . . packed us off to Wastwater with some booze at midnight'.[27]

There is only one recorded instance (also in 1955, when Gillian was home from Edinburgh) when she drove, somewhat reluctantly and certainly illegally, but it again illustrates the continued association of motoring with pleasure and romance at a time, and in a location, where travel by car had also become important for more mundane everyday transport (though this event could also be interpreted as a dominant male irresponsibly placing his female companion in a potentially dangerous situation):

> Norman, who had had a skin full of brandy invited Pooh [female friend] and I to go over Hard Knott with him. It was a magnificent sight – never have I seen it looking more forbidding & threatening. Coming down he said he was tired of driving so gave the wheel to me and down the pass I drove! Extremely amusing and several motorists were scared off the road. . . . I was terrified of coming across a policeman![28]

While in Edinburgh Gillian also had male friends with access to cars that were used almost entirely for pleasure: most everyday travel was on foot or by public transport. In 1957, Gillian had a regular boyfriend, but was tempted also to go out with another young man whose attractions included a smart, fast car. It seems he tried to impress Gillian with the speed of his driving, though this very nearly led to serious consequences:

> What a surprise. Just after breakfast the doorbell rang and there was KEN MARSH plus supersonic new green sports car – an A.C. Great fun. . . . Ken and I drove out to Duleton and spent the afternoon on the beach. What a fascinating rogue he is. I wish I had no morals! We drove back at high speed and were stopped by the police outside Portobello for speeding. They were very decent and let us off.[29]

Flying is the fastest form of transport, but air travel remained restricted to a very small elite (and those in the Air Force) before the 1950s, and in Britain it did not become a form of mass travel until the 1980s.[30] However, long before this, air travel could be particularly alluring. Gerald Gray Fitzmaurice (born 1901) was a young London lawyer and, in 1927, he visited an air pageant at Hendon. In his diary he recorded detailed information about different planes and wrote: 'It was an excellent show & makes me long to fly'.[31] Gillian Caldwell also enthused about a visit to Heathrow airport in 1953, and the following year Gillian herself experienced flying. As she had left school, and at the time was without work, her father arranged for her to go to Zurich to work for a few months for an English travel agent. She flew from Manchester (Ringway) airport on a '60 seat Swiss Constellation' and clearly enjoyed the experience, although she seemed to be as interested in the airport procedures as she was in the flight itself, with apparently little awareness of the speed at which she would have been travelling: 'I thoroughly enjoyed the flight and never noticed the take-off, though when we were coming in to land you could feel the plane losing height'.[32]

At all times from the 1840s to the 1950s, people were engaging with new and (for them) faster forms of transport. However, at the same time, it is important to remember that many everyday journeys continued to be undertaken by the oldest and slowest means of transport (walking); that travel was often multi-mode, with travel by train often preceded by a walk or a bus ride to the station; and that horse-drawn vehicles continued to compete for road space with motor vehicles for several decades in the twentieth century. In many rural areas, walking to access public transport was a necessity. For instance, the diary of Mary Anne Prout, written in 1882 when she lived in a small village near Truro in Cornwall, frequently recorded the comings and goings of relatives and visitors, who almost always had to walk several kilometres from the nearest railway station: 'Mr Henwood left about dinner time. Mrs Mitchel from Hayle came in just before he left she walked from Scorrier station this afternoon'.[33] Rural bus services in the 1920s could also be slow and unreliable, as recorded by Gerald Fitzmaurice when he was travelling in the south west of England: 'Annie and I left Charmouth together by the 8.55 bus to Axminster. It stopped so often en route that it undoubtedly would have missed the train if the train itself hadn't been half an hour late'.[34] Although in some instances speed and new forms of transport (bikes and cars) were associated with leisure and romance, the diaries contain just as many instances where it was the slow and leisurely experience of walking (especially at night) that was crucial to courting and romance, as in this extract from the diary of Ida Berry:

> It was a beautiful 'starlight' night and Norman took me for a walk through Spath Road etc. and behind the 'Park'. As we were coming down Northen Grove we saw a large 'Comet' fall from the sky, it looked lovely, just like a large rocket.[35]

At the start of this section, we suggested that newness (with regard to transport) is a relative concept. It is clear from the diary extracts that speed is also a relative concept. It depends very much on what transport speeds have been encountered before, and what one's expectations are. It also varies with physical exposure to the environment and the kinaesthetic experience of speed. Thus, on a bicycle, with wind in one's face and little protection from the elements, feelings of speed can be intense (as reported by Ida Berry). Much the same would be true in an early open-top car or on a motorbike. In a more enclosed space, feelings of speed are diminished, although sensations of speed can be generated in a car or on a train by watching the outside world flash by. In contrast, although flying is by far the fastest form of transport, sensations of speed are minimized owing to the passengers' insulation from the outside world and the lack of external markers against which to judge speed. The lack of personal control in a plane or on a train may also lessen perceptions of speed compared with travelling by car, where the driver has personal control (though the experience of a passenger may be subtly different). This section has focused on how travellers engaged with new (and potentially faster) forms of transport, and on the feelings and emotions sometimes associated with such journeys. However, it has ignored the main practical advantage of speed: the ability to get somewhere in a hurry.

The production of haste and stress in everyday travel

Hurry is almost always conditional on context: the need to travel more quickly than usual from one place to another is usually generated by external circumstances, often not in a traveller's control. Thus, a journey for which a traveller might initially have allocated ample time becomes rushed because of transport delays elsewhere in the system; family commitments might delay departure, thus leading to the need to hurry; or over-commitment might create a context in which it feels necessary to hurry all the time. Haste is thus often produced by circumstances independent of the transport mode available; hurried journeys can take place on foot, by bike, on public transport or in motor cars, and have occurred to some extent in all time periods. The title quote used for this chapter is taken from the diary of Annie Rudolph, a young woman who lived with her family in London in the 1920s, and it illustrates well the way in which changed family circumstances could create pressures that necessitated an unwelcome need to hurry. Annie was 17 at the time of writing and had previously lived a fairly carefree existence, attending college part-time and working with her father in their East End shop. However, following her mother's death, all this changed, and, as the oldest daughter at home, she was expected to take responsibility for running the home and caring for her younger siblings while continuing to work in the shop (her father relied on her to do most of the paperwork associated with the business). The journey from her home to the shop was about 5.6 km and was usually undertaken by tram. It could not be speeded up, and the changed circumstances meant that this became intolerable; she always felt rushed and unable to fulfil any of her duties properly, leading to her father's decision to move back to the East End. In the end, they did not move, as Annie's younger sister took on some of the load, but clearly for a while the journey was becoming intolerable and created significant additional stress in already difficult circumstances.

Although Annie's over-commitment and consequential need to rush from place to place was generated by family circumstances beyond their control, in other instances speed could be necessitated by the inefficiencies of the individual involved or of family and friends. Oversleeping before a journey, or simply neglecting to appreciate the length of time a journey might take, could produce considerable stress and the need to rush. This could occur in all time periods and across all social groups. John Lee travelled regularly by train and usually caught his preferred train without difficulty. However, there were some occasions on which he left insufficient time for the journey from home to the station, leading to the need to rush, not always successfully, as shown in the following example: 'Got up to go by the six o'clock train to Ripon, but I was about five minutes too late. I fortunately got to ride in a dray'.[36] His inability to catch the train, despite no doubt rushing, led to him having to travel by much older and slower means. Almost a century later, and with rather different transport options available, Gillian Caldwell also found herself rushing to catch a train, owing to the failure of the person she was relying on for a lift to the station to be ready on time. Living in a rural area with limited connections, for her journey to Edinburgh she persuaded another friend to drive to the station at great speed:

> Up early and Pooh [female friend] and I were all ready to go by 10 but Charlie certainly wasn't and by 10.15 I was getting worried so we piled into Ray's Morgan & he drove at a maniacal speed all the way to the station. We just made it. We were late into Edinburgh and thoroughly cheesed off.[37]

In this instance, having rushed for the train, the subsequent delay on the journey appears to have been especially frustrating. One final example comes from the diary of Annie Rudolph in 1923. She was travelling (alone) by train from her home in north London to stay with friends in Berkshire and was expecting to meet her elder sister who would accompany her across London. When her sister failed to turn up, Annie was forced to complete the journey unaided and at great speed, to just make the train on time. She concluded her detailed account with the words: 'Gee what a rush I was so puffed'.[38] Such situations are largely independent of the transport infrastructure itself.

Of course, there were also occasions where the need to rush between connections (or to wait a long period for another) was due to delays in transport caused either by technical failures or adverse weather,[39] but for the most part urban travel as reported by the diarists was relatively unproblematic; for instance, Ida Berry easily moved around Manchester in the first decade of the twentieth century and used a mixture of different transport systems that seemed to connect easily with each other and rarely necessitated haste. One example from 1905 is typical:

> After dinner we all went by train to Chorlton, and then walked to Stretford and then we got on the top of the bus, and went to Urmston . . . Coming home we had a cab from Stretford station to Chorlton station and caught the three minutes to seven train home.[40]

Waiting for transport is rarely explicitly mentioned as a problem in the diaries, though there must have been occasions when this occurred and, no doubt, caused frustration.

One source of occasional annoyance with the rail network was the way in which the lack of integrated ticketing, and the use of sometimes quite distant stations by different railway companies serving the same community, could necessitate hurried movement from one part of a settlement to another in order to complete a journey. It has been suggested that this was one of the major inefficiencies of the Victorian railway system, produced by the duplication of routes and competition between companies.[41] For instance, James Bennetts Williams's diary of 1883 gives an account of a journey by train from Cornwall to Liverpool, prior to sailing to South America where he worked for 4 years. He met up with a small group of other Cornish men, presumably all driven to seek work overseas as a result of the decline in the Cornish mining industry,[42] and noted that one had to change his ticket to travel with the group, presumably because the rail company he booked with specified a different route: 'Left St Agnes about 6 a.m. got to Truro about 7 and found several Boys who were going on to the same place. Charlie Batten came on to Truro from Chacewater Station and changed his ticket at Truro'.[43] On a train journey from Bristol to Charmouth, Gerald Fitzmaurice also complained about the complications of his change of trains at Yeovil: 'At Yeovil I had to take a taxi and go for 2½ miles across country to get to the Junction, the Southern Station (the other was GWR)'.[44]

In the early years of motoring (in theory, a fast form of transport), breakdowns were common and might necessitate a change of plans. After a dinner with friends in London prior to going to a dance in Banstead, Surrey, Gerald Fitzmaurice was 'greeted with the news that the car wouldn't function, or rather, its lights wouldn't. After attempts to hire a car and/or get a taxi to take us, it ended by our going on John Arthur's motor bike . . . he driving, Cecily in the side car and myself on the pillion, a cold night and a bumpy road and some 12 to 15 miles'.[45] Freda Smith's diary of 20 years earlier also makes frequent mention of car breakdowns: for example, 'Mother, U.W. and A.G. picked me up [at the station] and we started off. Beastly thing broke down. We had to walk [a] bit of the way home'.[46] In many of these examples, it was the difficulty of completing a journey – or the lack of speed – that was an issue, rather than the need for haste itself.

Conclusion

Caution must be exercised in drawing conclusions from the analysis of just nine diaries drawn from a wide range of times and places, although these are a subset of some 50 items of life-writing that so far have been consulted. They are not a source that can lead easily to generalizations. Rather, they should be seen as a collection of individual accounts that collectively shed some light on the experience of everyday travel in the past. It is certain that not all journeys were recorded in the same degree of detail, or indeed recorded at all, and it is likely that unusual events figure more prominently in these first-hand accounts than mundane trips. However, we suggest that the analysis of life-writing does provide some original insights into the experience of travel, and sheds light on the ways in which people engaged with increasingly speedy and complicated transport options in nineteenth- and twentieth-century Britain. We also argue that there are some themes that appear to be quite consistent over space and time. First, for most travellers from the 1840s to the 1950s, whatever their gender, location (urban or rural) or social background, everyday journeys were completed easily and seamlessly: travel was not something that the diarists saw as a problem and it was a routine part of their everyday life. Second, although the diarists were clearly aware of new and faster forms of transport, they did not always access them or deem them to be the most appropriate (see Jason Finch in this volume, on Ryecroft's resistance to the omnibus). For a wide range of reasons, including cost, convenience, accessibility and preference, older and slower forms of transport flourished alongside new technologies. Most architectures of travel – be they fast or slow – usually met the expectations of the diarists studied. Third, speed itself seems to have been valued as much for the kinaesthetic experience of travelling fast, especially in modes where the travellers were exposed to the elements, as it was for the ability to get somewhere quickly. Speed was not only convenient: it was also fun. Fourth, the need to hurry was most often generated by human failings and external circumstances, and was independent of the transport mode used and of its speed of travel, although even those not particularly seeking speed could become frustrated when things did not work smoothly. Finally, the individual accounts studied emphasize clearly the variety of human experiences and reactions to travel: no two journeys are ever precisely the same. They also remind us of the relative nature of both novelty and speed, because individual reactions to new and faster transport systems depended necessarily on both previous individual experiences and the expectations of travel that individuals had formed.

Notes

1 Anthony Townsend, 'Life in the Real-Time City: Mobile Telephones and Urban Metabolism', *Journal of Urban Technology* 7 (2000): 85–104; Barney Warf, *Time–Space Compression: Historical Geographies* (London: Routledge, 2008); Peter Kivisto, 'Time–Space Compression', in *The Wiley-Blackwell Encyclopedia of Globalization*, ed. George Ritzer (Oxford: Wiley-Blackwell, 2012).

2 Doreen Massey, 'Power-Geometry and a Progressive Sense of Place', in *Mapping the Futures: Local Cultures, Global Change*, ed. Jon Bird, Barry Curtis, Tim Putnam, George Robertson and Lisa Tickner (London: Routledge, 1993), 59–69; Gary Bridge, 'Mapping the Terrain of Time–Space Compression: Power Networks in Everyday Life', *Environment & Planning D: Society & Space* 15 (1997): 611–26.

3 Roland Robertson, 'Glocalization: Time–Space and Homogeneity–Heterogeneity', in *Global Modernities*, eds Michael Featherstone, Scott Lash and Roland Robertson (London: Sage, 1995), 25–44; John Agnew, 'The New Global Economy: Time–Space Compression, Geopolitics, and Global Uneven Development', J*ournal of World-Systems Research* 7 (2001): 133–54.

4 Jeremy Stein, 'Reflections on Time, Time–Space Compression and Technology in the Nineteenth Century', in *Timespace: Geographies of Temporality*, eds Jon May and Nigel Thrift (London: Routledge, 2001), 106–19; Warf, *Time–Space Compression.*

5 See, for example, Theodore Barker and Christopher Savage, *An Economic History of Transport in Britain* (Abingdon: Routledge, 1959); Harold James Dyos and Derek Aldcroft, eds, *British Transport: An Economic Survey from the Seventeenth Century to the Twentieth* (Leicester: Leicester University Press, 1969); Brian Robson, *Urban Growth: An Approach* (London: Routledge, 1973); Patrick Joyce, *Work, Society and Politics: The Culture of the Factory in Later Victorian England* (London: Taylor & Francis, 1980).

6 Dyos and Aldcroft, *British Transport*; Jack Simmons, *The Railways of Britain: An Historical Introduction* (London: Macmillan, 1968); Michael Freeman and Derek Aldcroft, eds, *Transport in Victorian Britain* (Manchester: Manchester University Press, 1988); Sean O'Connell, *The Car and British Society: Class, Gender and Motoring, 1896–1939* (Manchester: Manchester University Press, 1998); John Armstrong, 'From Shillibeer to Buchanan: Transport and the Urban Environment', in *The Cambridge Urban History of Britain 1840–1950*, ed. Martin Daunton (Cambridge: Cambridge University Press, 2000), 229–60; David Horton, Paul Rosen and Peter Cox, eds, *Cycling and Society* (Farnham: Ashgate, 2007); Gijs Mom, *Atlantic Automobilism: Emergence and Persistence of the Car, 1895–1940* (New York: Berghahn Books, 2014).

7 David Edgerton, *Shock of the Old: Technology and Global History since 1900* (London: Profile Books, 2006). See also Mackintosh in this volume.

8 Mimi Sheller and John Urry, 'The New Mobilities Paradigm', *Environment & Planning A* 38 (2006): 207–26. For overviews, see, for example: Tim Cresswell, *On the Move: Mobility in the Modern Western World* (New York: Routledge, 2006); John Urry, *Mobilities* (Cambridge: Polity, 2007); Peter Merriman, *Mobility, Space and Culture* (London: Routledge, 2012); Peter Adey, David Bissell, Kevin Hannam, Peter Merriman and Mimi Sheller, *The Routledge Handbook of Mobilities* (London: Routledge, 2014); James Faulconbridge and Alison Hui, 'Traces of a Mobile Field: Ten Years of Mobilities Research', *Mobilities* 11 (2016): 1–14; Mimi Sheller and John Urry, 'Mobilizing the New Mobilities Paradigm', *Applied Mobilities* 1 (2016): 10–25.

9 Benjamin Fincham, Mark McGuiness and Leslie Murray, *Mobile Methodologies* (Basingstoke: Palgrave, 2009); Monika Büscher, John Urry and Katian Witchger, eds, *Mobile Methods* (London: Routledge, 2010).

10 Jane Humphries, *Childhood and Child Labour in the British Industrial Revolution* (Cambridge: Cambridge University Press, 2010), 12–48; Emma Griffin, *Liberty's Dawn. A People's History of the Industrial Revolution* (New Haven, CT: Yale University Press, 2013).

11 Paul Thompson, *The Voice of the Past: Oral History* (Oxford: Oxford University Press, 1978); Robert Perks and Alistair Thomson, eds, *The Oral History Reader* (London: Routledge, 3rd edn, 2016).

12 Robert Fothergill, *Private Chronicles: A Study of English Diaries* (London: Oxford University Press, 1974); Philippe Lejeune, *On Diary* (Honolulu, HI: University of Hawai'i Press, 2009); Colin Pooley and Marilyn Pooley, 'Mrs Harvey Came Home

from Norwich . . . Her Pocket Picked at the Station and All Her Money Stolen', *Journal of Migration History* 1 (2015): 54–74.

13 Colin Pooley, Jean Turnbull and Mags Adams, *A Mobile Century? Changes in Everyday Mobility in Britain in the Twentieth Century* (Farnham: Ashgate, 2005).

14 Diary of John Lee, 1 April 1860, authors' collection.

15 Diary of John Lee, August 29 1860, authors' collection.

16 Diary of John Leeson, 2 August 1849. Bishopsgate Institute archive, London (GDP/8).

17 Diary of John Leeson, 23 December 1850. Bishopsgate Institute archive, London (GDP/8).

18 Horton et al., *Cycling and Society*.

19 Phillip Gordon Mackintosh and Glen Norcliffe, 'Men, Women and the Bicycle: Gender and Social Geography of Cycling in the Late Nineteenth-Century', in *Cycling and Society*, eds David Horton, Paul Rosen and Peter Cox (Farnham: Ashgate, 2007), 153–77; Phillip Gordon Mackintosh and Glen Norcliffe, 'Flaneurie on Bicycles: Acquiescence to Women in Public in the 1890s', *The Canadian Geographer/Le Géographe canadien* 50 (2006): 17–37. See also Dando in this volume.

20 Diary of Ida Berry, 11 April 1906. Bishopsgate Institute archive (GDP/28).

21 Diary of Ida Berry, 8 August 1906. Bishopsgate Institute archive (GDP/28).

22 Diary of Freda Smith, 18 February 1904. Bishopsgate Institute archive (GDP/99).

23 Diary of Freda Smith, 6 August 1904. Bishopsgate Institute archive (GDP/99).

24 O'Connell, *The Car and British Society*; Lynne Pearce, *Drivetime: Literary Excursions in Automotive Consciousness* (Edinburgh: Edinburgh University Press, 2016).

25 Diary of Freda Smith, 10 August 1907. Bishopsgate Institute archive (GDP/99).

26 *History of Road Safety, the Highway Code and the Driving Test*, www.gov.uk/government/publications/history-of-road-safety-and-the-driving-test/history-of-road-safety-the-highway-code-and-the-driving-test (accessed 7 October 2016).

27 Diary of Gillian Caldwell, 9 July 1955. Bishopsgate Institute archive (GDP/1).

28 Diary of Gillian Caldwell, 10 April 1955. Bishopsgate Institute archive (GDP/1).

29 Diary of Gillian Caldwell, 4 August 1957. Bishopsgate Institute archive (GDP/1).

30 Peter Lyth and Marc Dierikx, 'From Privilege to Popularity: The Growth of Leisure Air Travel since 1945', *The Journal of Transport History* 15 (1994): 97–116; Peter Lyth, 'Plane Crazy Brits: Aeromobility, Climate Change and the British Traveller', in *Transport Policy: Learning Lessons from History*, eds Colin Divall, Julian Hine and Colin Pooley (Farnham: Ashgate, 2016), 171–84.

31 Diary of Gerald Gray Fitzmaurice, 2 July 1927. Bishopsgate Institute archive (GDP/52).

32 Diary of Gillian Caldwell, 17 February 1954. Bishopsgate Institute archive (GDP/1).

33 Diary of Mary Anne Prout, 8 May 1882. Bishopsgate Institute archive (GDP/58).

34 Diary of Gerald Gray Fitzmaurice, 28 June 1926. Bishopsgate Institute archive (GDP/52).

35 Diary of Ida Berry, 26 February 1905. Bishopsgate Institute archive (GDP/28).

36 Diary of John Lee, 25 August 1859, authors' collection.

37 Diary of Gillian Caldwell, 12 April 1955. Bishopsgate Institute archive (GDP/1).

38 Diary of Annie Rudolph, 15 August 1923. Bishopsgate Institute archive (GDP/31).

39 Colin Pooley, 'Uncertain Mobilities: A View from the Past', *Transfers* 3 (2013): 26–44.

40 Diary of Ida Berry, 10 May 1905. Bishopsgate Institute archive (GDP/28).

41 Mark Casson, *The World's First Railway System: Enterprise, Competition, and Regulation on the Railway Network in Victorian Britain* (Oxford: Oxford University Press, 2009).

42 Allen Buckley, *The Cornish Mining Industry: A Brief History* (Redruth: Tor Mark Press, 1992).

43 Diary of James Bennetts Williams, 25 September 1883. Bishopsgate Institute archive (GDP/57).

44 Diary of Gerald Gray Fitzmaurice, 27 June 1926. Bishopsgate Institute archive (GDP/52).

45 Diary of Gerald Gray Fitzmaurice, 8 January 1926. Bishopsgate Institute archive (GDP/52).

46 Diary of Freda Smith, 8 July 1906. Bishopsgate Institute archive (GDP/99).

12 Epilogue

Mobilizing hurrysome historical geographies

Phillip Gordon Mackintosh, Deryck W. Holdsworth and Richard Dennis

As we conclude this volume on hurry's influence on and embeddedness in the geographies of the past, the changing hurried social and cultural world of the twenty-first century demands that we contemplate further the significance of hurry. Everyday twenty-first-century humans take hurry for granted. In many instances, an uncomplicated hurry frames and contextualizes the global digital world-view. A century on, E.M. Forster's 'hurry', which signified the angst- and pique-ridden rush of the modern city, is not necessarily ours – even when Forster's word occasionally fits. We propose a broader term – 'hurrysome' – which intimates a less emotive, and less moralizing, description of the hurriedness of contemporary life, and perhaps also the past; we have seen in this volume many instances when historical actors accepted with equanimity the hurry of their day.[1] Hurrysome as a more neutral term (but not wholly neutral; we give it some moral force, below) describes better, say, a millennial Shanghainese sitting in the back of a 'Didi' on her morning's ride to work, texting friends about late dinner and ordering new shoes on Taobao, hoping she can wear them that evening (Didi Chuxing is a Chinese ride-sharing service [and app], and Taobao is the Chinese 'Amazon', but often with same-day delivery). 'Hurry', in the Forster sense, may have no bearing on the circumstance, and yet hurrysome it is.

We also surmise that hurry obtains to more than mobility in the traditional sense. Rather, it seems to parallel, and perhaps even incite, the human pursuit of quality of life, convenience, comfort, power, security, consumption, and accumulation, however conceived, produced, and maintained. Hurry consorts not only with our curiosities in, but also our compulsion to improve, the human lot. Humans throughout their history have persistently sought to hurry social, cultural, geographical, and especially technological advantage, individually, tribally, or collectively. And urban modernity, that historical geographical concentration of peoples, social and material cultures and discourses, infrastructures and technologies, and surpluses and speculation, if it is anything, is the global expression of the hurrysome human condition – in environments that humans perceive will impede, hurt, or even destroy them without hurry's mediation.

Thus, hurry matters. As we've shown and implied throughout this volume, hurry is a historical impulse that may well connect viscerally to the human condition. It motivates, describes, and, seemingly, conditions human action. We hurry discursively because we think we *must* hurry – in a hurrysome world of our own construction. So, as we conclude *Architectures of Hurry*, it will be instructive to complicate the editors' own apprehensions of historical hurry as 'architectures' but, more precisely, as social and cultural geographies, by deconstructing two very different images of cities – a late-Victorian painting and an early twenty-first-century smartphone video – to make an additional point about hurry and urban modernity: their association with 'efficiency'.

Frederic Marlett Bell-Smith's *Lights of a City Street* (1894; Figure 12.1), a representation of Toronto, Canada, and its then famous intersection of King Street, 'long . . . the city's premier business address', and Yonge Street (its most recognizable street name), encapsulates modernity's preoccupation with hurry and efficiency.[2] It occupies a midpoint in 'street painting' between the distanced observation of impressionism and the anxieties and pleasures of early twentieth-century expressionism. Among the former, Gustave Caillebotte's *Paris Street: Rainy Day* (1877) depicts leisured flâneurs and flâneuses, untroubled by either crowds or technology, and Childe Hassam's *Rainy Day, Boston* (1885) is even less densely populated, and both are from the era of gaslight and horse traction. Bell-Smith presents an equally orderly 'factual' view, but of a much busier and technologically advanced street intersection, contrasting with more *modernist* representations, such as Carlo Carrà's *Piazza del Duomo* (1910) – a splash of electric and electrifying tangerine tramcars beset by a sea of dark-smudged

Figure 12.1 F.M. Bell-Smith (1846–1923), *Lights of a City Street*, 1894; oil on canvas. From the Hudson Bay Company Corporate Collection.

Source: Permission of the Hudson Bay Company.

would-be passengers – or the dizzying visual gymnastics of Dziga Vertov's *Man with a Movie Camera* (1929), also obsessed with converging, diverging, and apparently colliding and dividing electric trams.

In the foreground of Bell-Smith's painting, at 4.57 p.m. (by the clock on the south-east corner) and after an autumn shower, crowd the great promulgators of Victorian hurry: pedestrians. Here they walk on modern pavements: concrete (note the expansion joint in the sidewalk in the foreground), because wooden plank sidewalks literally impeded walking (see Mackintosh in this volume); and asphalt, an efficient accommodation of one of late-Victorian modernity's crucially hurrysome inventions: the pneumatic tyre. Curbs and gutters are plainly visible, their role not only to segregate street users and uses (although not at streetcar stops), but also to sluice water and sludge efficiently into newly constructed sewers (the one infrastructure Bell-Smith overlooks in his impressionist celebration of infrastructure). Streetcar tracks crisscross the roadway in the middle of the intersection. These tracks conveyed a newer representation of hurry in the *fin de siècle*: the electric streetcar – with beaming headlamp – is ostensibly the painter's reason for the painting's title, and it is hard not to imagine Bell-Smith equating such brightly modernized civility with bourgeois enlightenment. Electricity gave cities ubiquitous light, facilitating hurry at night in an age when people, carts and horses, bicycles, and automobiles regularly collided – in broad daylight – with streetcars. Electric lights on electric streetcars constituted a singular effort to make hurry by public transit efficient on late-Victorian streets infamous for their 'furious immobility', as J.B. Jackson described it.[3] It matters that, as the modern streetcars roll toward the viewer, the anti-modern horse and cart with two spectral carters move into the distance, while another pair of horses and a silhouetted carter on Yonge Street wait for all this new hurrysome traffic to pass – including bicycles, 'wheels' noted for hurrying in and around that furiously immobile traffic, while shunning oversubscribed streetcars on predominantly privatized street railway systems (crowded streetcars were a chief selling point of the bicycle in the 1890s). Bell-Smith places an umbrella, the ingenious tool of efficient urban hurrying afoot on concrete in the rain, at the centre of his painting.

Hurrysome street people abound: newsboys implore Bell-Smith himself for coins of admission to the era's most efficient and up-to-the-minute delivery system of information: the evening edition of the daily newspaper. And bourgeois women, who outnumber the visible men. Bourgeois women's beautifying presence on streets was actively encouraged by retailers and street beautifiers everywhere. This distinctly bourgeois feminist late-Victorian public propriety may well offer another meaning for 'lights' in Bell-Smith's Toronto.[4] We see both a traffic cop and a beat cop, to manage the crowding roadway and sidewalk; a woman riding a bicycle and a man walking one, the bicycle a *de rigueur* architecture of hurry in the 1890s; little children being led by women; and even a smirking adolescent lighting a cigarette (as if to hurry adulthood).

There are offices and banks, an indication of the efficient hurrying of capitalism, buildings that themselves will be hurried out of existence in the early 1900s.[5] Street lights grant daytime hurry efficient access to the potential of the city's nighttime.

Four utility poles magically transmit the energy founding Bell-Smith's luminescent city – because there are no unsightly wires in this romanticized vision of urban function. Yet wires must light the businesses on the north-east and south-east corners and conduct the spark above the oncoming streetcar. And, curiously, in an era of burgeoning city beautification discourse that denounced the ugliness of advertising (and wires), Bell-Smith includes legible building signage. Apart from 'Foster & Pender Carpets' ('largest retailing establishment of its kind in Canada'),[6] 'J.E. Ellis Co. Jeweller', and 'Renfrew & Co. Furriers', there are signs for 'Equitable Life Assurance' and 'London Guarantee Accident Co', both considerable insurers of urban hurriers. And, as if to suggest his representation of urban modernity was incomplete without all the modes of civilized hurry available to city people, Bell-Smith adds signs for the branch office of 'Cunard S.S. Line' and a portion of the signage for the 'Grand Trun[k Railway]' ticket office, forming the sage green wall on the right of the canvas. Both companies occupied primary places in the generation of architectures of hurry in the Dominion of Canada, the Grand Trunk its most dominant railway until bankruptcy in 1920, and the Cunard Steam Ship Line, a leading transatlantic passenger service from the 1840s onwards, and renowned as the operator of the ill-fated RMS *Lusitania* and of the fêted RMS *Mauretania*, the world's fastest transatlantic liner. And still there is another kind of sign: the spire of St. James Cathedral pierces the sky in the background, perhaps advertising a method of hurrying into heaven.

Bell-Smith's painting is a factual record of modernity. He includes the advertisements, the utility poles, and the cathedral spire, not as symbols but because they were there; but they were there because they were integral components of efficiently hurrying modernity. For the sake of completeness, we should also mention what his very literal representation necessarily omitted – what was close by but invisible from his viewpoint. The Leader Hotel, just out of sight down the block at 63 King East, offers an example of hurrysome hospitality (see Olson and Poutanen in this volume), which allows capitalism's urbanizing participants to whisk around the world, with a place to stay at each stop (and a place to eat, for the Thomas European Hotel with its famous English Chop House is just behind the painter's vantage, at 30 King West).

Although we can be confident Bell-Smith faithfully reproduced much of what he observed at the corner of King and Yonge in 1894, we should note his curious omissions in this remarkably sanitized, bourgeois, and hurrysome depiction of Toronto: the many impediments to efficient hurry that we know not only resided at that same corner, but also up and down Yonge Street, and along King Street East and King Street West: the numerous cart vendors, immigrant vegetable and fruit sellers, street musicians, shoeblacks, rowdies and dicers, loiterers and tobacco-juice expectorators, pencil and shoelace hawkers, throngs of working children (we do, of course, see very childlike 'newsies' in the foreground), people of colour, indigenous peoples, and numerous homeless men and women – those 'paupers', 'tramps', and 'idlers' who sojourned downtown and frequented its houses of Industry.[7] Missing too is an intimation of Toronto's unemployment and its general malaise in the 1890s, caused by a depression in North America between 1893 and 1898. Notwithstanding, *Lights of a City Street* is an eloquent homage to efficient modernity.

Contrast Bell-Smith's painting with 'Daily Life in India – Driving in Traffic in Delhi', taken by a Delhite, pseudonymously called 'Strong and Beyond', and shot from the passenger window of a car as it moves along a hurrysome New Delhi Street in 2014.[8] Promoting the wondrously 'ordinary daily routine in New Delhi's streets', with its 'crazy traffic' and '[h]orn-blowing, total mess', the videographer pronounces the experience, 'Incredible India!' Such videos in the era of social media are commonplace, but videoed with apparently similar aims: to illustrate the hurrysome spectacle of global southern urbanity.

And to Western eyes, spectacular it is. 'Driving in Traffic in Delhi', like many other videos on YouTube, animates Mike Davis's idea that modernity and the 'dynamics of Third World urbanization both recapitulate and confound the precedents of nineteenth- and early-twentieth-century Europe and North America'.[9] And while the gargantuan cities of the global south are hardly 'Third World', Davis's idea is apt notwithstanding. In this case, we watch furiously dense street activity as it occurs in contemporary New Delhi, but it mimics a famous image of traffic at Dearborn and Randolph Streets in Chicago, 1909, used as the cover of Gunther Barth's *City People*.[10] The overwhelming difference between the two cities resides in the populations – and staggering magnitude – of late-modern Indian urbanism itself: Chicago's 1909 population breached 2 million; New Delhi's nears 22 million. Yet scenes like the one in 'Driving in Traffic in Delhi' occur everywhere and every 'when' in twenty-first-century Indian metropolises, where the numbers of people and the intensity of the urban condition really do 'confound' the Western urban imagination. Another immediate difference is the infrastructure, arguably the one geography capable of indicating efficient modernity. Seemingly makeshift infrastructure follows the videographer: impossible snarls of street wires, crumbling concrete light standard bases and roadway medians, listing hydro poles, struggling street trees, goods heaped on the roadway – which itself looks surprisingly sound. Infrastructural decrepitude chaperones throngs of hurrysome pedestrians, bicycles and delivery bicycles and tricycles, scooters, motorcycles, jitneys and auto-rickshaws, carters and horses and cattle, and honking vehicles new and old – on thoroughfares dizzyingly arrayed with humble shops, stalls, and vendors selling, seemingly, everything. The dubious might exclaim, 'Yes, but it's a market. Of course it's crowded'. We and Delhites themselves would say no, because virtually every street is a market, and everywhere is full of the same Randolph Street-like traffic. Despite the inefficiencies and struggling modernization, the hurry is completely recognizable to Western eyes – except it is exponential, aspirational hurry, the taken-for-granted stuff of exhilarating and exhausting life in the streets of Delhi.

What are we to make of these contrasting examples of modern urban hurry, and how do they align with efficiency? Returning to *Lights of a City Street*, is the painting a harbinger of a wonderfully efficient and hurrysome urban future, or a woeful allegory of modern urban hubris *cum* blissful ignorance of bourgeois modernity's environmental consequences, climate change only the first on a long list of modernization-caused predicaments? What seems common sense to the committed modernizer and hurrysome efficiency seeker mirrors what French philosopher Jacques Ellul thinks is ultimately the negation of human freedom of the will, requiring 'all things human to be analyzed and integrated into orderly and manageable systems'.[11] Efficiency is a complex term (too much so to parse adequately in this brief epilogue) with many meanings and uses, technical, moral, or both; and its original use amounted to 'a kind of moral crusade . . . efficiency itself . . . a form of ethical superiority'.[12] But cities – as moderns have imagined, built, and lived in them – demand efficiency to maintain their populations (however successfully) irrespective of justifiable critique. In a life-and-death sense, efficiency and hurry are determinants of modern urban sustainability – the capacity to keep city people alive – even as efficiency and hurry simultaneously and ironically spawn processes, technologies, and capitalist impulses that undermine the planet's ability to support human and non-human life.

Efficiency, then, is the method and the goal of urban hurry – symptom and ailment of modernization. The development of modern cities as living spaces for billions of people has relied upon the ingenuity of efficiency- and hurry-minded problem-solvers over at least three centuries (or more, if classical Rome, with its public infrastructure, was the first modern city as we hinted in the Introduction).[13] The modern history of municipal physical and social infrastructure is a history of applied efficiency – technical and moral – with all the modern contradictions that attend it. Consider only the most obvious and crucial physical infrastructure on the typical twentieth-century residential street (which persists on ours, now with the inclusion of fibre-optic Internet cable): the roadway, utility poles and cables, light standards, and sewers. The municipal employment of sewers (as a public-health imperative to reduce human exposure to water-borne diseases on streets and in homes, but also as a capitalist fix for the conundrum of water-based mud impeding the flow of goods, services, information, or money) required the concomitant development of industrial mining and smelting to fabricate the network of clean water, effluent, and storm runoff pipes under the street, but visible on the surface as sewer grates and manholes. So too for the production of the paving equipment to lay the asphalt roadway, which depended on chemical and petro-chemical industries, not only to make the hardeners for the asphalt, but also the plastic insulation on telecommunications lines and high-voltage transmission wires running underground or underwater and into buildings; overhead cables are bare and use only the surrounding air as insulation.[14] Hydroelectric and coal- (and, later, oil-) fired power plants generated the electrical current in those cables threading towering, preservative-soaked, ever-present utility poles of pine and fir, and numerous industrial processes combined to put light on these poles in cities. All these efficient processes of mining, chemistry, logging, and heavy manufacturing – which still permit our cities to function – pose a clear and present danger to the environmental integrity of the planet, but without them cities and their multitudinous populations would collapse. And lest we think 'renewables' more virtuous, the same efficient and hurrysome environment-degrading processes create rechargeable batteries, solar panels, wind turbines, and the semiconductors that increasingly govern their use. Modernity, however efficient or hurrysome, demands of moderns difficult, sometimes impossible, choices every step of their uncertain journey – made even more unpredictable by the manner of the journeying.

So, from Bell-Smith's vantage point – the same vantage point as that of late-Victorian poverty, labour, housing, infrastructure, transportation, and public-health reformers, all cognizant of the threats to mortality and quality of life caused by cities eschewing efficiency – the answer to our question, above, must be the one affirming modernization. Indeed, Bell-Smith's painting anticipates the hurrysome modernized city with near-postmillennial enthusiasm.[15] Twenty-first-century critics of the consequences of such enthusiasm may be considerably less sanguine about modernity's future. But we can easily imagine today's reformers of the megalopolises, especially those of the global south, demanding an efficient urban future – from sanitation and garbage collection to property taxation.[16] City life in Delhi, without a future endowed with efficient and hurrysome urban reform – whatever that will mean or look like – amounts to urban environmental justice denied. Indeed, urban moderns tend to believe quality of life is impossible without the production of infrastructure so threatening to the ecology of the planet. Hence the sense of modern thrill in both examples above.

Deceptively static backcloths to the global performance of hurry are the buildings on each side of a city street (on the hurry of built space, see Holdsworth in this volume) – and here we return to our Shanghainese commuter, who we might imagine in two immobile, but ironically hurrysome, historical streetscapes. On the west side of the curvilinear, Thames-like Huangpu River, roughly bisecting Shanghai, she might move along 'the Bund' (in Puxi), which stands as a marker of international capital establishing a toehold in China, initially after the establishment of the treaty ports in the 1840s; a century later, the Bund had accumulated a collection of elegant modern office buildings, banks, and clubs, which present a fascinating Beaux Arts and art nouveau 'counter-skyline' to the more familiar postmodernist Shanghai skyline on the Pudong side of the Huangpu. Christopher Isherwood was crankily dismissive of this earlier array in 1939:

> Seen from the river the semi-skyscrapers of the Bund present, impressively, the façade of a great city. But it is only a façade. The spirit which dumped them upon this unhealthy mud-bank, thousands of miles from their kind, has been purely and brutally competitive. . . . Nowhere is there anything civic at all.[17]

Now, most of these office buildings have been transformed into hotels and condominiums. The centre of gravity for the Shanghai business district has shifted east across the river to Pudong, another district of Shanghai, but one that contains the dramatically contemporary Shanghai skyline, including its famous Oriental Pearl and Jin Mao Towers. Shanghai, the centre of the incomprehensibly populous Yangtze Delta metropolitan region (105 million people), is itself home to more than 24 million mostly highly modernized city people. A forest of skyscraper office buildings and hotels constitutes Pudong's Lujiazui Finance and Trade Zone, started in the 1990s. This towering landscape of hurry comprises an architectural 'survival of the fittest', subject to ever-decreasing life expectancy, as buildings no more than 10 or 20 years old are deemed obsolescent: too small to generate sufficient rental income to match burgeoning land values or too old to accommodate the latest technology. Nested here, too, is the Shanghai Stock Exchange, which helps shape the round-the-world financial markets that 'open' sequentially across a 24-hour cycle. Thus global financiers laud Frankfurt, Germany, and 'the never-ending day' of its banking time zone, which allows bankers access to Asian markets in the morning and the US markets in the evening.[18] Long gone is a risible financial world linked by pigeon or telegraph: now nanosecond trades flash on fibre-optic cables, hurrysome technologies and improvements that have largely emptied the once crowded pits of exchanges epitomized by the Chicago Board of Trade.[19]

The neighbours of the Shanghai Stock Exchange, at the end of Century Avenue and spread out along the curving Binjiang Avenue that follows the Huangpu, are no longer 'thousands of miles from their kind', when we think about other equally hurrysome skylines in Hong Kong, Singapore, or Kuala Lumpur – South East Asian cities where the colonial past has been swamped by modern capitalist hurry. The yardstick of urban form in these hyper-modern cities driven by hurried capital could easily be Dubai or Doha, Persian Gulf emirates seeking a post-oil future, or Panama City, Panama, acting as a refuge for Latin American money of many kinds, as much as Chicago, New York, or London. Ironically, the City of London itself is now being made over in the Pudong or Dubai manner, beginning with the 41-storey Swiss Re building (the 'Gherkin') at 30 St Mary Axe, and culminating (thus far) with the 95-story 'Shard' on the south bank of the Thames, contrasting with earlier efforts to contain the late twentieth-century business district in the more distant Canary Wharf on the Isle of Dogs.[20]

The point we are making, of course, is that hurry and urban capitalism do not merely associate when mutually appropriate; hurry is the substance of urban capitalist development. It is both the impulse of change in built and mobilized landscapes *and* the lens through which urban capitalists view valorized or devalorized buildings and landscapes in the context of capital circulation, whether in nineteenth-century industrial cities or twenty-first-century global cities.[21] Today's global financiers, with 'expanded ambitions of reinvestment in the inner city', continue to find, as investors throughout the modern era have found, the city a convenient place to circulate their capital, whether profiting from infrastructure expansion in the nineteenth century or from the redevelopment of rail yards and port lands in the twentieth and twenty-first centuries (think Canary Wharf or King's Cross Goods Yard in London, or the Hudson Yards in New York).[22] Watching even smaller cities, such as Toronto or Seattle or Frankfurt – let alone the nearly 150 Chinese cities with populations of more than 1 million – change in similar ways and at the same hurrysome rate as their more famous and heavily populous cousins suggests that hurry is an imperative for developers, financiers, and states.[23] Especially municipal governments: as we write, city councils across North America slaver at the prospect of their cities attracting

Amazon.com Inc's 'second headquarters, a massive complex the company says comes with up to 50,000 jobs and more than $5-billion (U.S.) in investment over the next 15 years'.[24] Canada's top-tier cities – Toronto, Vancouver, Montreal, and Calgary – but also some smaller cities (Ottawa, Winnipeg, Halifax, Kitchener-Waterloo), along with American cities – Dallas, Denver, Pittsburgh, Baltimore, Boston, New York, Nashville, Cincinnati, and Houston – hope to inject both hurry and serious global city-ness into their post-2008 municipal economic recovery. And it cannot come fast enough for Winnipeg, Manitoba, which, along with the provincial government, launched '"Team Manitoba" . . . to woo Amazon into planting roots in Winnipeg' – because winning the Amazon 'lottery' is much like winning an Olympic bid, with similar intensive urban development consequences, favourable and otherwise.[25]

All this hurrysome urban capitalism pushes our imaginations in still another direction: to ruminate on the effect 'Silicon Valley' is having on cities and city people, by reducing or eliminating labour, labour opportunities, and once-upon-a-time well-paid wages, and even human physical energy, with energized, digitized, algorithmic technologies and robots. In this wonderfully convenient but paradoxical circumstance, we get email, texting, and social media communications that threaten the extinction of letter carriers, postal workers, and nationalized post offices, which rely on parcel/goods delivery to stay solvent; industrial robots that continue to eliminate and/or make redundant industrial labourers; ATMs supplanting bank tellers (and increasingly bank buildings); Bitcoin, Paypal, and credit/debit chip cards with their ubiquitous 'PayPass', 'payWave', or 'Flash' – 'tap' – terminals, all new expressions of the digital economy and its digital currencies that eschew physical money, its production, and handling (and, in the case of Bitcoin, the elimination of third-party guarantees, tracking, and fees); self-checkout and the automation of retail stores that undervalue human cashiers and, in turn, decrease access to precarious part-time and full-time work for women and marginalized people; online retailing reducing traditional retail infrastructure and labour overheads, and online food and retail delivery services that grant more time for leisure and, alas, work; the emergence of self-drive vehicles that will dislodge human operators and, potentially, the human causes of traffic generation and its disruption of the flow of people, goods, services, and money – but will also disaffect employed 'drivers' everywhere; the use of smartphone GIS mapping to provide trip navigation, making traffic reports, reporters, and car radios and listeners superfluous – hence the rise of subscription 'satellite radio', scrubbed of news, traffic, and even commercials, and the people whose job it was to provide all three; automated public transit kiosks that vend transit passes, placed conveniently aside from the hurrysome entrance and egress of passengers, and require little paid labour to supervise and maintain; app-driven 'news' and 'information' sites that underpay their journalists, making both economically and democratically poorer for the bargain; digital dating and classified advertisement services superseding newspaper personal and classified ads, and the workers in charge of them. In this context, we long for the reassurance of a few 'clipped words' and 'potted expressions'.

Then there is the emergence of the so-called flexible 'gig economy' with its sharing services, chief among them ride sharing, offering cheap and immediate app access to the entire city, but subsidized by the lower wages and vehicle ownership and maintenance of and by the drivers themselves; automobile and bike sharing services, popular with metropolitan and condo-dwelling millennials, that are nudging car rental companies and car rental employment toward obsolescence; online hospitality 'sharing' services, allowing individuals to rent their homes/properties to tourists, not just threatening the precarious labour of hotel housekeepers, but worsening the already deleterious vacancy rates of global cities; the decline of the 'Hollywood' model of entertainment and the rise of entertainment and information sharing, from YouTube or Tumblr to Reddit or Instagram, sites where sharers ironically pay ISPs for the privilege of uploading, without remuneration, personal and 'borrowed' content (YouTube does have a payment model for content with consistently large viewerships); this also includes music 'sharing', where digitized and copyrighted recorded music is uploaded to the web without permission, not only subverting the traditional economic model, but also forcing music makers to invent other models of payment for their efforts.[26] And because virtually everyone carries a smartphone – with its sweeping array of service plans and fees – all these gigabytes and terabytes of information and entertainment hurry along with us, accessed through our myriad apps or 'cloud' storage. When any of these technologies malfunction, untold numbers of precarious 'call' and 'chat' centre workers around the globe are paid low wages to help us troubleshoot our problems.[27]

We cannot yet know how all this hurrysome digital, algorithmic, and robotic change will play out, but geographers suggest their presence nevertheless creates new spatial relations, with unforeseeable social and urban outcomes.[28] Moderns have rightly suspected disruptive change for centuries – and certainly Marx and Engels's observation of capitalism's 'need of a constantly expanding market for its products chases the bourgeoisie over the whole surface of the globe. It must nestle everywhere, settle everywhere, establish connections everywhere' has the whiff of hurrysome change about it and, worse, change that broad populations of humans have not sought but had imposed on them.[29] Yes, some of these digital changes create work opportunities where before there was none, but employment in this new digital world of so-called 'creative classes' and 'creative cities' is desperately uneven, requiring skill sets those losing grip of deindustrialized workplaces will never have (which, in the early twenty-first century, is the stuff of labour-based class segregation).[30] Indeed, at the Atlas Café (in Williamsburg, in infamously gentrifying and hipster-blooming Brooklyn, New York) – or 'laptopistan' according to one satirical observer – laptop jockeys take up most of the seats, deep in their own world and creative economy. Freelance journalist, textbook editor, adjunct professor, editor of a custom press, freelance astrophysicist, documentary maker, screenwriter, novelist, product designer, corporate human resources consultant, doctoral student, choreographer, biotech startup, online educator, architect, event planner, database designer, app developer.[31] Laptops, notebooks, smartphones/androids, and, for a few, smartwatches effect the ultimate hurrysome redundancy of designated places of work in a knowledge-based global economy flush with neo-liberalized entrepreneurs.

Yet with this brave new digital world comes the difficult idea that life in an increasingly hurrysome algorithmic urban world is beginning to shape how we 'know' both our cities and ourselves, a condition that will have dramatic consequences, with 'many futures at stake' for working people.[32] The broader point, however, is that contemporary hurry walks abreast with hurrysome digital technologies to produce the social disruptions and distortions, algorithmic and otherwise, plaguing the modern globe, of which quality of labour-life is paramount – because increasingly, in financialized and exclusionary global cities, quality-of-life wages are the entry point for residential tenure.[33] Apparently, *geographies of hurry are geographies of instabilities*, and further research beyond hurry and mobilities will need to investigate the affinities between hurry, cities, and smartphones/androids and apps, software and algorithms, artificial intelligence, and robots.

A generation ago, futurologists proclaimed the 'death of geography', because the annihilation of space by time seemed to be complete. In practice, the world witnessed a rise in localism and, more contentiously, tribal nationalism, as people and places sought to reaffirm their distinct identities in a globalized world. But today's hurrysome economics and technology threaten a more insidious 'death of geography'. In a world of instantaneous digital transactions, manufactures, and social interaction, there is no longer any differentiation between where we live and work, pursue our leisure activities, shop, worship, or learn. All of them can be wherever we happen to be and at the time of our choosing. We carry our geography with us – digitally.[34] If, previously, hurry connoted the emotions, anxieties, and thrills of those experiencing the hurrying, perhaps it is time to ask (as critics now are) whether the robots, terminals, and other devices interacting, supposedly on our behalf, are also subject to the exhilaration and debilitation of hurry.[35] When is fast too fast, for machines, for humans, for nature?

Notes

1 *The English Dialect Dictionary* offers 'hurrysome' as both an adjectival and adverbial form of Forster's intimation of 'hurry' (*The English Dialect Dictionary Vol. II—F-M*, ed. Joseph Wright [London: Henry Frowde; New York: G.P. Putnam's Sons, 1900], 290).

2 Gunter Gad and Deryck W. Holdsworth, 'Streetscape and Society: The Changing Built Environment of King Street', in *Patterns of the Past: Interpreting Ontario's History*, eds Roger Hall, William Westfall, and Laurel Sefton MacDowell (Toronto and Oxford: Dundurn Press, 1988), 174. For further discussion of Bell-Smith's painting, see Richard Dennis, *Cities in Modernity: Representations and Productions of Metropolitan Space 1840–1930* (New York: Cambridge University Press, 2008), 173–5; and Rae Fleming, 'The Trolley Takes Command, 1892 to 1894', *Urban History Review/Revue d'histoire urbaine* 19 (1991): 218–25.

3 John Brinckerhoff Jackson, *American Space: The Centennial Years 1865–1876* (New York: W.W. Norton, 1972), 204.

4 On this broader point, see Mary Blanchard, 'Boundaries and the Victorian Body: Aesthetic Fashion in Gilded Age America', *American Historical Review* 100 (1995): 21–50; and Mona Domosh, 'The "Women of New York": A Fashionable Moral Geography', *Environment & Planning D: Society & Space* 19 (2001): 573–92.

5 A salient point, here, is that Bell-Smith's painting – relatively shortly after its creation – represented an ephemeral moment in the historical geography of this intersection. By 1914, the corner of King and Yonge was dominated by banks and skyscraper offices. On the south-west corner, the five-storey 1879 Dominion Bank Building, with its mansard roof (not visible in the painting), was replaced by a 12-storey skyscraper – the bank surprisingly demolished after 35 years. On the south-east corner sat the 15-storey CPR Building. On the north-west corner was the Lawlor Building, 1897. On the north-east corner was the Royal Bank, with the Bank of Nova Scotia nearby along King Street. Both these now preeminent Canadian banks were originally based in Halifax, Nova Scotia; they relocated to Toronto to better profit from proximity to the emerging twentieth-century epicentre of rail, resources, and industries. See Gad and Holdsworth, 'Streetscape and Society'.

6 See the Foster and Pender advertisement, 'Advertising Department', *Canadian Magazine* 1 (October 1898): xxiv.

7 Phillip Gordon Mackintosh, *Newspaper City: The Liberal Press and Toronto's Street Surfaces, 1860–1935* (Toronto and Buffalo: University of Toronto Press, 2017), 182–6. Research on the presence of First Nations in Toronto's Victorian streets is in its infancy (see Victoria Jane Freeman, *'Toronto has no history': Indigeneity, Settler Colonialism, and Historical Memory in Canada's Largest City* (unpublished PhD thesis, Department of History, University of Toronto, 2010); and Jasmine Chorley, 'Disappearing into White Space: Indigenous Toronto, 1900–1914', *ACTIVEHISTORY.CA* [accessed 28 September 2017] http://activehistory.ca/papers/disappearing-into-white-space-indigenous-toronto-1900-1914/#21).

8 'Daily Life in India – Driving in Traffic in Delhi', *YouTube* (accessed 24 August 2017), www.youtube.com/watch?v=mDLCd70iTeE.

9 Mike Davis, *Planet of Slums* (London and New York: Verso, 2006), 11. Examples of such videos perhaps include one that 'went viral' in the West a few years ago: 'India Driving'. The drivers', motorcyclists', and pedestrians' care and patience navigating a street without traffic controls are mesmerising, and may cause thoughtful viewers to reconsider regulations on hurry ('India Driving', *YouTube* [accessed 24 August 2017], www.youtube.com/watch?v=RjrEQaG5jPMovercrowde). Still another might be one of overcrowded passenger trains in modern Nepal ('The MOST OVERCROWDED TRAINS in the World!', *YouTube* [accessed 24 August 2017], www.youtube.com/watch?v=bfMfXuNRRkY) recapitulating and confounding the overcrowding of North American streetcars in the mid and late nineteenth century. These Nepalese train riders would likely not recognize the overcrowding on, for example, the New York Street Railway in 1867 as pathological urbanism (see 'The New York Street Railway: What Is It Coming to?' and 'Street-Car Salad', *Harpers Weekly Magazine* 11 (23 March 1867): 188; and Glen Holt, 'The Changing Perception of Urban Pathology: An Essay on the Development of Mass Transit in the United States', in *Cities in American History*, eds Kenneth Jackson and Stanley Schultz (New York: Alfred A. Knopf, 1972), 324–43.

10 Gunther Barth, *City People: The Rise of Modern City Culture in Nineteenth-century America* (Oxford, New York, Toronto, and Melbourne: Oxford University Press, 1980).

11 Jennifer Karns Alexander, *The Mantra of Efficiency: From Waterwheel to Social Control* (Baltimore: Johns Hopkins University Press, 2008), 1.

12 Ibid., 3–4; Suzanne Raitt, 'The Rhetoric of Efficiency in Early Modernism', *Modernism/Modernity* 13 (2006): 836.

13 Miles Ogborn (*Spaces of Modernity: London's Geographies, 1680–1780* [New York: Guilford Press, 1998], 2) suggests we can find 'spaces of modernity' at and in specific historical times and places, enabling us to think of the modern beyond its 'grand' theorizations.

14 'Insulator (electricity)', *Wikipedia* (accessed 17 September 2017), https://en.wikipedia.org/wiki/Insulator_(electricity), and 'Overhead power line', *Wikipedia* (accessed 17 September 2017), https://en.wikipedia.org/wiki/Overhead_power_line.

15 A ubiquitous turn-of-the-twentieth-century liberal evangelical Protestant idea that cities must be perfected before the second coming of Jesus Christ can occur. On post-millennialism and urban reform in North America, see Phillip Gordon Mackintosh and Clyde Forsberg, '"Co-agent of the Millennium": City Planning, Urban Reform and Christian Eschatology in North America, 1890–1920', *Annals of the Association of American Geographers* 103 (2013): 727–47.

16 We can see this in, for example, contemporary reports of urban reform in Delhi: Viba Sharma, 'Swachh Rankings: All Delhi Municipal Bodies Fare Badly', *Hindustan Times*, 5 May 2017 (accessed 17 September 2017), www.hindustantimes.com/delhi-news/swachh-rankings-all-delhi-municipal-bodies-fare-badly/story-oAce3sG0uu6nrsIQm-r1u8K.html; Moushumi Das Gupta, 'Most Municipalities Lag in Basic Urban Reforms, says Govt Report', *Hindustan Times*, 2 July 2017 (accessed 17 September 2017), www.hindustantimes.com/india-news/most-municipalities-lag-in-basic-urban-reforms-says-govt-report/story-aeylSCxkvPyVdrgaUax39J.html.

17 W.H. Auden and Christopher Isherwood, *Journey to a War* (New York: Random House, 1939), 240–52. Quoted in Jeffrey N. Wassertrom, 'Metropolis East, a Review of Stella Dong, *Shanghai, 1842–1949: The Rise and Fall of a Decadent City*', William Morrow, *The American Scholar* 69:2 (2000): 147–50. Also in Jeffrey N. Wasserstrom, 'A Big Ben with Chinese Characteristics: The Customs House as Urban Icon in Old and New Shanghai', *Urban History* 33:1 (2006): 65–84.

18 Joe Miller, 'Frankfurt is Winning the Battle for Brexit Spoils', *BBC News*, 29 August 2017 (accessed 29 August 2017), www.bbc.com/news/business-41026575.

19 'End of Era: Trading Pits Close', *Wall Street Journal*, www.youtube.com/watch?v=aluuekJIhWI; see also 'CME to shutter trading pits after 167 years', www.youtube.com/watch?v=NqZqcqo1SLA.

20 An interesting novelistic take on the new London, which uses the travails of architect Matthew Halland to describe both speculative office towers in Canary Wharf and the gentrification of a Georgian and Victorian fabric in the old City, is Penelope Lively, *City of the Mind* (Harmondsworth: Penguin, 1992).

21 Rachel Weber, 'Extracting Value from the City: Neoliberalism and Urban Redevelopment', *Antipode* 34 (2002): 519–40; Mark Davidson, 'Gentrification as Global Habitat: A Process of Class Formation or Corporate Creation?', *Transactions of the Institute of British Geographers* 32 (2007): 490–506.

22 Loretta Lees and David Ley, 'Introduction to Special Issue on Gentrification and Public Policy', *Urban Studies* 45 (2008): 2382.

23 See, for example, Zachary Slobig, 'The Violence of Frankfurt's Frantic Construction Boom', *Wired*, 23 March 2015 (accessed 17 September 2017), www.wired.com/2015/03/violence-frankfurts-frantic-construction-boom/; and Meike Fischer, *Auflösung* (accessed 17 September 2017), http://meike-fischer.de/lightfigure/dokumentation/aufloesung/.

24 Jeff Gray and Sean Silcoff, 'Canadian Cities Jump at Chance to Play Host To Amazon', *Globe and Mail*, 7 September 2017 (accessed 19 September 2017), https://beta.theglobeandmail.com/report-on-business/international-business/us-business/amazon-plans-second-headquarters-in-n-america/article36194091/?ref=http://www.theglobeandmail.com&.

25 See Robert Oliver, 'Toronto's Olympic Aspirations: A Bid for the Waterfront', *Urban Geography* 32 (2011): 767–87; Nathan Edelson, 'Inclusivity as an Olympic Event at the 2010 Vancouver Winter Games', *Urban Geography* 32 (2011): 804–22.

26 Digital hurry also affects sound and traditional ways of hearing; see Damon Krukowski, *The New Analog: Listening and Reconnecting in a Digital World* (New York and London: New Press, 2017).

27 Beyond this, high-tech globalism is shifting to India, with one-third of IBM's total workforce (about 130,000 employees) now residing in India, a circumstance 'vital to

keeping down costs at IBM' (Vindu Goel, 'IBM Now Has More Employees in India than in the US', *New York Times*, 28 September 2017 [accessed 28 September, 2017], www.nytimes.com/2017/09/28/technology/ibm-india.html?hpw&rref=technology&action=click&pgtype=Homepage&module=well-region&region=bottom-well&WT.nav=bottom-well&_r=0).

28 See Vincent del Casino, 'Social Geographies II: Robots', *Progress in Human Geography* 40 (2016): 846–55.

29 Karl Marx and Frederick Engels, *Communist Manifesto* (Chicago: Charles H. Kerr, 1910), 17.

30 Jamie Peck, 'Struggling with the Creative Class', *International Journal of Urban and Regional Research* 29 (2005): 740–70; Steve Herbert and Elizabeth Brown, 'Conceptions of Space and Crime in the Punitive Neoliberal City', *Antipode* 38 (2006): 755–77.

31 David Sax, 'Destination: Laptopistan', *NYT.com*, 3 December 2010 (accessed 25 August 2017), www.nytimes.com/2010/12/05/nyregion/05laptop.html?pagewanted=1&_r=1

32 del Casino, 'Social Geographies II: Robots', 851.

33 See Manuel Aalbers, ed., *Subprime Cities: The Political Economy of Mortgage Markets* (Malden, MA, and Oxford: Wiley-Blackwell, 2012).

34 See James Ash, Rob Kitchin, and Agnieszka Leszczynski, 'Digital Turn, Digital Geography?', *SSRN*, 14 October 2015 (accessed 28 September 2017), https://ssrn.com/abstract=2674257. This curious situation is anticipated by E.M. Forster in a story written the year before *Howards End*, in *The Machine Stops* (1909). In Forster's dystopian future, people do not move around at all: they live underground, so that they never encounter the natural world; everything they need is delivered to them, and they communicate by a kind of Skype.

35 For example, Bill Gates, Stephen Hawking, and Elon Musk fear the hurry by which we are developing artificial intelligence. See Kevin Rawlinson, 'Microsoft's Bill Gates Insists AI Is a Threat', *BBC News*, 29 January 2015 (accessed 28 September 2017), www.bbc.com/news/31047780.

Bibliography

Historical literature

Albion, Robert G. *The Rise of New York Port, 1815–1860*. Boston: Northeastern University Press, 1984[1939].

Auden, W. H. and Christopher Isherwood. *Journey to a War*. New York: Random House, 1939.

Baudelaire, Charles. 'Perte d'Auréole', in *Le spleen de Paris: Ou les cinquante petit poemes en prose de Charles Baudelaire*, 152. Paris: Chez Emile-Paul, 1869.

Baudelaire, Charles. 'Loss of Halo', in *Paris Spleen*, Louise Varése, trans., 94. New York: New Directions, 1970[1869].

Beard, George. *American Nervousness*. New York: Putnam, 1881.

Benjamin, Walter. *The Arcades Project*, translated by Howard Eiland and Kevin McLaughlin. Cambridge, MA: Harvard University Press, 1999.

Blackwood, Algernon. *Episodes Before Thirty*. London, New York, Toronto and Melbourne: Cassell, 1923.

Borrett, George Tuthill. *Letters from Canada and the United States*. London: J.E. Adlard, 1865.

Boston Transit Commission. *Statement of the Subway Commission*. Boston: Boston Transit Commission, 1894.

Carhart, E.R. 'The New York Produce Exchange', *The Annals of the American Academy of Political & Social Science* 38 (1911): 206–21.

Caymari, Bernardo. *Propuesta para la Construcción en la Ciudad de Buenos Aires de un Tranvía Eléctrico Elevado Metropolitano*. Buenos Aires: Gunche, Wiebeck, Turtl, 1896.

Century Road Club of America. *Century Road Club Manual*. Terre Haute, IN: Moore & Langen, 1898.

Clarke, D.E. Martin, ed and trans. *The Hávamál: With Selections from Other Poems of the Edda, Illustrating the Wisdom of the North in Heathen Times*. Cambridge: Cambridge University Press, 1923.

Control Committee of Buenos Aires Transport. *New Principles in Urban Transportation Economy*. Buenos Aires: Ministry of the Interior, 1941.

Cornish, C.J. *Wild England of To-Day and the Wild Life in It*. London: Thomas Nelson, 1895.

Davidson, Lillias. *The Handbook for Lady Cyclists*. London: H. Nisbet and Co., 1896.

Defoe, Daniel. *The Anatomy of Exchange-Alley: or, a System of Stock-Jobbing*. London: E. Smith near the Exchange-Alley, 1719.

Dickens, Charles. *Bleak House*, edited by Norman Page. Harmondsworth, UK: Penguin, 1971[1853].

Dickens, Charles. *Nicholas Nickleby*, edited by Paul Schlicke. Oxford: Oxford University Press, 1990.

Eaton, Earl. 'Two on a Tandem', in *Lyra Cyclus or the Bards and the Bicycle*, edited by Edmond Redmond, 32. Rochester NY: n/a, 1897.

Eliot, T.S. *Notes towards the Definition of Culture*. London: Faber & Faber, 1948.

Erskine, Fanny. *Lady Cycling: What to Wear & How to Ride*. London: British Library Publishing Division, 1897.

Estrada, Ezequiel Martínez. *La Cabeza de Goliat. Microscopía de Buenos Aires*. Buenos Aires: Centro Editor de América Latina, 1968[1947].

Follett, Helen. 'A Honeymoon on Wheels', *Outing* 29 (October 1896): 3–7.

Forster, E.M. *Howards End*. London: Penguin, 2000[1910].

Gissing, George. *The Private Papers of Henry Ryecroft*. London: Constable, 1921[1903].

Gissing, George. *Workers in the Dawn*, edited by Debbie Harrison. Brighton: Victorian Secrets, 2010[1880].

Gissing, George. *Thyrza*, edited by Pierre Coustillas. Brighton: Victorian Secrets, 2013[1887].

Gordon, W.J. *The Horse-World of London*. London: Religious Tract Society, 1893.

Grahame, Gordon Hill. *Short Days Ago*. Toronto: Macmillan, 1972.

Howells, William Dean. *A Hazard of New Fortunes*. Oxford: Oxford University Press, 1990[1890].

Howells, William Dean. *Through the Eye of the Needle*. New York: Harper, 1907.

Johnson, Joshua. *Joshua Johnson's Letterbook, 1771–1774: Letters from a Merchant in London to his Partners in Maryland*, edited by Jacob M. Price (London: London Record Society, 1979), *British History Online* (accessed 12 September 2017), www.british-history.ac.uk/london-record-soc/vol15.

Lloyd, Reginald. *Twentieth Century Impressions of Argentina. Its History, People, Commerce, Industries, and Resources*. London: Lloyd's Greater Britain, 1911.

London, Jack. *The People of the Abyss*. London: The Macmillan Company, 1903.

Louis, Chevalier de Jaucourt. 'Hospitalité', in *Encyclopédie ou Dictionnaire raisonné des sciences, des arts et des métiers, par une société de gens de lettres*. Vol. 8, edited by Denis Diderot and Jean d'Alembert, 316. Paris: 1765.

Marx, Karl and Frederick Engels. *Communist Manifesto*. Chicago: Charles H. Kerr, 1910.

Moore, H.C. *Omnibuses and Cabs: Their Origin and History*. London: Chapman & Hall, 1902.

Municipalidad de Buenos Aires. *Censo general de población, edificación, comercio é industrias de la ciudad de Buenos Aires*. Buenos Aires: Municipalidad de Buenos Aires, 1910.

Musil, Robert. *The Man Without Qualities*. London: Picador, 2017[1930].

Nietzsche, Friedrich. *The Gay Science*, edited by Bernard Williams. Cambridge: Cambridge University Press, 2012[1882].

Norris, Frank. *The Pit*. London, Edinburgh, and New York: Thomas Nelson, 1903.

Paolera, Carlos della. *Buenos Aires y sus Problemas Urbanos*. Buenos Aires: Olkos, 1977[1937].

Pedestrians' Association. *Ninth Annual Report of the Pedestrians' Association*. London: Pedestrians' Association, 1937.

Pedestrians' Association. *Tenth Annual Report of the Pedestrians' Association*. London: Pedestrians' Association, 1938.

Pennell, Elizabeth. 'Cycling', in *Ladies in the Field*, edited by V. Greville, 257–65. New York: D. Appleton and Co., 1894.

Pennell, Joseph and Elizabeth Robins Pennell. *Our Sentimental Journey through France and Italy*. London: T. Fisher Unwin, 1893.

Picton, J.A. *Memorials of Liverpool Historical and Topographical*, Vol II. London: Longmans, Green, 1875.

Porter, Luther. *Cycling for Health and Pleasure: An Indispensable Guide to the Successful Use of the Wheel*. New York: Dodd, Mead, 1895.

Pratt, Charles. *The American Bicycler: A Manual for the Observer, the Learner, and the Expert*. Boston: Houghton, Osgood, 1879.

Redfern, Percy. *The New History of the CWS*. London: Dent, 1938.

Redmond, Edmond. *Lyra Cyclus or The Bards and the Bicycle*. Rochester, NY: n/a, 1897.

Rhys, Captain Horton. *A Theatrical Trip for a Wager! Through Canada and the United States*. Vancouver: Alcuin Society, 1966.

Robertson, Anthony B. *Report of Cases Argued and Determined in the Superior Court of New York, Volume II*. Albany, NY: W.C. Little, 1867.

SCA. 'Las Grandes Obras (El Subterráneo)', *Revista de la Sociedad Central de Arquitectos* 112 (1917): 78.

Sheppard, Edmund E. *Toronto by Gaslight: Nighthawks of a Great City*. Toronto: Edmund E. Sheppard, 1884.

Simmel, Georg. 'The Metropolis and Mental Life', in *Classic Essays on the Culture of Cities*, edited by Richard Sennett, 47–60. Englewood Cliffs, NJ: Prentice Hall, 1969[1903].

Simmel, Georg. *The Philosophy of Money*, 3rd edn, translated by David Frisby. New York and London: Routledge, 2011[1900].

Smith, Dexter. 1886. *Cyclopedia of Boston and Vicinity*. Boston, MA: Cashin & Smith.

Sterne, Lawrence. *A Sentimental Journey Through France and Italy*. Reprinted in New York: Golden Cockerel Press, 1928[1768].

Sullivan, Algernon. 'Speech', in *Ceremonies on Leaving the Old and Opening the New Produce Exchange, May 5th and 6th, 1884*. New York: The Art Interchange Press, 1884.

Swinnerton, Frank. *George Gissing: A Critical Study*. London: Martin Secker, 1912.

Tillson, George. *Street Pavements and Paving Materials – A Manual of City Pavements: The Methods and Materials of Their Construction*. New York: John Wiley, 1900.

Tripp, Alker. *Road Traffic and Its Control*. London: E. Arnold, 1938.

Veblen, Thorstein. *The Theory of the Leisure Class*. New York: Macmillan, 1899.

Ward, Maria. *Bicycling for Ladies: The Common Sense of Bicycling*. New York: Brentano's, 1896.

Wickett, S. Morley. 'Municipal Government of Toronto', in *Municipal Government in Canada*, edited by S. Morley Wickett, 37–58. Toronto: Librarian of the University of Toronto, 1907.

Willard, Frances. *Wheel within a Wheel: How I Learned to Ride the Bicycle*. London: Hutchison, 1895.

Woolf, Virginia. *Mrs Dalloway*. Oxford: Oxford University Press, 1992[1925].

Woolf, Virginia. 'George Gissing', in *The Common Reader Volume II*, edited by Andrew McNeillie, 220–25. London: Vintage, 2003[1932].

Wright, Joseph, ed. *The English Dialect Dictionary Vol. II – F–M* (London: Henry Frowde; New York: G.P. Putnam's Sons, 1900).

Secondary literature

Aalbers, Manuel, ed. *Subprime Cities: The Political Economy of Mortgage Markets*. Malden, MA, and Oxford: Wiley-Blackwell, 2012.

Ackley, Clifford, ed. *British Prints from the Machine Age: Rhythms of Modern Life 1914–1939*. London: Thames & Hudson, 2008.

Adey, Peter, David Bissell, Kevin Hannam, Peter Merriman, and Mimi Sheller. *The Routledge Handbook of Mobilities*. London: Routledge, 2014.

Agnew, John. 'The New Global Economy: Time–Space Compression, Geopolitics, and Global Uneven Development', *Journal of World-systems Research* 7 (2001): 133–54.

Alexander, Jennifer Karns. *The Mantra of Efficiency: From Waterwheel to Social Control*. Johns Hopkins University Press, 2008.

Alford, Terry. *Fortune's Fool: The Life of John Wilkes Booth*. New York: Oxford University Press, 2016.

Allen, M.D. '"Feeble Idyllicism": Gissing's Critique of *Oliver Twist* and *Ryecroft*', *Gissing Journal* 43 (July 2007): 26–32.

Allen, M.D. 'Bleak House and The Emancipated', *Gissing Journal* 43 (October 2007): 17–27.

Alto, Sërpa. 'Commercial Travel and Hospitality in the Kings' Sagas', *Mirator* 10 (2009): 31–43.

Armstrong, John. 'From Shillibeer to Buchanan: Transport and the Urban Environment', in *The Cambridge Urban History of Britain 1840–1950*, edited by Martin Daunton, 229–60. Cambridge: Cambridge University Press, 2000.

Aronson, Sidney. 'The Sociology of the Bicycle', *Social Forces* 30 (1952): 305–12.

Ascoli, David. *The Queen's Peace: The Origins and Development of the Metropolitan Police 1829–1979*. London: Hamish Hamilton, 1979.

Ash, James, Rob Kitchin, and Agnieszka Leszczynski. 'Digital Turn, Digital Geography?' *SSRN* (14 October 2015; accessed 28 September 2017), https://ssrn.com/abstract=2674257.

Barker, Theodore and Michael Robbins. *A History of London Transport*. London: Allen & Unwin, 1963, 1974.

Barker, Theodore and Christopher Savage. *An Economic History of Transport in Britain*. Abingdon: Routledge, 1959.

Barth, Gunther. *City People: The Rise of Modern City Culture in Nineteenth-Century America*. Oxford, New York, Toronto, and Melbourne: Oxford University Press, 1980.

Beard, Mary. *SPQR: A History of Ancient Rome*. New York: W.W. Norton, 2015.

Berg, Maggie and Barbara Seeber. *The Slow Professor: Challenging the Culture of Speed in the Academy*. Toronto, Buffalo, and London: University of Toronto Press, 2016.

Berman, Marshall. *All That Is Solid Melts into Air: The Experience of Modernity*. New York: Simon & Schuster, 1982.

Bernard, Andreas. *Lifted: A Cultural History of the Elevator*. New York: New York University Press, 2014.

Black, Mary. *Old New York in Early Photographs, 1853–1901, 196 Prints from the Collection of New-York Historical Society*. New York: Dover, 1973.

Blanchard, Mary. 'Boundaries and the Victorian Body: Aesthetic Fashion in Gilded Age America', *American Historical Review* 100 (1995): 21–50.

Bluestone, Daniel. *Constructing Chicago*. New Haven, CT: Yale University Press, 1991.

Bobrick, Benson. *Labyrinths of Iron: A History of the World's Subways*. New York: Henry Holt, 1994.

Boehm, Katharina and Josephine McDonagh. 'Urban Mobility: New Maps of Victorian London', *Journal of Victorian Culture* 15 (2010): 184–200.

Bogardus, Ralph. 'The Reorientation of Paradise: Modern Mass Media and Narratives of Desire in the Making of American Consumer Culture', *American Literary History* 10 (1998): 508–23.

Boyer, Christine. *Dreaming the Rational City: The Myth of American City Planning.* Cambridge, MA: MIT Press, 1983.

Bradley, Simon and Nikolaus Pevsner. *London 6: Westminster.* New Haven, CT: Yale University Press, 2005.

Braudel, Fernand. *The Wheels of Commerce: Civilization and Capitalism, 15th–18th Century, Volume 2.* New York: Harper & Row, 1982.

Bridge, Gary. 'Mapping the Terrain of Time–Space Compression: Power Networks in Everyday Life', *Environment & Planning D: Society & Space* 15 (1997): 611–26.

Brown, Marilyn R. *Degas and the Business of Art: A Cotton Office in New Orleans.* University Park, PA: Pennsylvania State University Press, 1994.

Bruce, J. Graeme and C.H. Curtis. *The London Motor Bus: Its Origins and Development.* London: London Transport, 1973.

Buchanan, Dave, ed. *A Canterbury Pilgrimage/An Italian Pilgrimage by Joseph Pennell and Elizabeth Robins Pennell.* Edmonton, AB: University of Alberta Press, 2015.

Buchanan, Dave. 'Pilgrims on Wheels: The Pennells, F.W. Bockett, and Literary Cycle Travels', in *Culture on Two Wheels: The Bicycle in Literature and Film*, edited by Jeremy Withers and Daniel Shea, 19–40. Lincoln, NE: University of Nebraska Press, 2016.

Buckley, Allen. *The Cornish Mining Industry: A Brief History.* Redruth: Tor Mark Press, 1992.

Burch, Stuart. 'An Unfolding Signifier: London's Baltic Exchange in Tallinn', *Journal of Baltic Studies* 39 (2008): 451–73.

Büscher, Monika, John Urry, and Katian Witchger, eds. *Mobile Methods.* London: Routledge, 2010.

Buzard, James. *The Beaten Track: European Tourism, Literature, and the Ways to 'Culture', 1800–1918.* Oxford: Oxford University Press, 1983.

Buzard, James. *Disorienting Fiction: The Autoethnographic Work of Nineteenth-Century British Novels.* Princeton: Princeton University Press, 2005.

Cameron, Frank. *Bicycling in Seattle 1879–1904.* Seattle, WA: F. Cameron, 1982.

Casson, Mark. *The World's First Railway System: Enterprise, Competition, and Regulation on the Railway Network in Victorian Britain.* Oxford: Oxford University Press, 2009.

Chauncey, George. *Gay New York: Gender, Urban Culture, and the Making of the Gay Male World, 1890–1940.* New York: Basic Books, 1994.

Chorley, Jasmine. 'Disappearing into White Space: Indigenous Toronto, 1900–1914', *ACTIVEHISTORY.CA* (accessed 28 September 2017), http://activehistory.ca/papers/disappearing-into-white-space-indigenous-toronto-1900-1914/#21.

Clark, Norman. *Deliver Us from Evil: An Interpretation of American Prohibition.* New York: Norton, 1976.

Clayton, Barbara. *A Penelopean Poetics: Reweaving the Feminine in Homer's Odyssey.* Lanham: Lexington Books, 2003.

Cresswell, Tim. *On the Move: Mobility in the Modern Western World.* New York: Routledge, 2006.

Cresswell, Tim. 'Mobilities II: Still', *Progress in Human Geography* 36 (2012): 645–53.

Cresswell, Tim and Tanu Priya Uteng. 'Gendered Mobilities: Towards a Holistic Understanding', in *Gendered Mobilities*, edited by Tanu Priya Uteng and Tim Cresswell, 1–12. Farnham, UK: Ashgate, 2008.

Cronon, William. *Nature's Metropolis: Chicago and the Great West*. New York: W.W. Norton, 1991.

Croot, Patricia E.C., ed. *A History of the County of Middlesex Volume XIII: The City of Westminster Part I*. Woodbridge, UK: Boydell & Brewer for the Institute of Historical Research, 2009 (Victoria County Histories).

Cushing, George M. Jr. *Great Buildings of Boston: A Photographic Guide*. New York: Dover, 1982.

Dando, Christina. 'Riding the Wheel: Selling American Women Mobility and Geographic Knowledge', *Acme: An International E-Journal for Critical Geographies* 6 (2007): 174–210.

Davidson, Mark. 'Gentrification as Global Habitat: A Process of Class Formation or Corporate Creation?', *Transactions of the Institute of British Geographers* 32 (2007): 490–506.

Davies, Philip. *Lost London: 1870–1945*. Croxley Green: English Heritage, 2009.

Davis, Mike. *Planet of Slums*. London and New York: Verso, 2006.

Day, John R. *The Story of the London Bus*. London: London Transport, 1973.

de Certeau, Michel. *The Practice of Everyday Life*. Berkeley: University of California Press, 1984.

del Casino, Vincent. 'Social Geographies II: Robots', *Progress in Human Geography* 40 (2016): 846–55.

De Luca, Giuseppe. 'Infrastructure Financing in Medieval Europe: On and Beyond "Roman Ways"', in *Infrastructure Finance in Europe: Insights Into the History of Water, Transport, and Telecommunications*, edited by Youssef Cassis, Giuseppe De Luca, and Massimo Florio, 39–60. Oxford: Oxford University Press, 2016.

Dennis, Richard. *Cities in Modernity: Representations and Productions of Metropolitan Space 1840–1930*. New York: Cambridge University Press, 2008.

Dennis, Richard. 'The Architecture of Hurry', in *Cityscapes in History: Creating the Urban Experience*, edited by Katrina Gulliver and Heléna Tóth, 115–36. Farnham, UK: Ashgate, 2014.

Derrida, Jacques. *'De l'Hospitalité' Anne Dufourmantelle invite Jacques Derrida à répondre*. Paris: Calmann-Lévy, 1997.

Dikeç, Mustafa and Carlos Lopez Galviz. '"The Modern Atlas": Compressed Air and Cities c.1850–1930', *Journal of Historical Geography* 53 (2016): 11–27.

Divall, Colin, ed. *Cultural Histories of Sociabilities, Spaces and Mobilities*. Abingdon: Routledge, 2016.

Domosh, Mona. 'The "Women of New York": A Fashionable Moral Geography', *Environment & Planning D: Society & Space* 19 (2001): 573–92.

Dufaux, François and Sherry Olson. 'Reconstruire Montréal, rebâtir sa fortune', *Revue de la Bibliothèque et archives nationales du Québec* 1 (2009): 44–57.

Duffy, Enda. *The Speed Handbook: Velocity, Pleasure, Modernism*. Durham, NC: Duke University Press, 2009.

Dufresne, Sylvie. 'Le Carnaval d'hiver de Montréal, 1803–1889', *Revue d'Histoire Urbaine* 11 (1983): 25–45.

Dunford, Michael and Weidong Liu, eds. *The Geographical Transformation of China*. Abingdon, UK: Routledge, 2017.

Dyos, Harold James. 'The Slums of Victorian London', in *Exploring the Urban Past: Essays in Urban History by H.J. Dyos*, edited by David Cannadine and David Reeder, 129–53. Cambridge: Cambridge University Press, 1982.

Dyos, Harold James and Derek Aldcroft, eds. *British Transport: An Economic Survey from the Seventeenth Century to the Twentieth*. Leicester: Leicester University Press, 1969.

Edelson, Nathan. 'Inclusivity as an Olympic Event at the 2010 Vancouver Winter Games', *Urban Geography* 32 (2011): 804–22.

Edensor, Tim. 'Rhythm and Arrhythmia', in *The Routledge Handbook of Mobilities*, edited by Peter Adey, David Bissell, Kevin Hannam, Peter Merriman, and Mimi Sheller (New York: Routledge, 2014).

Edgerton, David. *Shock of the Old: Technology and Global History since 1900*. London: Profile Books, 2006.

Edwards, Paul. 'Infrastructure and Modernity: Force, Time and Social Organization in the History of Sociotechnical Systems', in *Modernity and Technology*, edited by Thomas Misa, Philip Brey, and Andrew Feenberg, 185–225. Cambridge and London: MIT Press, 2003.

Eesfehani, Amir Moghaddaas. 'The Bicycle's Long Way in China: The Appropriation of Cycling as a Foreign Cultural Technique, 1860–1940', *Cycle History 13: Proceedings of the 13th International Cycle History Conference*, edited by Nick Clayton and Andrew Ritchie, 94–102. San Francisco: Cycle Publishing, 2003.

Ellegård, Kajsa and Bertil Vilhelmson. 'Home as a Pocket of Local Order: Everyday Activities and the Friction of Distance', *Geografiska Annaler* 86B (2004): 281–96.

Ellis, Aytoun. *The Penny Universities: A History of the Coffee-house*. London: Decker & Warburg, 1956.

Emsley, Clive. '"Mother, What Did Policemen Do When There Weren't Any Motors?" The Law, the Police and the Regulation of Motor Traffic in England, 1900–1939', *The Historical Journal* 36 (1993): 357–81.

Evans, Elizabeth F. 'We are Photographers, Not Mountebanks! Spectacle, Commercial Space, and the New Public Woman', in *Amy Levy: Critical Essays*, edited by Naomi Hetherington and Nadia Valman, 25–46. Athens, OH: Ohio University Press, 2010.

Faulconbridge, James and Alison Hui. 'Traces of a Mobile Field: Ten Years of Mobilities Research', *Mobilities* 11 (2016): 1–14.

Feng, Suwei and Qiang Li. 'Car Ownership Control in Chinese Mega Cities: Shanghai, Beijing and Guangzhou', *Journeys* (September 2013): 40–49.

Fenske, Gail and Deryck W. Holdsworth. 'Corporate Identity and the New York Office Building, 1895–1915', in *The Landscape of Modernity: Essays on New York City, 1900–1940*, edited by David Ward and Oliver Zunz, 129–59. New York: Russell Sage, 1992.

Finch, Jason. *E.M. Forster and English Place: A Literary Topography*. Turku, Finland: Åbo Akademi Press, 2011.

Finch, Jason. *Deep Locational Criticism: Imaginative Place in Literary Research and Teaching*. Amsterdam: Benjamins, 2016.

Fincham, Benjamin, Mark McGuiness, and Leslie Murray. *Mobile Methodologies*. Basingstoke, UK: Palgrave, 2009.

Finison, Lorenz. *Boston's Cycling Craze, 1880–1900: A Story of Race, Sport, and Society*. Amherst and Boston, MA: University of Massachusetts Press, 2014.

Fleming, Rae. 'The Trolley Takes Command, 1892 to 1894', *Urban History Review/Revue d'histoire urbaine* 19 (1991): 218–25.

Fogelson, Robert. *Bourgeois Nightmares: Suburbia, 1870–1930*. New Haven, CT, and London: Yale University Press, 2005.

Foglesong, Richard. *Planning the Capitalist City: The Colonial Era to the 1920s*. Princeton: Princeton University Press, 1986.

Fothergill, Robert. *Private Chronicles: A Study of English Diaries*. London: Oxford University Press, 1974.

Fraser-Stephen, Elspet, *Two Centuries in the London Coal Trade: The Story of Charringtons*. London: n.p., 1952.

Freeman, Michael. *Railways and the Victorian Imagination*. New Haven, CT: Yale University Press, 1999.

Freeman, Michael and Derek Aldcroft, eds. *Transport in Victorian Britain*. Manchester: Manchester University Press, 1988.

Freeman, Victoria Jane. 'Toronto has no history': Indigeneity, Settler Colonialism, and Historical Memory in Canada's Largest City. Unpublished PhD thesis, Department of History, University of Toronto, 2010.

Freund, David. *Colored Property: State Policy and White Racial Politics in Suburban America*. Chicago: University of Chicago Press, 2007.

Gad, Gunter and Deryck W. Holdsworth. 'Streetscape and Society: The Changing Built Environment of King Street', in *Patterns of the Past: Interpreting Ontario's History*, edited by Roger Hall, William Westfall, and Laurel Sefton MacDowell, 174–205. Toronto and Oxford: Dundurn Press, 1988.

Galinou, Mireille and John Hayes. *London in Paint*. London: Museum of London, 1996.

Gao, Boyang, Weidong Liu, and Michael Dunford. 'State Land Policy, Land Markets and Geographies of Manufacturing: The Case of Beijing, China', *Land Use Policy* 36 (2014): 1–12.

Garvey, Ellen. 'Reframing the Bicycle: Advertising-Supported Magazines and Scorching Women', *American Quarterly* 47 (1995): 66–101.

Gilbert, David and Claire Hancock. 'New York City and the Transatlantic Imagination: French and English Tourism and the Spectacle of the Modern Metropolis, 1893–1939', *Journal of Urban History* 33 (2006): 77–107.

Girouard, Mark. *Cities and People: A Social and Architectural History*. New Haven, CT: Yale University Press, 1985.

Glaisyer, Natasha. 'Merchants at the Royal Exchange, 1660–1720', in *The Royal Exchange*, edited by A. Saunders, 199–205. London: London Topographical Society, 1997.

Gleick, James. *The Information*. New York: Vintage, 2011.

Gorelik, Adrián. *La Grilla y el Parque. Espacio público y cultura urbana en Buenos Aires 1887–1936*. Buenos Aires: Universidad Nacional de Quilmes, 1998.

Gorelik, Adrián. 'A Metropolis in the Pampas: Buenos Aires 1890–1940', *in Cruelty and Utopia: Cities and Landscapes of Latin America*, edited by Jean-François Lejeune, 146–59. New York: Princeton Architectural Press, 2005.

Gould, Peter. 'Dynamic Structures of Geographic Space', in *Collapsing Space and Time: Geographic Aspects of Communication and Information*, edited by S.D. Brunn and T.R. Leinbach, 3–30. London: HarperCollins, 1991.

Goy, Richard R. *The Building of Renaissance Venice: Patrons, Architects, and Builders*. New Haven, CT: Yale University Press, 2006.

Green, David R. and Alastair Owens. 'Geographies of Wealth: Real Estate and Personal Property Ownership in England and Wales, 1870–1902', *Economic History Review* 66 (2013): 848–72.

Gregory, Derek. 'The Friction of Distance? Information Circulation and the Mails in Early Nineteenth-Century England', *Journal of Historical Geography* 13 (1987): 130–54.

Griffin, Emma. *Liberty's Dawn. A People's History of the Industrial Revolution*. New Haven, CT: Yale University Press, 2013.

Gruschetsky, Valeria. 'Ingeniería vial y diseño urbano en el proyecto de la Avenida General Paz. Buenos Aires en los años treinta.' Paper presented at the Jornadas de Becarios y Tesistas, Universidad Nacional de Quilmes, Prov. de Buenos Aires, Argentina, 2015.

Gudis, Catherine. 'Driving Consumption', *History & Technology* 26 (2010): 369–78.

Guroff, Margaret. 'How We Roll', *Raritan* 35 (2016): 93–115.

Hall, Tom and Robin James Smith. 'Stop and Go: A Field Study of Pedestrian Practice, Immobility and Urban Outreach Work', *Mobilities* 8 (2013): 272–92.

Hanson, Susan. 'Gender and Mobility: New Approaches for Informing Sustainability', *Gender, Place & Culture* 17 (2010): 5–23.

Harris, Richard and Robert Lewis. 'The Geography of North American Cities and Suburbs: A New Synthesis', *Journal of Urban History* 27 (2001): 262–92.

Harvey, David. *The Urbanization of Capital: Studies in the History and Theory of Capitalist Urbanization*. Baltimore: Johns Hopkins University Press, 1985.

Harvey, David. 'Between Space and Time: Reflections on the Geographical Imagination', *Annals of the Association of American Geographers* 80 (1990): 418–34.

Harvey, David. 'Time–Space Compression and the Postmodern Condition', in *Modernity: Critical Concepts – Volume IV: After Modernity*, edited by Malcolm Waters, 98–118. New York and London: Routledge, 1999.

Harvey, David. *Paris, Capital of Modernity*. New York: Routledge, 2003.

Harvey, David. *The Enigma of Capital: And the Crises of Capitalism*. London: Profile, 2010.

Harvey, David. *Seventeen Contradictions and the End of Capitalism*. Oxford and New York: Oxford University Press, 2014.

Herbert, Steve and Elizabeth Brown. 'Conceptions of Space and Crime in the Punitive Neoliberal City', *Antipode* 38 (2006): 755–77.

Hillier, Bill. *Space Is a Machine*. Cambridge: Cambridge University Press, 1996.

Hobhouse, Hermione, ed. *Survey of London Volume XLIII: Poplar, Blackwall and the Isle of Dogs*. London: Athlone Press for the Royal Commission on the Historical Monuments of England, 1994.

Holdsworth, Deryck W. 'Morphological Change in Lower Manhattan, New York, 1893–1920', in *Urban Landscapes: International Perspectives*, edited by J.W.R. Whitehand and P.J. Larkham, 114–29. London: Routledge, 1992.

Holt, Glen. 'The Changing Perception of Urban Pathology: An Essay on the Development of Mass Transit in the United States', in *Cities in American History*, edited by Kenneth Jackson and Stanley Schultz, 324–43. New York: Alfred A. Knopf, 1972.

Horton, David, Paul Rosen, and Peter Cox, eds. *Cycling and Society*. Farnham, UK: Ashgate, 2007.

Hughes, Quentin. *Seaport: Architecture and Townscape in Liverpool*. Liverpool: Bluecoat Press, 1993.

Humphries, Jane. *Childhood and Child Labour in the British Industrial Revolution*. Cambridge: Cambridge University Press, 2010.

Hunt, Tristam. *Ten Cities that Made an Empire*. London: Allen Lane, 2014.

Illich, Ivan. *Tools for Conviviality*. New York: Harper & Row, 1973.

Ingleby, Richard, Jonathan Black, David Cohen, and Gordon Cooke. *C.R.W. Nevinson: The Twentieth Century*. London: Merrell Holberton, 1999.

Jackson, John Brinckerhoff. *American Space: The Centennial Years 1865–1876*. New York: W.W. Norton, 1972.

Jacobson, Matthew Frye. *Whiteness of a Different Color: European Immigrants and the Alchemy of Race*. Cambridge, MA: Harvard University Press, 1998.

Jessup, Lynda, ed. *Antimodernism and the Artistic Experience: Policing the Boundaries of Modernity*. Toronto, Buffalo, and London: University of Toronto Press, 2001.

Joy, David. *A Regional History of the Railways of Great Britain: Volume VIII South and West Yorkshire*. Newton Abbot, UK: David & Charles, 1975.

Joyce, Patrick. *Work, Society and Politics: The Culture of the Factory in Later Victorian England*. London: Taylor & Francis, 1980.

Keating, Peter. *The Working Classes in Victorian Fiction*. London: Routledge, 1979.

Kern, Stephen. *The Culture of Time and Space, 1880–1918*. Cambridge, MA: Harvard University Press, 1983.

Keunen, Bart and Luc De Droogh. 'The Socioeconomic Outsider: Labour and the Poor', in *The Cambridge Companion to the City in Literature*, edited by Kevin R. McNamara, 99–113. Cambridge: Cambridge University Press, 2014.

Kivisto, Peter. 'Time–Space Compression', in *The Wiley-Blackwell Encyclopedia of Globalization*, edited by George Ritzer. Oxford: Wiley-Blackwell, 2012.

Krukowski, Damon. *The New Analog: Listening and Reconnecting in a Digital World*. New York and London: New Press, 2017.

Kynaston, David. *The City of London Volume II, Golden Years 1890–1914*. London: Chatto & Windus, 1995.

Labalme, Patricia H. and Laura White, eds. *Venice, Cità Excelentissima: Selections from the Renaissance Diaries of Marin Sanudo*, translated by Linda L. Carroll. Baltimore: Johns Hopkins University Press, 2008.

Landau, Sarah and Carl Condit. *Rise of the New York Skyscraper, 1865–1913*. New Haven, CT: Yale University Press, 1996.

Langan, Mary and Bill Schwarz, eds. *Crises in the British State 1880–1930*. London: Hutchinson, 1985.

Laperrière, Guy. 'Le congrès eucharistique de Montréal en 1910: une affirmation du catholicisme montréalais', *SCHEC, Études d'histoire religieuse* 77 (2011): 21–39.

Latchford, A.L. and H. Pollins. *London General: The Story of the London Bus, 1856–1956*. London: London Transport, 1956.

Law, Michael John. *The Experience of Suburban Modernity: How Private Transport Changed Interwar London*. Manchester: Manchester University Press, 2014.

Lawrence, David, ed. *Omnibus: A Social History of the London Bus*. London: London Transport Museum, 2014.

Laybourn, Keith and David Taylor. *Policing in England and Wales, 1918–39: The Fed, Flying Squads and Forensics*. Basingstoke, UK: Palgrave Macmillan, 2011.

Lears, T.J. Jackson. *No Place of Grace: Antimodernism and the Transformation of American Culture, 1880–1920*. Chicago and London: University of Chicago Press, 1994.

Lees, Loretta and David Ley. 'Introduction to Special Issue on Gentrification and Public Policy', *Urban Studies* 45 (2008): 2379–84.

Lefebvre, Henri. *Towards an Architecture of Enjoyment*, edited by Łukasz Stanek. Minneapolis: University of Minnesota Press, 2014.

Lejeune, Philippe. *On Diary*. Honolulu, HI: University of Hawai'i Press, 2009.

Lemon, James. *Liberal Dreams and Nature's Limits: Great Cities of North America Since 1600*. Toronto: Oxford University Press, 1996.

Lewis, Michael. *Flash Boys: A Wall Street Revolt*. New York: Norton, 2014.

Li, Jinhua. 'Beijing Bicycle: Desire, Identity and the Wheels', in *Culture on Two Wheels: The Bicycle in Literature and Film*, edited by Jeremy Withers and Daniel P. Shewa, 281–99. Lincoln, NE: University of Nebraska Press, 2016.

Lin, George. *Developing China: Land, Politics and Social Conditions*. London: Routledge, 2009.

Lipartito, Kenneth J. 'The New York Cotton Exchange and the Development of the Cotton Futures Market', *The Business History Review* 57 (1983): 50–72.

Liu, Hongguang and Weidong Liu. 'Decomposition of Energy-Induced CO2 Emissions in Industry of China', *Progress in Geography* 2 (2009): 285–92.

Liu, Weidong and Peter Dicken. 'Transnational Corporations and "Obligated Embeddedness": Foreign Direct Investment in China's Automobile Industry', *Environment & Planning A* 38 (2006): 1229–47.

Liu, Weidong and Henry Yeung. 'China's Dynamic Industrial Sector: The Automobile Industry', *Eurasian Geography & Economics* 49 (2008): 523–48.

Lively, Penelope. *City of the Mind*. Harmondsworth, UK: Penguin, 1992.

Longhurst, James. 'The Sidepath Not Taken: Bicycles, Taxes, and the Rhetoric of the Public Good in the 1890s', *The Journal of Policy History* 25 (2013): 557–86.

Lyth, Peter. 'Plane Crazy Brits: Aeromobility, Climate Change and the British Traveller', in *Transport Policy: Learning Lessons from History*, edited by Colin Divall, Julian Hine, and Colin Pooley, 171–84. Farnham, UK: Ashgate, 2016.

Lyth, Peter and Marc Dierikx. 'From Privilege to Popularity: The Growth of Leisure Air Travel since 1945', *The Journal of Transport History* 15 (1994): 97–116.

McClintock, Anne. *Imperial Leather: Race, Gender and Sexuality in the Colonial Contest*. New York and London: Routledge, 1995.

McCrossen, Alexis. *Marking Modern Times: A History of Clocks, Watches, and Other Timekeepers in American Life*. Chicago and London: University of Chicago Press, 2013.

McDonagh, Josephine. 'Space, Mobility, and the Novel: "The Spirit of Place is a Great Reality"', in *A Concise Companion to Realism*, edited by Matthew Beaumont, 50–67. Oxford: Blackwell, 2010.

McDowell, Linda, Adina Batnitzky, and Sarah Dyer. 'Division, Segmentation, and Interpellation: The Embodied Labors of Migrant Workers in a Greater London Hotel', *Economic Geography* 83 (2007): 1–25.

McKay, Ian. *The Quest of the Folk: Antimodernism and Cultural Selection in Twentieth Century Nova Scotia*. Montreal and Kingston: McGill–Queen's University Press, 1994.

Mackintosh, Phillip Gordon. '"The Development of Higher Urban Life" and the Geographic Imagination: Beauty, Art, and Moral Environmentalism in Toronto, 1900–1920', *Journal of Historical Geography* 31 (2005): 688–722.

Mackintosh, Phillip Gordon. 'A Bourgeois Geography of Domestic Cycling: Using Public Space Responsibly in Toronto and Niagara-on-the-Lake, 1890–1900', *Journal of Historical Sociology* 20 (2007): 126–57.

Mackintosh, Phillip Gordon. 'The "Occult Relation between Man and the Vegetable": Transcendentalism, Immigrants, and Park Planning in Toronto, c.1900', in *Rethinking the Great White North: Race, Nature and the Historical Geographies of Whiteness in Canada*, edited by Andrew Baldwin, Laura Cameron and Audrey Kobayashi, 85–106. Vancouver: UBC Press, 2011.

Mackintosh, Phillip Gordon. *Newspaper City: The Liberal Press and Toronto's Street Surfaces, 1860–1935*. Toronto and Buffalo: University of Toronto Press, 2017.

Mackintosh, Phillip Gordon and Clyde Forsberg. '"Co-agent of the Millennium": City Planning, Urban Reform and Christian Eschatology in North America, 1890–1920', *Annals of the Association of American Geographers* 103 (2013): 727–47.

Mackintosh, Phillip Gordon and Glen Norcliffe. 'Flâneurie on Bicycles: Acquiescence to Women in Public in the 1890s', *The Canadian Geographer* 50 (2006) 17–37.

Mackintosh, Phillip Gordon and Glen Norcliffe. 'Men, Women and the Bicycle: Gender and the Social Geography of Cycling in the Late Nineteenth-Century', in *Cycling and Society*, edited by David Horton, Paul Rosen, and Peter Cox, 153–77. Farnham, UK: Ashgate, 2007.

McShane, Clay. 'Transforming the Use of Urban Space: A Look at the Revolution in Street Pavements, 1880–1924', *Journal of Urban History* 5 (1979): 279–307.

McShane, Clay. *Down the Asphalt Path: The Automobile and the American City*. New York: Columbia University Press, 1994.

Macy, Sue. *Wheels of Change: How Women Rode the Bicycle to Freedom (With a Few Flat Tires Along the Way)*. Washington, DC: National Geographic Society, 2011.

Massey, Doreen. 'Power-Geometry and a Progressive Sense of Place', in *Mapping the Futures: Local Cultures, Global Change*, edited by Jon Bird, Barry Curtis, Tim Putnam, George Robertson, and Lisa Tickner. London: Routledge, 1993.

Mathieson, Charlotte. '"A Moving and a Moving On": Mobility, Space, and the Nation in Charles Dickens's *Bleak House*', *English* 61 (2012): 395–405.

Mathieu, S.J. 'North of the Colour Line: Sleeping Car Porters and the Battle against Jim Crow on Canadian Rails, 1880–1920', *Labour/Le Travail* 47 (2001): 9–42.

Mei, Lixia and Jici Wang. 'The Changing Geography of Chinese Bicycle Industry: The Case of Tianjin Bicycle Cluster and Its Evolutional Trajectory.' Paper presented at the Annual General Meeting, Association of American Geographers, Boston, MA, April 22–7, 2008.

Merriman, Peter. 'Mobility', in *International Encyclopedia of Human Geography: Volume 7*, edited by Rob Kitchin and Nigel Thrift, 134–43. London: Elsevier, 2009.

Merriman, Peter. *Mobility, Space and Culture*. London: Routledge, 2012.

Milbrandt, Roger. 'How Poor Was George Gissing? A Study of Gissing's Income between 1877 and 1888', *Gissing Journal* 43 (October 2007): 1–17.

Misa, Thomas. 'The Compelling Tangle of Modernity and Technology', in *Modernity and Technology*, edited by Thomas Misa, Philip Brey, and Andrew Feenberg, 1–30. Cambridge and London: MIT Press, 2003.

Misa, Thomas, Philip Brey, and Andrew Feenberg, eds. *Modernity and Technology*. Cambridge and London: MIT Press, 2003.

Molz, J. Germann. 'Representing Pace in Tourism Mobilities: Staycations, Slow Travel and The Amazing Race', *Journal of Tourism & Cultural Change* 7 (2009): 270–86.

Mom, Gijs. *Atlantic Automobilism: Emergence and Persistence of the Car, 1895–1940*. New York: Berghahn Books, 2014.

Moran, Joe. 'Crossing the Road in Britain 1931–1976', *Historical Journal* 49 (2006): 477–96.

Murray, James M. *Bruges. Cradle of Capitalism, 1280–1390*. Cambridge: Cambridge University Press, 2005.

Nash, John. 'Arnold Bennett and Home Management', *English Literature in Transition, 1880–1920* 59 (2016): 210–33.

Nead, Lynda. *Victorian Babylon: People, Streets and Images in Nineteenth-Century London*. New Haven, CT: Yale University Press, 2000.

Nightingale, Carl. *Segregation: A Global History of Divided Cities*. Chicago: University of Chicago Press, 2012.

Norcliffe, Glen. *The Ride to Modernity: The Bicycle in Canada, 1869–1900*. Toronto: University of Toronto Press, 2001.

Norcliffe, Glen. *Critical Geographies of Cycling: History, Political Economy and Culture*. London: Routledge, 2016.

Nord, Deborah Epstein. *Walking the Victorian Streets: Women, Representation and the City*. Ithaca, NY: Cornell University Press, 1995.

Norton, Peter. 'Street Rivals: Jaywalking and the Invention of the Motor Age Street', *Technology & Culture* 48 (2007): 331–59.

Norton, Peter. *Fighting Traffic: The Dawn of the Motor Age in the American City*. Cambridge, MA: MIT Press, 2008.

O'Connell, Sean. *The Car and British Society: Class, Gender and Motoring, 1896–1939*. Manchester: Manchester University Press, 1998.

Ogborn, Miles. *Spaces of Modernity: London's Geographies, 1680–1780*. New York: Guilford Press, 1998.

Oliver, Robert. 'Toronto's Olympic Aspirations: A Bid for the Waterfront', *Urban Geography* 32 (2011): 767–87.

Olson, Sherry. 'Feathering Her Nest in Nineteenth-Century Montreal', *Social History/Histoire Sociale* 33 (2000): 1–35.

Olson, Sherry. 'Silver and Hotcakes and Beer, Irish Montreal in the 1840s', *Canadian Ethnic Studies*, 45 (2013): 179–201.

O'Mahony, Barry. 'The Role of the Hospitality Industry in Cultural Assimilation: A Case Study from Colonial Australia', in *Hospitality: A Social Lens*, edited by Conrad Lashley, Paul Lynch, and Alison J. Morrison, 89–100. Amsterdam: Elsevier, 2007.

O'Malley, Michael. *Keeping Watch: A History of American Time*. New York: Viking, 1990.

Owens, Alastair, Nigel Jeffries, Karen Wehner, and Rupert Featherby. 'Fragments of the Modern City: Material Culture and the Rhythms of Everyday Life in Victorian London', *Journal of Victorian Culture* 15 (2010): 212–25.

Pan, Haixiao, Qing Shen, and Song Xue. 'Intermodal Transfer between Bicycles and Rail Transit in Shanghai, China', *Transportation Research Record* 2144 (2010): 181–8.

Parsons, Deborah. *Streetwalking the Metropolis: Women, the City and Modernity*. Oxford: Oxford University Press, 2000.

Pearce, Lynne. *Drivetime: Literary Excursions in Automotive Consciousness*. Edinburgh: Edinburgh University Press, 2016.

Peck, Jamie, 'Struggling with the Creative Class', *International Journal of Urban & Regional Research* 29 (2005): 740–70.

Perks, Robert and Alistair Thomson, eds. *The Oral History Reader*, 3rd edn. London: Routledge, 2016.

Petty, Ross. 'Peddling the Bicycle in the 1890s: Mass Marketing Shifts Into High Gear', *Journal of Macromarketing* 15 (1995): 32–46.

Petty, Ross. 'Women and the Wheel: The Bicycle's Impact on Women', in Cycle History: Proceedings of the 7th International Cycle History Conference, edited by Rob van der Plas, 112–33. San Francisco: Rob van der Plas, 1997.

Petty, Ross. 'Bicycling in Minneapolis in the Early 20th Century', *Minnesota History* 62 (2010): 84–95.

Pevsner, Nikolaus. *A History of Building Types*. Princeton: Princeton University Press, 1979.

Piketty, Thomas. *Le Capital au XXIe siècle*. Paris: Seuil, 2013.

Plowden, William. *The Motor Car and Politics in Britain 1896–1970*. London: Bodley Head, 1971.

Poole, Adrian. *Gissing in Context*. Totowa, NJ: Rowman & Littlefield, 1975.

Pooley, Colin. 'Uncertain Mobilities: A View from the Past', *Transfers* 3 (2013): 26–44.

Pooley, Colin and Marilyn Pooley. 'Mrs Harvey Came Home from Norwich . . . Her Pocket Picked at the Station and All Her Money Stolen', *Journal of Migration History* 1 (2015): 54–74.

Pooley, Colin, Jean Turnbull, and Mags Adams. *A Mobile Century? Changes in Everyday Mobility in Britain in the Twentieth Century*. Farnham, UK: Ashgate, 2005.

Potter, Eliza. *A Hairdresser's Experience in High Life*, edited by Xiomara Santamarina. Chapel Hill: University of North Carolina Press, 2009.

Poutanen, Mary Anne. *Beyond Brutal Passions: Prostitution in Early Nineteenth-Century Montreal*. Montreal: McGill-Queen's University Press, 2015.

Poutanen, Mary Anne. 'Due Attention Has Been Paid to All Rules: Women, Tavern Licences, and Social Regulation in Montreal, 1840–1860', *Histoire sociale/Social History* 50 (2017): 43–68.

Prévost, Michel. *La Belle époque de Caledonia Springs*. Hull, QC: Lettres Plus, 1997.

Price, Jacob M. 'Directions for the Conduct of Merchant's Counting House, 1766', *Business History* 28 (1986): 134–50.

Pugh, Martin. *State and Society: A Social and Political History of Britain 1870–1997*, 2nd edn. London: Arnold, 1999.

Raitt, Suzanne. 'The Rhetoric of Efficiency in Early Modernism', *Modernism/Modernity* 13 (2006): 835–51.

Reece, Steve. *The Stranger's Welcome: Oral Theory and the Aesthetics of the Homeric Hospitality Scene*. Ann Arbor: University of Michigan Press, 1993.

Reed, John. *London Buses: A Brief History*. Harrow, UK: Capital Transport, 2007.

Reid, Carlton. *Roads Were Not Built for Cars: How Cyclists were the First to Push for Good Roads & Became the Pioneers of Motoring*. Washington, DC: Island Press, 2015.

Revill, George. 'Histories', in *The Routledge Handbook of Mobilities*, edited by Peter Adey, David Bissell, Kevin Hannam, Peter Merriman, and Mimi Sheller, 506–16. London: Routledge, 2014.

Rich, Paul. 'Doctrines of Racial Segregation in Britain: 1900–1944', *New Community: Journal of the Commission for Racial Equality* 12 (1984): 75–88.

Robertson, Roland. 'Glocalization: Time–Space and Homogeneity–Heterogeneity', in *Global Modernities*, edited by Michael Featherstone, Scott Lash, and Roland Robertson, 25–44. London: Sage, 1995.

Robson, Brian. *Urban Growth: An Approach*. London: Routledge, 1973.

Roediger, David. *Working Toward Whiteness: How America's Immigrants Became White: The Strange Journey from Ellis Island to the Suburbs*. New York: Basic Books, 2005.

Roman, Gretta. 'The Reach of the Pit: Negotiating the Multiple Spheres of the Chicago Board of Trade Building in the Nineteenth Century.' Unpublished doctoral dissertation, Department of Art History, Penn State University, 2015.

Roomet, Louise B. 'Vermont as a Resort Area in the Nineteenth Century', *Vermont History* 44 (1976): 1–13.

Rosa, Harmut and William E. Scheuerman, eds. *High-Speed Society: Social Acceleration, Power and Modernity*. University Park, PA: Pennsylvania State University Press, 2009.

Rothstein, Morton. 'Centralizing Firms and Spreading Markets: The World of International Grain Traders, 1846–1914', *Business and Economic History* 2nd series, 17 (1988): 103–13.

Rudy, Jarrett. 'Do You Have the Time? Modernity, Democracy, and the Beginnings of Daylight Saving Time in Montreal, 1907–1928', *Canadian Historical Review* 93 (2012): 531–54.

Rumilly, Robert. *Histoire de la province de Québec, Vol. XV*. Montréal: Bernard Valiquette, 1945.

Sargent, Charles. *The Spatial Evolution of Greater Buenos Aires, Argentina, 1870–1930*. Tempe: Center for Latin American Studies, Arizona State University, 1974.

Saunders, Max. *Self-Impression: Life-Writing, Autobiografiction, and the Forms of Modern Literature*. Oxford: Oxford University Press, 2010.

Schivelbusch, Wolfgang. *The Railway Journey*. Leamington Spa, UK: Berg, 1986.

Schnapp, Jeffrey. 'Crash (Speed as Engine of Individuation)', *Modernism/Modernity* 6 (1999): 1–49.

Schultz, Stanley. *Constructing Urban Culture: American Cities and City Planning, 1800–1920*. Philadelphia: Temple University Press, 1989.

Scobie, James. *Buenos Aires: Plaza to Suburb, 1870–1910*. New York: Oxford University Press, 1974.

Semple, Janet. *Bentham's Prison: A Study of the Panopticon Penitentiary*. Oxford: Clarendon Press, 1993.

Sennett, Richard. *The Fall of Public Man*. Cambridge: Cambridge University Press, 1974.

Sennett, Richard. *Flesh and Stone: The Body and the City in Western Civilization*. London: Faber, 1994.

Seung, Sebastian. *Connectome: How the Brain's Wiring Makes Us Who We Are*. Boston: Houghton Mifflin Harcourt, 2012.

Sharples, Joseph and John Stonard. *Built on Commerce: Liverpool's Central Business District*. Swindon, UK: English Heritage, 2008.

Sheehy, Barry. *Montreal City of Secrets: Confederate Operations in Montreal during the American Civil War*. Montreal: Baraka Books, 2017.

Sheller, Mimi. 'Moving with John Urry', *Theory, Culture & Society* (2016), www.theoryculturesociety.org/moving-with-john-urry-by-mimi-sheller/.

Sheller, Mimi and John Urry. 'The City and the Car', *International Journal of Urban & Regional Research* 24 (2000): 737–57.

Sheller, Mimi and John Urry. 'The New Mobilities Paradigm', *Environment & Planning A* 38 (2006): 207–26.

Sheller, Mimi and John Urry. 'Mobilizing the New Mobilities Paradigm', *Applied Mobilities* 1 (2016): 10–25.

Shumsky, Neil Larry, ed. *The Physical City: Public Space and the Infrastructure, Volume 2*. New York and London: Garland, 1997.

Simmons, Jack. *The Railways of Britain: An Historical Introduction*. London: Macmillan, 1968.

Singh, Dhan Zunino. 'Towards a Cultural History of Underground Railways', *Mobility in History* 4 (2012), 106–12.

Singh, Dhan Zunino. 'Meaningful Mobilities: The Experience of Underground Travel in the Buenos Aires Subte (1913–1944)', *Journal of Transport History* 35 (2014): 97–113.

Singh, Dhan Zunino. 'The Circulation and Reception of Mobility Technologies: The Construction of Buenos Aires's Underground Railways', in *Peripheral Flows: A Historical Perspective on Mobilities between Cores and Fringes*, edited by Simone Fari and Massimo Moraglio, 128–53. Newcastle upon Tyne, UK: Cambridge Scholar Press, 2016.

Smith, Susan. *The Politics of 'Race' and Residence: Citizenship, Segregation and White Supremacy in Britain*. Cambridge: Polity Press, 1989.

Somers, Dale. 'A City on Wheels: The Bicycle Era in New Orleans', *Louisiana History: The Journal of the Louisiana Historical Association* 8 (1967): 236.

Soppelsa, Peter. 'Intersections: Technology, Mobility, and Geography', *Technology & Culture* 52 (2011): 673–7.

Stein, Jeremy. 'Reflections on Time, Time–Space Compression and Technology in the Nineteenth Century', in *Timespace: Geographies of Temporality*, edited by Jon May and Nigel Thrift, 106–19. London: Routledge, 2001.

Sternass, Jon. *First Resorts: Pursuing Pleasure at Saratoga Springs, Newport, and Coney Island*. Baltimore: Johns Hopkins University Press, 2003.

Summerson, John. *The Architecture of Victorian London*. Charlottesville, VA: University of Virginia Press, 1976.

Summerson, John. 'The Victorian Rebuilding of the City of London', *The London Journal* 3 (1977): 163–85.

Swafford, Kevin. 'Mourning, Pleasure and the Aesthetic Ideal in *The Private Papers of Henry Ryecroft*', *Gissing Journal* 38 (July 2002): 1–13.

Taylor, William and Thomas Bender. 'Culture and Architecture: Aesthetic Tensions in the Shaping of New York', in *In Pursuit of Gotham: Culture and Commerce in New York*, edited by William Taylor, 51–69. New York: Oxford University Press, 1992.

Thompson, F.M.L. *Hampstead: Building a Borough, 1650–1974*. London: Routledge, 1974.

Thompson, Paul. *The Voice of the Past: Oral History*. Oxford: Oxford University Press, 1978.

Thornton, Agathe. 'The Homecomings of the Achaeans', in *People and Themes in Homer's Odyssey*. Dunedin, NZ: University of Otago in association with Methuen, London, 1970.

Tindall, Jane. *The Born Exile: George Gissing*. New York: Harcourt Brace Jovanovich, 1974.

Tobin, Gary. 'The Bicycle Boom of the 1890s: The Development of Private Transportation and the Birth of the Modern Tourist', *Journal of Popular Culture* 7 (Spring 1974): 838–49.

Tomlinson, John. *The Culture of Speed: The Coming of Immediacy*. London: Sage, 2007.

Torres, Horacio. 'Evolución de los Procesos de Estructuración Espacial Urbana. El Caso de Buenos Aires', *Desarrollo Económico* 15 (1975): 281–306.

Townsend, Anthony. 'Life in the Real-time City: Mobile Telephones and Urban Metabolism', *Journal of Urban Technology* 7 (2000): 85–104.

Tuan, Yi-Fu. *Cosmos and Hearth: A Cosmopolite's Viewpoint*. Minneapolis: University of Minnesota Press, 1996.

Tung, Charles M. 'Baddest Modernism: The Scales and Lines of Inhuman Time', *Modernism/Modernity* 23 (2016): 515–38.

Urry, John. *Mobilities*. Cambridge, UK and Malden, MA: Polity, 2007.

Uteng, Tanu Prya and Tim Cresswell, eds. *Gendered Mobilities*. Farnham, UK: Ashgate, 2008.

van der Wijk, Wim. *Beijing Bicycle Strategy and Policy: The Way Back to 'Bicycle Capital of the World'*. Royal Haskoning DHV Engineering, Ecofys, China Academy of Transport Science, and Asian Development Bank, 2016.

Walkowitz, Judith R. *City of Dreadful Delight: Narratives of Sexual Danger in Victorian London*. Chicago: University of Chicago Press, 1992.

Wang, Qiuning. 'A Shrinking Path for Bicycles: A Historical Review of Bicycle Use in Beijing.' Unpublished MA thesis, Department of Community and Regional Planning, University of British Columbia, 2012.

Warf, Barney. *Time–Space Compression: Historical Geographies*. London: Routledge, 2008.

Wasserstrom, Jeffrey. 'Metropolis East: A Review of Stella Dong, *Shanghai: The Rise and Fall of a Decadent City, 1842–1949*', *The American Scholar* 69 (2000): 147–50.

Wasserstrom, Jeffrey. 'A Big Ben with Chinese Characteristics: The Customs House as Urban Icon in Old and New Shanghai', *Urban History* 33 (2006): 65–84.

Weaver, C., D. Fyfe, A. Robinson, D. Holdsworth, D. Peuquet, and A.M. MacEachren. 'Visual Analysis of Historic Hotel Visitation Patterns', *IEEE Transactions on Visualization & Computer Graphics* 12 (2006): 35–42.

Weaver, C., D. Fyfe, A. Robinson, D. Holdsworth, D. Peuquet, and A.M. MacEachren. 'Visual Exploration and Analysis of Historic Hotel Visits', *Information Visualization* 6 (2007): 89–103.

Weber, Rachel. 'Extracting Value from the City: Neoliberalism and Urban Redevelopment', *Antipode* 34 (2002): 519–40.

Weinert, Jonathan, Chaktan Ma, and Christopher Cherry. 'The Transition to Electric Bikes in China: History and Key Reasons for Rapid Growth', *Transportation* 34 (2007): 301–18.

White, Jerry. *London in the Twentieth Century: A City and Its People*. London: Vintage, 2001.

White, Jerry. *London in the Nineteenth Century: A Human Awful Wonder of God*. London: Vintage, 2008.

Whitehall, Bryant. *London Coffeehouses: A Reference Book of Coffee Houses of the Seventeenth, Eighteenth and Nineteenth Centuries*. London: George Allen & Unwin, 1963.

Wilder, Craig Steven. *A Covenant with Color: Race and Social Power in Brooklyn 1636–1990*. New York: Columbia University Press, 2000.

Williams, Raymond. *Keywords: A Vocabulary of Culture and Society*. New York: Oxford University Press, 1983.

Wilson, David A. 'The Fenians in Montreal, 1862–68: Invasion, Intrigue, and Assassination', *Éire-Ireland* 38 (2003): 109–33.

Winks, Robin. *Canada and the United States: The Civil War Years*. Montreal: McGill-Queen's University Press, 1998.

Winter, James. *London's Teeming Streets 1830–1914*. London: Routledge, 1993.

Wise, Sarah. 'The Eclectic Hall, Headquarters of Soho Radicalism', *History Workshop Journal* 83 (2017): 289–300.

Wolff, Janet. 'The Invisible "*Flâneuse*": Women and the Literature of Modernity', *Theory, Culture & Society* 2–3 (1985): 37–46.

Wolff, Janet. 'Gender and the Haunting of Cities (or, the Retirement of the Flâneur)', in *The Invisible Flâneuse: Gender, Public Space and Visual Culture in Nineteenth-Century Paris*, edited by Aruna D'Souza and Tom McDonough, 18–32. Manchester: Manchester University Press, 2006.

Wosk, Julie. *Women and the Machine: Representations from the Spinning Wheel to the Electronic Age*. Baltimore and London: Johns Hopkins University Press, 2001.

Yang, Jun, Ying Liu, Ping Qin, and Antung Liu. 'Review of Beijing's Vehicle Registration Lottery: Short-term Effects on Vehicle Growth and Fuel Consumption', *Energy Policy* 75 (2014): 157–66.

Zacharias, John. 'Bicycle in Shanghai: Movement Patterns, Cyclist Attitudes and the Impact of Traffic Separation', *Transport Reviews* 22 (2002): 309–22.

Zacharias, John and Bingjie Zhang. 'Local Distribution and Collection for Environmental and Social Sustainability – Tricycles in Central Beijing', *Journal of Transport Geography* 49 (2015): 9–15.

Zacharias, John and Jian Ming Zhang. 'Estimating the Shift from Bicycle to Metro in Tianjin', *International Development Planning Review* 30 (2008): 93–111.

Zhao, Pengjun. 'Sustainable Urban Expansion and Transportation in a Growing Megacity: Consequences of Urban Sprawl for Mobility on the Urban Fringe of Beijing', *Habitat International* 34 (2010): 236–43.

Index